The Plant-Feeding Gall Midges of North America

THE PLANT-FEEDING GALL MIDGES OF NORTH AMERICA

Raymond J. Gagné

Systematic Entomology Laboratory
Plant Sciences Institute
Agricultural Research Service
U.S. Department of Agriculture

Comstock Publishing Associates *a division of*
Cornell University Press *Ithaca and London*

First published 1989 by Cornell University Press.

Printed in the United States of America.
Color plates printed in Hong Kong.

The paper in this book is acid-free and meets the guidelines for permanence and durability of the Committee on Production Guidelines for Book Longevity of the Council on Library Resources.

Library of Congress Cataloging-in-Publication Data

Gagné, Raymond J.
The plant-feeding gall midges of North America.

Bibliography: p.
Includes index.
1. Gall midges—North America. 2. Gall midges—North America—Host plants. 3. Insects—North America. 4. Insects—North America—Host plants. I. Title.
QL537.C33G34 1989 595.77′1 88-47921
ISBN 0-8014-1918-2

To Sally

Contents

Acknowledgments

All the drawings in this book were executed by Deborah Leather Roney. Preparation of these illustrations and publication of the color plates were made possible in part through the generous support of the Hershey Foods Corporation Fund, Hershey, Pennsylvania, and the Forest Insect and Disease Research and Forest Pest Management Funds of the U.S. Forest Service, U.S. Department of Agriculture.

Many friends and colleagues helped me with this book. They supplied much of the information and many of the records used here, they assisted with research and production of the manuscript, and they reviewed the manuscript. I am most grateful to the following for their goodwill and generosity: Alicia A. Amis, Agricultural Research Service, Byron, Georgia; James E. Appleby, Illinois Natural History Survey, Champaign; Paul H. Arnaud, Jr., California Academy of Sciences, San Francisco; Jeffrey K. Barnes, New York State Biological Survey, Albany; William F. Barr, University of Idaho, Moscow; Lekh R. Batra and Suzanne W. T. Batra, Agricultural Research Service, Beltsville, Maryland; Frederick D. Bennett, University of Florida, Gainesville; Theodore L. Bissell, Hyattsville, Maryland; Paul E. Boldt, Agricultural Research Service, Temple, Texas; Howard P. Boyd, Vincentown, New Jersey; J. Wayne Brewer, Auburn University, Auburn, Alabama; Richard L. Brown, Mississippi State University, Starkville; David A. Carmean, Oregon State University, Corvallis; Lance G. Clark, New York State Agricultural Experiment Station, Geneva; Edwin F. Cook, University of Minnesota, St. Paul; Jack R. Coulson, Agricultural Research Service, Beltsville, Maryland; James A. Digiulio, Oregon State University, Corvallis; Françoise Dreger-Jauffret, Institut de Botanique, Strasbourg; Arnold T. Drooz, Cary, North Carolina; Robert Duncan, Canadian Forestry Service, British Columbia; A. Lee Eavy, University of Maine, Orono; Bernard H. Ebel, U.S. Forest

Service, Athens, Georgia; Douglas C. Ferguson, Agricultural Research Service, Washington D.C.; G. Wilson Fernandes, Arizona State University, Flagstaff; Richard H. Foote, Agricultural Research Service, Washington D.C.; O. T. Forrester, Animal and Plant Health Inspection Service, Otis Air National Guard Base, Massachusetts; Gordon W. Frankie, University of California, Berkeley; Rudolph T. Franklin, University of Georgia, Athens; Roy F. Fritz, Sun City, Arizona; Saul I. Frommer, University of California, Riverside; Malcolm M. Furniss, University of Idaho, Moscow; Irène M. Gagné, Meriden, Connecticut; Janice M. Gillespie, Cranford, New Jersey; Richard D. Goeden, University of California, Riverside; Eric E. Grissell, Agricultural Research Service, Washington, D.C.; Dale H. Habeck, University of Florida, Gainesville; Paul E. Hansen, Oregon State University, Corvallis; Keith M. Harris, Commonwealth Institute of Entomology, London; J. H. Hatchett, Kansas State University, Manhattan; Bradford A. Hawkins, University of California, Riverside; Alan F. Hedlin, Victoria, British Columbia; K. L. Hibbard, Fort Pierce, Florida; Stephen D. Hight, Agricultural Research Service, Beltsville, Maryland; David Johnson, University of Florida, Gainesville; Robert G. Jones, Animal and Plant Health Inspection Service, Florence, South Carolina; Lloyd Knutson, Agricultural Research Service, Beltsville, Maryland; Marian Kohn, Seattle, Washington; Vera A. Krischik, University of Maryland, College Park; the late Jean L. Laffoon, Iowa State University, Ames; Hiram G. Larew, Agricultural Research Service, Beltsville, Maryland; John D. Lattin, Oregon State University, Corvallis; J. Frank McAlpine, Agriculture Canada, Ottawa; Miles McEvoy, Cornell University, Ithaca, New York; Max W. McFadden, U.S. Forest Service, Rosslyn, Virginia; Phuangthong Malikul, Agricultural Research Service, Washington, D.C.; Frank W. Mead, Florida Department of Agriculture, Gainesville; Edward P. Merkel, U.S. Forest Service, Olustee, Florida; Douglass R. Miller, Agricultural Research Service, Beltsville, Maryland; Jeffrey C. Miller, Oregon State University, Corvallis; Robert C. Morris, Leland, Mississippi; John C. Moser, U.S. Forest Service, Pineville, Louisiana; John W. Neal, Jr., Agricultural Research Service, Beltsville, Maryland; Nolan H. Newton, Raleigh, North Carolina; W. C. Nijveldt, Wagenigen, The Netherlands; Allen L. Norrbom, Agricultural Research Service, Washington, D.C.; Eben A. Osgood, Jr., University of Maine, Orono; William A. Palmer, Temple, Texas; Jerry A. Payne, Agricultural Research Service, Byron, Georgia; Robert W. Pemberton, Agricultural Research Service, Albany, California; Robert V. Peterson, Agricultural Research Service, Washington, D.C.; John D. Plakidas, Frederick, Maryland; Elizabeth A. Pullar, Quincy, Massachusetts; William H. Robinson, Virginia Polytechnic Institute and State University, Blacksburg; Emily A.

Rock, Wayne General and Technical College, Orrville, Ohio; Charlie E. Rogers, Agricultural Research Service, Tifton, Georgia; Odette Rohfritsch, Institut de Botanique, Strasbourg; Hans Roskam, Rijksuniversiteit Leiden, The Netherlands; Curtis W. Sabrosky, Agricultural Research Service, Washington, D.C.; Timothy D. Schowalter, Oregon State University, Corvallis; John T. Schulz, North Dakota State University, Fargo; Joseph D. Shorthouse, Laurentian University, Sudbury, Ontario; the late Floyd F. Smith, Agricultural Research Service, Beltsville, Maryland; Mario Solinas, Universita degli Studi di Perugia, Perugia, Italy; Carl E. Stegmaier, Tallahassee, Florida; George C. Steyskal, Agricultural Research Service, Washington, D.C.; Manya B. Stoetzel, Agricultural Research Service, Beltsville, Maryland; Edvard Sylvén, Swedish Museum of Natural History, Stockholm; F. Christian Thompson, Agricultural Research Service, Washington, D.C.; Karl R. Valley, Pennsylvania Department of Agriculture, Harrisburg; James K. Wangberg, University of Wyoming, Laramie; Gwendolyn L. Waring, Arizona State University, Flagstaff; Alfred G. Wheeler, Jr., Pennsylvania Department of Agriculture, Harrisburg; Robert C. Wilkinson, Jr., University of Florida, Gainesville; William M. Woods, Western Australian Department of Agriculture, South Perth; Allen L. Wuensche, Wilson, Texas; Harry O. Yates III, U.S. Forest Service, Athens, Georgia; Allen M. Young, Milwaukee Public Museum, Wisconsin; Junichi Yukawa, University of Kagoshima, Kagoshima, Japan.

RAYMOND J. GAGNÉ

Washington, D.C.

Introduction

Although gall midges are responsible for a significant proportion of insect-related crop damage, the most recent guide to plant-associated cecidomyiids of North America and their damage appeared in 1940 (Felt 1940). As a measure of what was known then, only one gall midge, an unidentified species, had been reported on Douglas-fir. Since then, eight species have been described from Douglas-fir cones and needles, and another, undescribed species has been found feeding on sap in twigs. My main objectives in writing the present book were to list all of the gall midges associated with plants in North America and to review the pertinent literature in the hope of encouraging and assisting further study of these insects.

Gall midges belong to the family Cecidomyiidae in the insect order Diptera, or two-winged flies. They have four separate stages in their life cycle: the egg, larva, pupa, and adult. Adults are tiny, delicate flies usually only 2–3 mm long. They are short-lived, typically living only long enough to mate and lay eggs. All or most feeding is done by the larvae, which are usually less than 5 mm long and spindleform, legless, and white, yellow, or red. They feed by sucking plant juices, which are broken down by salivary secretions and then ingested. The larval stage may last from less than two weeks to more than two years. The long-lived forms spend most of their life cycle in diapause, a period of suspended animation.

Within this general life pattern, cecidomyiids are tremendously varied both anatomically and biologically. Gall midges feed on many plants and on all kinds of plant tissue. They are found on diverse groups of gymnosperms and angiosperms, and a few even live on ferns. Many species are predators of other arthropods, and some that feed on plants have secondarily developed symbiotic relationships with fungi. This book treats all North American cecidomyiids associated with plants, whether they are primary plant-feeders or secondary inhabitants of galls or other plant damage. Although *gall midges* is the common name most often used for these insects, the common name *cecidomyiids* is preferable for species that are not primary plant-feeders. I use both terms interchangeably for the gall makers.

More than 1,200 named species of Cecidomyiidae are known from North America, of which about 900 are known or presumed to be associated with plants. The remaining species have the primitive habit of feeding on fungus growing in decaying organic matter and are not treated further here except briefly in chapter 3. Many additional cecidomyiids, most of them found on plants, are known but still unnamed. In this book, 732 examples of galls or other plant damage and 891 species of cecidomyiids are recorded. Of the gall midges treated here, more than one-third, 298 species, are not identified beyond family level (67) or genus (231). Most of these unidentified species are probably new to science and provide a measure of the amount of work still to be done on the family. Most of them are assigned here to one of four genera: *Asphondylia* (44 species), *Contarinia* (59), *Dasineura* (33), and *Neolasioptera* (40).

Available information on damage, biology, hosts, and distribution is given for each species, and general information is provided for each genus. An illustrated generic key to larvae is presented, along with a key to galls and plant damage. Because this book is intended for a general audience, technical terms have been held to a minimum, and a glossary of those technical terms deemed necessary is provided. Some knowledge of biology and basic entomology is assumed of the reader, who may need to refer to a book that gives the scientific names of plants.

1 Biology

Gall midge biology is diverse, particularly in the strategies evolved for diapause and for leaving the host. One reason for the diversity is that gall midges have evolved as plant-feeders several times, and thus developed different adaptations to serve similar purposes. All known plant-feeding gall midges mate and lay eggs. Other lines of Cecidomyiidae exhibit asexual reproduction and paedogenesis (Wyatt 1967), but those groups are not considered further here. This chapter reviews what is known of gall midge biology, including the many ways in which the plant-feeding forms have adapted to cope with different hosts and environments.

Larval Searching and Settling

Eggs hatch within a few days of deposition; they are not known to diapause. Larvae that hatch from eggs not laid directly into plant tissue must crawl to an appropriate feeding site. Many die during this period. Days after some larvae of *Dasineura porrecta* (Felt) have become enveloped in galls on elm samaras, others are still crawling over the developing flowers, searching for a feeding site. Some are already dead and shriveled. Larvae that hatch from eggs deposited within host tissue can of course begin feeding directly. Larvae that live in simple galls such as leaf rolls are often gregarious. Those that live in complex galls tend to be in individual cells.

Once settled on the proper host tissue, the larvae begin to feed. At this time, the larva of many species that form complex galls changes shape and comes to lie in close contact with the plant tissue (Marchal 1897, Rohfritsch and Shorthouse 1982). The gall then begins to grow around the larva. If such a larva is removed experimentally, it can no longer move again in search of another site (Rohfritsch and Shorthouse 1982). *Dasi-*

neura balsamicola (Lintner), an inquiline often found in galls formed by *Paradiplosis tumifex* Gagné, lays its eggs at about the same time the gall maker does. Newly hatched larvae of the inquiline search for the gall-making larvae that have already begun to settle. The inquiline then remains close to the gall maker while the plant tissue grows around both (Akar and Osgood 1987).

Stem-infesting species reportedly bore into tissue. Wilson and Heaton (1987) observed larvae of *Neolasioptera brevis* Gagné penetrate the soft tissue of new shoots while gyrating their bodies from a position perpendicular to the stem. Within 8–10 hours, larvae made their way into the shoot, leaving a cylindrical channel behind them. The channel is detectable externally by a small closed scar. The larvae later make use of the channels to escape the galls, a strategy also reported for other, more succulent, closed galls (Johnson 1952, Wilson and Heaton 1987).

The Indian *Lasioptera cephalandrae* (Mani) "burrows" between the base of a petiole and the stem within 20–30 minutes (Haridass 1987). It enters initially by "gnawing" the epidermal cells of the stem, and then forces its way between plant cells. First-instar larvae seemingly lack anatomical adaptations for burrowing or mouthparts for gnawing, so possibly the slender, piercing mandibles hold fast to the plant while the rapidly developing plant tissue surrounds the larva.

Leaving the Host

Some cecidomyiid larvae leave the host when full-grown; others pupate in the gall and leave only as adults. Leaving the host in the larval stage is probably the primitive habit and is common to species that live in flower heads or in succulent galls such as leaf rolls that begin to decay after larval feeding ceases. Larvae that leave the gall to pupate always have a spatula, an elongate, sclerotized, epidermal structure on the venter of the prothorax (see fig. 19 in chapter 4). This structure is often lost in groups that pupate in the host.

Larvae That Leave the Gall to Pupate

Such larvae are usually in simple leaf folds with a longitudinal slit or other opening, or live freely in flower heads, in fruit, or on the underside of leaves. When full-grown they simply crawl out of the gall and drop to the soil. Once on the ground the larvae move to a suitable site and then dig into the soil with the spatula. The primitive role of the spatula is probably for digging through soil (Pitcher 1957). When a larva begins to dig into soil, its head is retracted so that the spatula forms the leading edge of the body. The spatula serves to push grains of soil out of the way to facilitate burrowing (Milne 1961). Larvae of some species may burrow up to 5 cm in depth (Vincinaux and Biernaux 1974), but the depth probably varies with the character of the soil. The spatula is used also in locomotion, allowing larvae to vault several centimeters at a time once they have dropped to the soil. A larva arches its body by bringing the posterior end across the venter to catch on the posterior end of the spatula. The body then becomes taut, the connection between the two parts of the body is released, and the larva springs away. Jumping larvae are positively phototactic, and their behavior appears to be adapted for their emergence from the gall and their movement for only a short distance on the ground (Tokuhisa et al. 1979).

Larvae That Pupate in the Gall

Many unrelated groups pupate in their galls, so this condition has evidently evolved many times. Two main strategies occur here: in one, the full-grown larva uses its spatula to cut a channel through which the pupa can later escape; in the other, the pupa cuts its way out to the surface with the help of sclerotized, pointed structures at the anterior end of the body (see figs. 2, 3 in chapter 2). Groups that use the first strategy always have a well-developed larval spatula; those that use the second have, with the exception of *Asphondylia*, lost the spatula or have a weakly developed one.

In most Alycaulini and Lasiopterini, which live primarily in woody stem galls, full-grown larvae use the spatula to cut a cylindrical passage from their feeding tunnels almost to the exterior, leaving only a final thin layer of bark. Larvae of some species cut completely out to the exterior but then seal the opening with silk and retreat into the tunnel (Wilson and Heaton 1987). Larvae then enter diapause or pupate directly. Eventual passage through the thin remaining bark or silk seal is made by the pupa, which pushes through the remaining weak covering. Interestingly, many gregarious *Neolasioptera* and *Lasioptera* species in woody stems form a common exit hole instead of one per larva. One case is known of alycauline pupae that were unsuccessful in escaping from a grass stem gall because the larvae had prepared no escape tunnel (Frohne 1939).

Larvae of other groups that pupate in their galls, e.g., most *Rhopalomyia*, have secondarily lost the spatula. The task of digging through the sometimes woody tissue is taken over by the pupa, which has well-developed antennal horns and other armature suitable for the purpose. *Rhopalomyia* pupae from the hardest galls on sagebrush have the most conspicuous head development (Jones et al. 1983).

The function of the well-developed spatula found on most *Asphondylia* larvae is unknown. The larva does not prepare an escape channel. The mature pupa cuts a channel by rotating its body around a longitudinal axis, using its antennal horns to cut its way out of the gall. Strong abdominal spines help the pupa gain purchase along the channel that forms.

A larva in a complex gall usually feeds with its head toward the interior of the plant; when full-grown, the larva reverses its alignment so that the head is pointed toward the exterior. When the pupa develops, it will have to move only anteriorly to exit from the gall. Whether the larva or pupa cuts a channel, the mature pupa exits only about two-thirds of the way to the exterior. The pressure of the tight passage then holds the pupa steady while the adult breaks out of the pupal skin (color plate 2D, E).

Diapause

Cecidomyiids have evolved many ways to survive unsuitably dry or cold periods, particularly when their host plants are herbaceous annuals. These strategies serve to synchronize the adult's emergence closely with its host's cycle so that the proper plant tissue is available for the newly hatching larvae.

Larval Diapause

In the temperate zone, diapause most commonly occurs in overwintering, full-grown larvae. Larvae of *Dasineura porrecta*, the elm samara gall midge, drop to the ground in mid-April in Maryland and burrow into the soil, where they remain as larvae until late the following winter. Adults emerge from the ground just before elm flower buds open. Most *Neolasioptera* and *Lasioptera* species overwinter in their stem galls, also as fully developed larvae. Larvae of a *Neolasioptera* sp. diapause in flower galls until late summer the following year when their host, *Baccharis* (Asteraceae), comes into bloom. But a few *Asphondylia* and *Rhopalomyia* species overwinter as first-instar larvae in plant tissue that will develop into the following year's buds. The first instar of *Asphondylia ilicicola* Foote lies dormant for 10 months; that of the Japanese *Pseudasphondylia neolitseae* Yukawa lies dormant for up to 20 months (Takasu and Yukawa 1984).

Pupal Diapause

The only known North American record of pupal diapause is for *Contarinia negundinis* (Gillette) (Wilson 1966). Pupal diapause has been recorded elsewhere for *Resseliella lavandulae* (Barnes) (Coutin 1982) in France and *Hasegawaia sasacola* Monzen in Japan (Sunose 1978). The last may overwinter either as a full-grown larva or as a pupa.

Onset of Diapause

Species with more than one generation per year, e.g., *Dasineura pyri* (Bouché) (Coutin 1959) and *Asteromyia carbonifera* (Osten Sacken), should need to enter diapause only in the last summer generation. But part of each generation of some multivoltine cecidomyiids will enter and remain in diapause until the following spring (Coutin 1959, Johnson 1952, Readshaw 1966, Vallotton 1969). The number of diapausing larvae increases with successive generations through the season (Readshaw 1966). The spring generation is a composite one, consisting not only of the last generation of the previous year, but also of some individuals from each of the previous year's other generations (Johnson 1952).

Diapause in hot climates may be brought on by drought. In sub-Saharan Africa, larvae of *Geromyia penniseti* Coutin and Harris are quiescent for 8–9 months in florets of millet; they pupate when rain eventually falls (Coutin and Harris 1969).

In addition to diapause, some species may also undergo a short-term dormancy in response to drought. Larvae of *Contarinia nasturtii* (Kieffer) may become dormant during a drought, but growth resumes after a rainfall (Readshaw 1966, Vallotton 1969).

Cocoons

Larvae of many species spin cocoons in which they eventually pupate. This happens whether the larvae pupate in host tissue or in soil. The silk is spun out of the mouthparts, and a cocoon can be completed in 12–24 hours (Bishop 1954). Considerable variation occurs in cocoon making. In some cases only part of a population makes a cocoon, at least in laboratory colonies (Bishop 1954). In nature this variation may be a response to the environment. For example, some full-grown larvae of *Resseliella dizygomyzae* (Barnes) on willow in Europe overwinter in their burrows in the host stem and do not spin a cocoon, while others drop to the ground and form a cocoon (Urban and Skuhravá 1982).

Larvae of the spring generation of the European *Contarinia nasturtii* and *Contarinia pisi* (Winnertz) have different fates when they enter the soil (Coutin 1959). They may make ovoid cocoons just below the surface of the ground from which they emerge as adults 10–48 days later, or they may form spherical cocoons deeper down, in which they will diapause. When dormancy in the spherical cocoons is broken, larvae break out, move to the soil surface, reenter the soil, and make ovoid cocoons in which they pupate directly (Readshaw 1966). Some groups, notably the Asphondyliini, have apparently lost the need and ability to make cocoons (Möhn 1961).

Extended Diapause

Diapause in full-grown larvae of some species can last several years. Most diapausing larvae will emerge the year after dropping to the ground, but small numbers may not emerge for two or more years. Adults of *Sitodiplosis mosellana* (Géhin) were reared for 13 successive years from one collection of larvae, the numbers of emerging adults fewer in each succeeding year (Barnes 1956). Some females of the European *Contarinia pisi* have some offspring that diapause longer than others, so this ability is probably genetically controlled (Vallotton 1969). *Contarinia pyrivora* (Riley) adults emerged for three years from one brood of larvae (Coutin 1959). *Lasioptera camelliae* Ohno and Yukawa on camellia in Japan exhibits extended diapause in its second-instar larva. The species usually has one full generation each year, but some individuals may take 2–3 years to complete their life cycle in host tissue (Ohno and Yukawa 1984).

Extended diapause can benefit a species by insuring survival of some individuals through lean years or unusual weather. *Contarinia oregonensis* Foote in Douglas-fir cones may have an advantage in extended diapause, inasmuch as the cone crop is not always good each year (Hedlin 1964a). The Japanese *Hasagawaia sasacola* Monzen, another species with staggered emergence, is limited not only by the number of bamboo buds available each year, but also by competition with a lepidopteran for that same resource (Sunose 1978).

Pseudasphondylia neolitseae normally takes one year to complete development, but a small, variable portion of the population undergoes prolonged diapause and takes two years to develop (Takasu and Yukawa 1984). Galls grow to the same size whether or not larvae are in diapause, but larvae remain in diapause as first instars until autumn of the second year, when they develop rapidly to third instars by winter. A parasitoid does not oviposit in galls containing only first instars, so part of a population escapes parasitism.

Life Cycle Strategies

The duration of a species' life cycle may vary in one locale or in different parts of its range, due to differing conditions. Larvae of *Taxomyia taxi* (Inchbald) in England may take one or two years to develop depending on whether they become embedded in active or dormant buds (Redfern 1975). Larvae hatch from eggs in June and will remain first instars for either 2–3 months through August of the year of hatching or 14–15 months through July of the following year. The "artichoke" bud galls of both the one- and two-year life cycles are generally similar except that the dormant bud galls are larger. The progeny of a female can give rise to either or both types of life cycle, so the population is always mixed.

Dasineura oleae (Löw) on olives in Syria differs in the number of generations per year depending on whether it lives along the coast or in the cooler interior (Katlabi and Coutin 1979). Along the coast, adults fly in February and March and lay eggs on buds and inflorescences. The generation in flowers develops quickly and produces another in April on newly emerging leaves; the generation in buds and the second generation in leaves do not develop as quickly, but enter a summer diapause and continue their development through the winter, to emerge as adults the following February-March. In the interior of Syria, this species emerges in spring, too late for flowers, and lays eggs on new leaves and leaf buds. Second instars diapause until winter. Part of the population completes its growth and emerges as adults the following spring, but the remainder enters diapause that lasts through the following winter also.

Adult Emergence

Emergence must be well timed to the host's biology. *Dasineura porrecta* pupates in late winter, when the ground may still be frozen, to emerge the first week of March in Maryland, when elm flower buds are burgeoning. What precipitates pupation of *D. porrecta* is unknown. In the sorghum midge, *Contarinia sorghicola* (Coquillett), temperature and moisture appear to be more important than photoperiod in breaking diapause (Baxendale and Teetes 1983).

Mechanics

When the adult is fully developed and ready to emerge, the pupa wriggles through the narrow end of the ovoid cocoon, if there is one. If the cocoon is underground, the pupa wriggles all the way to the soil surface before the adult breaks out (Bishop 1954). The pupal thorax splits behind the antenna and along the midline, forming a T-shaped exit, through which the adult emerges. When the legs are freed, they push against the pupal skin to help pull the remainder of the body out. After the abdomen is free, the legs and wings expand rapidly to their full size and harden.

Duration

Individual emergence takes a few minutes to several hours, depending on humidity and temperature (Barnes 1930a). Most adults emerge at dawn or dusk, when conditions are usually more humid than at other times of day, but some emerge at midday (Yukawa and Miyamoto 1979), others at night (Yukawa and Sunose 1976). Emergence of a population can last a week, but most individuals will emerge on the same day. Adults usually emerge over a period of hours, beginning in small numbers and peaking at some point. *Masakimyia pustulae* Yukawa and Sunose, however, has two separate peaks, one at dawn, the other at dusk (Sunose 1983b). Male emergence usually peaks slightly before female emergence (Teetes 1985, Yukawa et al. 1976, Yukawa and Sunose 1976). This allows for optimal sclerotization of male body parts before the females emerge. In West Africa, both *Geromyia penniseti* and *Contarinia sorghi* (Harris) lay eggs in florets of the same grass. Adults of *G. penniseti* are most active around midnight; adults of *C. sorghi* emerge in the morning (Coutin and Harris 1974).

Sex Ratio

Although both males and females of any given species are usually present in approximately equal numbers, individual females of many species give rise to progeny of a single sex (Barnes 1931b). But daughter offspring of all-female families can produce all males, all females, or both males and females. The reason for this, at least in the Hessian fly, *Mayetiola destructor* (Say), is that paternally derived chromosomes are eliminated during spermatogenesis. The end effect is that females produce two kinds of gametes, female or male, but males produce only one kind (Gallun and Hatchett 1969).

Flight

Males generally remain near the site where they emerged and where females will emerge soon after. At least one exception to this pattern is

Contarinia pseudotsugae Condrashoff. Here both sexes follow the trunk of the tree to the crown, where females rest on foliage and males swarm around the tips of branches (Condrashoff 1962).

Swarms

True swarms, that is aggregates of males from which individuals leave to mate with females and later rejoin the swarm, have been reported for some cecidomyiids (Chiang 1968, Baylac 1986, Yukawa et al. 1976). Adults of many plant-feeding gall midges may emerge simultaneously in such large numbers that any flying activity may appear to represent a true swarm: 2000 individuals of *Contarinia pseudotsugae* emerge per square foot under Douglas-fir trees (Condrashoff 1962). Swarming may be temperature-dependent (Yukawa et al. 1976) and occurs near the site where the males emerge.

Courtship and Mating

Courtship

Some species apparently display no courtship prior to mating. The Hessian fly merely approaches the female from the rear, attaches, and mates in the side to side position (McKay and Hatchett 1984). This is true also for *Asphondylia morivorella* (Naito) in Japan. A male leaves its swarm and attaches immediately to a female (Sunose 1983a). But simple precopulatory activity involving wing vibration by males just before copulation has been observed in several plant-feeding gall midges (Barnes 1939, Yukawa and Miyamoto 1979), including an undescribed *Contarinia* from sour gum.

The most complex courtship reported for Cecidomyiidae is that of a *Lestodiplosis* species, a predator of a scale insect in France (Baylac 1986). The male approaches a female from behind and the female exserts the end of her abdomen. The male vibrates his wings at varying speeds for different periods and also vibrates his antennae intermittently. Eventually the male pivots to face the opposite direction from the female and simultaneously grasps the end of her abdomen with his genitalia. Predaceous cecidomyiids may require more complex mating behavior than do plant-feeders to elicit a response from the female: these species emerge intermittently over the length of a season and may be less numerous in one locality at any given time than a plant-feeder that emerges in great numbers almost simultaneously.

Sexual Attraction

Unmated females are extremely attractive to males. Newly emerged females can often be seen exserting and waving their ovipositors, and a drop of fluid, presumably containing a pheromone, may be seen at the tip. Waving the extruded ovipositor prior to mating is a widespread habit in Cecidomyiinae (Harris 1979, McKay and Hatchett 1984, Miller and Borden 1981, Yukawa and Miyamoto 1979), but at least in some species it seems to require the presence of the male (Barnes 1939, Baylac 1986). Virgin females of the European *Dasineura tetensi* (Rübsaamen) extrude their ovipositors when placed with males of the same species. The males react by vibrating their antennae and moving their wings before attempting to mate with the females (Barnes 1939). When males of another European species, *Dasineura plicatrix* (Loew), and females of *D. tetensi* were placed together, neither sex reacted; when males of both species were enclosed with virgin females of *D. tetensi*, the females extruded their ovipositors but only male *D. tetensi* reacted.

Number of Matings

Most reports indicate that females mate only once. After mating, a female retracts the ovipositor, possibly cutting off pheromone release, and is evidently no longer attractive to males. She then either flies off directly or rests under a leaf for a short while (Yukawa and Sunose 1976) or up to two hours (Readshaw 1966). The only reported exception to the rule that females mate only once is *Rhopalomyia hirtipes* (Osten Sacken), which reportedly mates with several males (Spence 1969).

Host Seeking and Oviposition

Males do not usually need to fly far from the emergence site, but females, especially if emerging from the ground, may have to travel a considerable distance to find the host. Gall midges are nonetheless relatively weak fliers. A grass-infesting gall midge in England was seen to fly above grasses in mild weather, but flew among them if it was windy (Jones 1940). *Rhopalomyia hirtipes* is egg-heavy after mating and flies clumsily until it has laid some eggs (Spence 1969).

Host Attraction

Females are evidently attracted by host chemicals. *Dasineura brassicae*

(Winnertz) females in Sweden fly passively with the wind until leeward of a host plant, and then they actively fly toward the host (Sylvén 1970, 1979). Females, but not males, of that species are attracted to sinigrin, a naturally occurring chemical in *Brassica*, the host plant (Pettersson 1976). Females of *Aphidoletes aphidimyza* (Rondani), a predator of aphids, are more attracted to brussels sprouts than to other plants if they were reared on that host (Mansour 1975). The Hessian fly selects particular varieties of wheat at least partially on the basis of physical characteristics, such as hairiness (Morrill 1982).

Host Specificity

Some species have decided preferences for certain hosts, although they will sometimes lay eggs on related plants on which the larvae do not grow well (Åhman 1981, 1985, Morrill 1982, Pitcher 1955a, Stokes 1953). *Dasineura brassicae* will lay more eggs on some *Brassica* spp. than on others. Full-grown larvae from eggs laid on other than favored *Brassica* spp. weigh less at maturity, and their resulting females are less fecund than those on the preferred hosts (Åhman 1985, 1986). *Dasineura brassicae* may even lay eggs on plants where larvae will not develop (Åhman 1986).

Most gall midges are monophagous, restricted to one host, or oligophagous, restricted to a few related hosts. A few are polyphagous and feed on hosts that are not closely related, but these generally cause simple bud swellings. Among generalist plant-feeders are *Prodiplosis* spp. and some *Contarinia* spp., which feed on plants of several families (Gagné 1986b). *Contarinia maculipennis* Felt, reportedly a West Indian immigrant in Hawaii, has been recorded from buds of such diverse plants as tomatoes, jasmine, *Brassica*, and hibiscus (Jensen 1946).

Oviposition

Species lay their eggs in characteristic patterns. Eggs may be laid singly, in regular rows (Wilson and Heaton 1987), or in batches, either on the surfaces or in crevices of the host. They may also be laid directly into plant tissue if the female is equipped with a piercing ovipositor. Females must manage to lay their eggs near the tissue that will be infested. Having alighted on a plant, a female finds her way to the oviposition site by antennating and palpating the plant surface (Hallberg and Åhman 1987). When a suitable oviposition site is found, the female backs into the site by feeling with the tip of her ovipositor (Hallberg and Åhman 1987, Marchal 1897). The Hessian fly must lay eggs on the upper surface of a leaf, because the newly hatched larvae will usually crawl downward until tightly ensconced between two leaves, where they will begin to feed (Morrill 1982). Larvae from eggs laid on the underside of an outer leaf would crawl downward to the ground and die.

Environmental factors play a role in where eggs are laid. *Dasineura swainei* Felt lays eggs chiefly on the sunny side of open-grown spruces (Clark 1952). The European *Dasineura brassicae*, which lays its eggs in pods, depends on feeding punctures made by the turnip seed weevil (Åhman 1986). As the density of the weevil increases, so does that of the gall midge (Sylvén and Svenson 1975). *Lasioptera berlesiana* Paoli, a Mediterranean species that lives in olives, inserts its eggs in oviposition punctures made by tephritid flies (Solinas 1967).

In most plant-feeding gall midges, females have long, tapered, telescoped, and pliable ovipositors. These allow the eggs to be inserted in such narrow places as among bud scales or between closely fitted parts of burgeoning buds. Females of many other species have hard and stiff ovipositors modified for inserting eggs directly into plant tissue.

Mortality

Adults of plant-feeding cecidomyiids are short-lived. Most probably survive less than a day or two, although some individuals may live longer in the laboratory (Hedlin 1961). Females usually die soon after they deposit their eggs (Möhn 1961, Teetes 1985). Larvae, much longer lived than adults, are subject to many more environmental dangers as well as risk from parasitoids and predators (Redfern and Cameron 1978, Sone 1986a).

Parasitoids

Parasitoids of cecidomyiids are mostly Hymenoptera, especially of the families Platygasteridae, Pteromalidae, and Torymidae, but less commonly Ceraphronidae, Encyrtidae, Eupelmidae, and Eulophidae, and, more rarely, Braconidae and Ichneumonidae. Parasitoids may be either generalists with a wide host range or specialists on one or more species of gall midges (Hawkins and Goeden 1984). Hymenoptera and other inhabitants of cecidomyiid galls have been well surveyed. A few such studies are by Askew and Ruse 1974, Baxendale et al. 1983, Coutin and Katlabi 1986, Frankie et al. 1987, Hawkins and Goeden 1984, Neiswander

1962, Parnell 1963, Roskam 1979, 1986, Soné 1986b, Sunose 1984, Yukawa 1983, and Yukawa et al. 1981. Parasitism rates can be very high: a pteromalid parasitized 24–45% of a population of Japanese *Asphondylia morivorella* (Sunose 1983a), and two species of *Platygaster* killed 50% of a *Contarinia coloradensis* Felt population (Brewer and Johnson 1977).

Asteromyia carbonifera (Osten Sacken) lays eggs in small to large clutches that result proportionately in one to many larval cells per gall. In galls that have two or more larvae, not all larvae are parasitized, because the parasitoid visiting the gall does not oviposit in each of the cells. The possibility of death through parasitism increases in galls with only one larva, because parasitoids visit galls randomly, whatever the size. Yet female *A. carbonifera* reared from galls with a single larva are usually larger and bear more eggs (Weis 1982a).

Some parasitoids, particularly egg-larva parasitoids, have extremely close relationships with their hosts. *Platygaster pini* Fouts (Platygasteridae) lays its eggs in those of *Janetiella coloradensis* Felt on pinyon pine needles, but the parasitoid does not begin growing until the gall midge larva is fully developed in spring of the following year. The *Platygaster* then develops quickly to appear as an adult shortly after the unparasitized host normally emerges (Brewer 1971). Roskam (1986) shows similar cases of European *Platygaster* spp. on birch gall midges.

Predators

Among the major predators are cecidomyiids of the genus *Lestodiplosis*, which prey on cecidomyiids in simple galls or flower heads (Neal 1988, Neiswander 1962, Nijveldt 1969). Other Diptera predators include larvae of Phoridae that feed in galls of gregarious cecidomyiids (Robinson 1971) and adults of Empididae that catch gall midges in flight (Stark and Wetzel 1987). Hemiptera of the families Reduviidae, Nabidae, and Miridae may be seen feeding on pupae breaking out of their galls, but they also feed on larvae and pupae through gall tissue (Jones et al. 1983). Beetles, ants, harvestmen, and spiders also feed on gall midges.

Vertebrates can be important predators. Some branches of an elm that suffered a heavy infestation of *Dasineura ulmea* Felt had every bud, infested or not, torn apart by birds searching for larvae. Needle galls of *Paradiplosis tumifex* appear to be pecked apart by birds (Struble and Osgood 1976). In Europe, birds break apart stems of reed (*Phragmites*) to pick out the gall midge larvae (Skuhravý 1981). Rodents will eat succulent galls that are close to the ground (Spence 1969).

Fungi also take a toll on larvae. An epidemic of *Entomophthora brevinucleata* (Zygomycetes: Entomophthoraceae) killed many gall midges on grasses in Europe in 1983 (Keller and Wilding 1985).

Environmental Hazards

In addition to parasitoids and predators, accidents and weather can affect survival: eggs are laid on the wrong host or an unsuitable part of the plant, or they do not adhere; larvae die before reaching susceptible tissue; adults emerging from the ground in late winter encounter strong winds that drive them out of range of their hosts. Not least of all, plant toxins can kill gall midge eggs and larvae (Åhman 1986).

Intra- and Interspecific Competition

Several larvae of *Janetiella* sp. on pinyon pine enter a particular fascicle bundle, but only one larva per needle fascicle reaches maturity (Gillet and Brewer 1983). Crowding of *Contarinia oregonensis* in Douglas-fir cone scales, although each larva lives isolated in its individual cell, results in smaller adults and, presumably, reduced fecundity; and whole cone scales with several cells died when larval density was too high (Miller 1984). Competition with other arthropods for the same tissue has been noted between mites and *Taxomyia taxi* on yew in England (Redfern and Cameron 1978) and between a lepidopteran and *Hasegawaia sasacola* on bamboo (Sunose 1978) in Japan.

Inquilinous Cecidomyiidae

An inquiline is an organism that can regularly be found in certain galls but is not the gall-forming agent. Inquilines of gall midges include cecidomyiids and other arthropods. Some of the cecidomyiid inquilines are benign, particularly if they live in the outer leaves of rosette galls on such plants as willow and spirea and never influence gall development or come into contact with the gall maker at the center. Other inquilines not only compete with the gall maker for space and food but can even modify the gall. *Dasineura balsamicola*, an inquiline in balsam needle galls formed by *Paradiplosis tumifex*, is always fatal to the gall maker (Osgood and Gagné 1978, Shorthouse and West 1986). Galled needles normally drop off the tree in autumn soon after full-grown *P. tumifex* larvae leave; if the galls are infested with the inquiline, the needles remain on the tree past the date when they would otherwise fall, giving the slower growing *D. balsamicola* time to develop (Osgood and Gagné 1978). A *Contarinia* species in Europe colonizes flower galls of *Paragephyraulus diplotaxis* Solinas when larvae of the latter begin to leave the galls. The *Contarinia* then takes control of the gall, causing it to change shape and continue producing food (Solinas and Bucci 1982).

Asphondylia galls are sometimes attacked by *Tetrastichus* wasps (Eulophidae), which encapsulate themselves in spherical cells within the gall tissue. The enlarging cells gradually constrict and kill the *Asphondylia* larva (Hawkins and Goeden 1982).

The inquilinous habit is widespread among the Cecidomyiidae. *Dasineura* and *Contarinia* both include inquilines, but the majority of species in those genera are gall makers. Some genera are wholly inquilinous. All *Trotteria* spp. are specialists in galls made by *Asphondylia* spp. They indirectly kill the gall maker through competition unless the latter is almost fully developed (Parnell 1964). Many inquilines are generalist fungus feeders that arrive late in the gall cycle, often after the gall maker has departed.

Gall Midge and Fungus Associations

Some gall midges have a symbiotic association with fungi. In some cases the association is obligatory because the fungi serve as food for the larvae; in others the fungi are incorporated in the gall but are apparently benign. In North America, fungi are found in galls made by all *Asphondylia* and by most Lasiopterini and Alycaulini (Gagné 1986c).

Symbiotic Fungi as Food

The fungi are evidently obligatory in all galls of *Asphondylia* and are the food of those gall midges (Meyer 1952). The biology of the holly berry gall midge, *Asphondylia ilicicola*, is outlined here as a representative example of this symbiosis. In spring one or more eggs are inserted into a flower ovary. All the berries appear normal until winter, when those that are infested with the gall midge fail to turn red. Usually no carpels develop normally and at least one carpel cavity per fruit will contain a single first-instar *Asphondylia* larva that will remain in that stage through the winter. Accompanying the larva in each cavity is a small, white patch of fungus hardly larger in diameter than the length of the larva. Neither the larva nor the fungus grows noticeably during the winter. In spring, the fungus grows to completely line the gall cavity with a dense, white mycelium. The cavity, originally unevenly lobed, becomes evenly ellipsoidal and enlarges to accommodate the growing larva. The rapidly growing larva continuously rotates within the chamber to feed evenly upon the substrate. In such galls the fungal mycelium forms a thick mat connected lightly to the plant tissue by intercellular haustoria (Batra and Lichtwardt 1963, Neger 1908).

The plant tissue surrounding the fungus lacks nutritive cells, the layer of cytoplasmically dense cells that appears next to the larval chamber in most galls (Meyer 1952). It therefore appears that the *Asphondylia* larvae feed on the fungus. The manner by which the fungus is carried from gall to gall is unknown, but it is always present with the egg (Puzanowa-Malysheva 1935). One hypothesis is that the female scoops up fungal spores through a shoveling action of the hind end of her abdomen and carries them in the ventral pouch that lies between the seventh and eighth abdominal segments (Borkent and Bissett 1985). Spores can be found, however, in the ventral pouch of other genera of the tribe Asphondyliini, such as *Schizomyia* and *Polystepha*, which are not known to have fungal symbionts in their galls.

At least some Lasiopterini also apparently feed on fungi in galls. Meyer (1952) found that galls of the European *Lasioptera rubi* (Schrank) on *Rubus* also lacked the characteristic layer of nutritive plant cells. Other Old World *Lasioptera* spp. are known to feed on fungi. Larvae of *Lasioptera cephalandrae* (Mani) were successfully reared on in vitro cultures of the fungus normally found in their galls (Haridass 1987). The European *Lasioptera berlesiana* feeds mainly on its fungal associate that grows in the tunnel in due course, but here the larva first feeds on the egg or first-instar larva of an olive-infesting fruit fly (Coutin and Katlabi 1986, Solinas 1967). The cecidomyiid is also cannibalistic if another of its own kind is in the tunnel (Solinas 1967). Although *L. berlesiana* is always found in nature in tunnels begun as tephritid punctures, larvae of the gall midge will survive on the fungus alone if it is artificially introduced in the fruit (Solinas 1967). The oviposition holes of the fruit fly serve mainly as a point of entry for the ovipositor of the gall midge. Another European *Lasioptera, L. donacis* Coutin and Faivre-Amiot, feeds on fungi growing in old galleries of a chloropid fly in leaf sheaths of *Phragmites* (Coutin and Faivre-Amiot 1981).

Fungi as Part of the Gall Structure

Fungal associates are not always present in Alycaulini and Lasiopterini, and, when present, the fungus is apparently not always taken as food. An example is the blister leaf gall of *Asteromyia carbonifera* on goldenrod. The larva is stationary while it is growing, and feeds on one spot, which remains green until the larva is full-grown. A white mycelium almost completely surrounds the larva. When the larva is full-grown, the fungus turns black and forms a hard cast around the larva (Batra 1964, Weis 1982a). This cast can protect the gall midge from parasitoids (Weis 1982b). A close relative of that gall midge, *Asteromyia modesta* (Felt), forms a blister gall on leaves of *Conyza canadensis* and *Erigeron* spp., but has no fungal symbiont.

Elsewhere in Alycaulini and Lasiopterini there are cases of a closely associated pair of species, one with a fungal associate and one without. Examples are the two species of *Lasioptera* on *Ephedra* in western North America and several species of *Lasioptera* (s.l.) on *Phragmites* in eastern Europe (Skuhravá and Skuhravý 1981). But a fungal associate is present more often than not in lasiopterine and alycauline galls. The method of inoculation or transferral of the fungus from gall to gall appears to be through fecal contamination. Haridass (1987) has found the fungal associate of *Lasioptera cephalandrae* in feces deposited during egg laying. Gravid females deposit the fungal inoculum at the same time as the accessory secretions used to attach the eggs to the plant. Fungal spores have been found on the exterior of the ovipositors of Alycaulini and Lasiopterini (Borkent and Bissett 1985), but those spores were possibly picked up only incidentally during the course of egg laying.

Gall Midges as Pests

Two of the most serious pest gall midges in North America, the Hessian fly and the sorghum midge, are introduced species. But some native gall midges are also pests, particularly in crop monocultures. *Contarinia schulzi* Gagné, the sunflower-head gall midge, went unnoticed until sunflower was grown in vast acreages in the Midwest (Schulz 1973). Other native gall midge pests are *Contarinia oregonensis* in cones of Douglas-fir, *Cecidomyia bisetosa* Gagné in cones of slash pines in Forida, and an undescribed species of *Cecidomyia* in spruce seedling plantations in British Columbia. Under natural conditions more than enough seed from these conifers would survive for propagation of the plants, but seed plantations are not run for the benefit of cone insects. Gall midge pests of ornamentals are important in a different way because the appearance of each individual plant is critical. If holly is being grown for sprays of green leaves and red berries for Christmas and the berries remain green because of an infestation by *Asphondylia ilicicola*, the value of the plant is considerably diminished.

Control of cecidomyiids, particularly chemical control, requires knowledge of their biology (Baskin and Brewer 1983, DeBoo and LaPlante 1975). Insecticides have mixed effects because gall midges and their parasitoids, which are in flight at the same time, are indiscriminately killed (Curry et al. 1983). Pheromones might be an effective control measure and have been used as a population monitoring device (Miller and Borden 1981).

The control of many crop pest midges is discussed in Barnes's (1946–1956) series of books, *Gall Midges of Economic Importance*, volumes 1–7. Rotating crops, destroying stubble and volunteer hosts, and planting late are partially effective measures against various species (Barnes 1956).

The most effective control appears to be the development of resistant plant varieties. Control of the first generation of a multivoltine species is important to reduce population build-up. Barnes (1931a) advocated producing early varieties of some hosts, which would then flower before peak emergence of the gall midges ovipositing on them. This method is effective with the sorghum midge, but requires the additional expense of irrigation if planting is very early (Teetes 1985).

Wheat varieties resistant to the Hessian fly were reported as long ago as 1785 (Painter 1951). The Hessian fly apparently does not discriminate among wheat varieties while laying eggs; the limiting factor is the ability of the larvae to feed (Painter 1951).

Varieties of willow are differentially susceptible to various species of *Rabdophaga* (Barnes 1930b, Smith et al. 1983), and there are remarkable differences in infestation among individual trees of various species (Chen and Appleby 1984b). Nonhost *Brassica* spp. of *Dasineura brassicae* in Europe have chemical defenses that might be bred into preferred-host *Brassica* spp. (Åhman 1986). Conifers have resistant qualities that could be used to limit attack of gall midges. Shoots of ponderosa pine with sticky surfaces appear more susceptible to *Cecidomyia* spp. than those with smooth surfaces (Eaton and Yuill 1960). Development of Douglas-fir varieties that leaf out after the main flight of the needle-infesting *Contarinia pseudotsugae* is complete might reduce the need for repeated sprayings in spring (Mitchell and Nagel 1969).

2 Anatomy

This chapter gives an overview of cecidomyiid anatomy. More specific anatomical structures are discussed under various group names in chapter 4. Anatomical terminology follows the system used in McAlpine et al. (1981). Cecidomyiids are distinctive among Diptera, yet are fairly simple insects. The larvae are wormlike with greatly reduced heads; the adults have very simple wing venation and are relatively weak flyers.

Egg

Eggs are ovoid to elongate-ovoid. In some species, the narrow end is drawn out to a point (Wilson 1966). Eggs may be white, yellow, orange, or red. In some species, they may be colorless initially and become darker as they mature (Roskam 1977).

Larva

Larvae are legless, flat-cylindrical, and tapered at both ends. Examples from various genera are illustrated in figures 19–31 in chapter 4. Larvae may be white, yellow, orange, or red, depending on the species or age. Those of the Japanese *Masakimyia pustulae* Yukawa and Sunose in *Euonymus* leaf galls may be white or yellow, a difference that is genetically determined (Sunose 1979): lighter colored larvae are found in southern Japan, yellow ones throughout the country.

Cecidomyiid larvae have three instars, which can be distinguished by measurements of the head capsules and other anatomical differences (Gagné and Hatchett, in press, Osgood and Gagné 1978, Pitcher 1955b,

Redfern 1975, Roskam 1977, 1979, and Solinas 1965a–c). In reports of four instars the prediapausing third instar is counted as a fourth (e.g., Metcalfe 1933), or a late second or an early third instar that shows a partially developed spatula is counted as an additional instar (Abbas 1965, Haridass 1987). These reports were not supported by head capsule measurements. As the pupal tissues begin to develop in the third instar, the larva does change shape slightly. This is sometimes called a prepupal stage, but no molt occurs until the pupa breaks through the larval skin. There are many detailed descriptions of larvae (see, e.g., Otter 1938, Principi 1959, Roskam 1977, 1979, and Solinas 1965a–c).

The largest growth increment for most species is in the third instar. An exception is the Hessian fly (Gagné and Hatchett, in press). Its third larval instar does not feed, but lies fully developed inside the second-instar skin. The pupa then develops, still within the second larval instar skin. This retention of the second-instar skin as a puparium appears several times in this family, as, e.g., in an undescribed species from *Purshia* (q.v.) and another new species from Japanese maple seeds. The puparium is analogous to that found in higher Diptera.

The larval head is conical. It is eyeless but has sensory plates, papillae, and short, conical antennae. Movable mouthparts are composed of a pair of vertically articulated mandibles and a pair of maxillary plates. To feed, the anterior part of the head capsule is closely appressed to the host tissue, the mandibles abrade or grasp the host, salivary secretions are expressed, and strong pharyngeal dilator muscles effect ingestion through the esophagus (Refai et al. 1956, Solinas 1968). The head capsule is modified for various modes of life; predators have long antennae, as do crawling first-instar larvae of many plant-feeding species. Mandibles are used for scraping, grasping, and piercing, depending on the mode of life of the larva (Solinas 1971).

Behind the head are a neck segment, three thoracic segments, and nine abdominal segments, including the terminal or caudal segment (see fig. 19 in chapter 4). Second- and third-instar larvae have spiracles on the first thoracic and first through eighth abdominal segments. First instars have a variable number of spiracles: none (Parnell 1964), the eighth abdominal (Gagné and Hatchett, in press), or the first thoracic and the eighth abdominal (Roskam 1977, 1979). The third instar (rarely, the second also) has a spatula, which is a sclerotized, usually elongate, dermal structure on the venter of the prothorax. The spatula has become variously shaped and may be secondarily reduced or lost in some groups.

The integument is smooth, spiny, or rough. Raised, horizontal welts used in locomotion occur on the anterior part of the abdominal segments, even on many larvae that are otherwise smooth. Some species have ventral pseudopods or dorsal and lateral lobes. The body has a number of papillae, which are regularly placed, sensory organs, with or without setae. Some papillae are lost on sedentary species, particularly those that live in complex galls; such losses are correlated more with larval habits than with natural affinities (Sylvén 1975). The basic papillar pattern is shown in figure 19. A few changes in the terminology of the papillae from that used by Möhn (1955), Mamaev and Krivosheina (1965), and Roskam (1979) are made here for simplicity and to reflect apparent homology between thoracic and abdominal papillae: in this book inner pleural papillae of the thorax and hind ventral papillae of the abdomen are ventral papillae; and sternal papillae of the thorax and anterior ventral papillae of the abdomen are sternal papillae.

Terminal papillae are the most often modified. In particular, most Cecidomyiini have one pair with short, stubby, slightly recurved setae, which are very different from the remaining, setiform papillae on that segment. The stubby pair is used by many species in vaulting. The caudal end of the body is bent forward ventrally and those papillae hook into the fold at the posterior end of the spatula. The body is held under tension and the hold released to flip the larva up to several centimeters away.

The digestive tract is simpler in plant-feeding gall midges than in primitive fungus-feeding ones (Mamaev 1968). Salivary glands are large and variable in structure. The midgut may be constricted into two distinct parts. Two Malpighian tubules are present at the end of the midgut. For further details on internal anatomy, see Slepyan (1960) and Mamaev (1968). See also Stuart and Hatchett (1987) concerning the development of the salivary glands of the Hessian fly.

Pupa

The pupa is a stage of usually short duration between the larval and adult stages. The only comprehensive study of pupae for any large group of gall midges was done by Möhn (1961) for the tribe Asphondyliini, although pupae of many species in other groups have been described. Pupae usually have a large, elongate prothoracic spiracle and short abdominal spiracles. The antennal bases are the anteriormost parts of a pupa and are usually strongly sclerotized to effect escape from soil, cocoons, or galls. Figures 1–6 show representative pupae of gall midges; figure 7 shows a Hessian fly (*Mayetiola destructor*) puparium, which is the sclerotized second-instar skin in which the nonfeeding third-instar larva and the pupa successively develop.

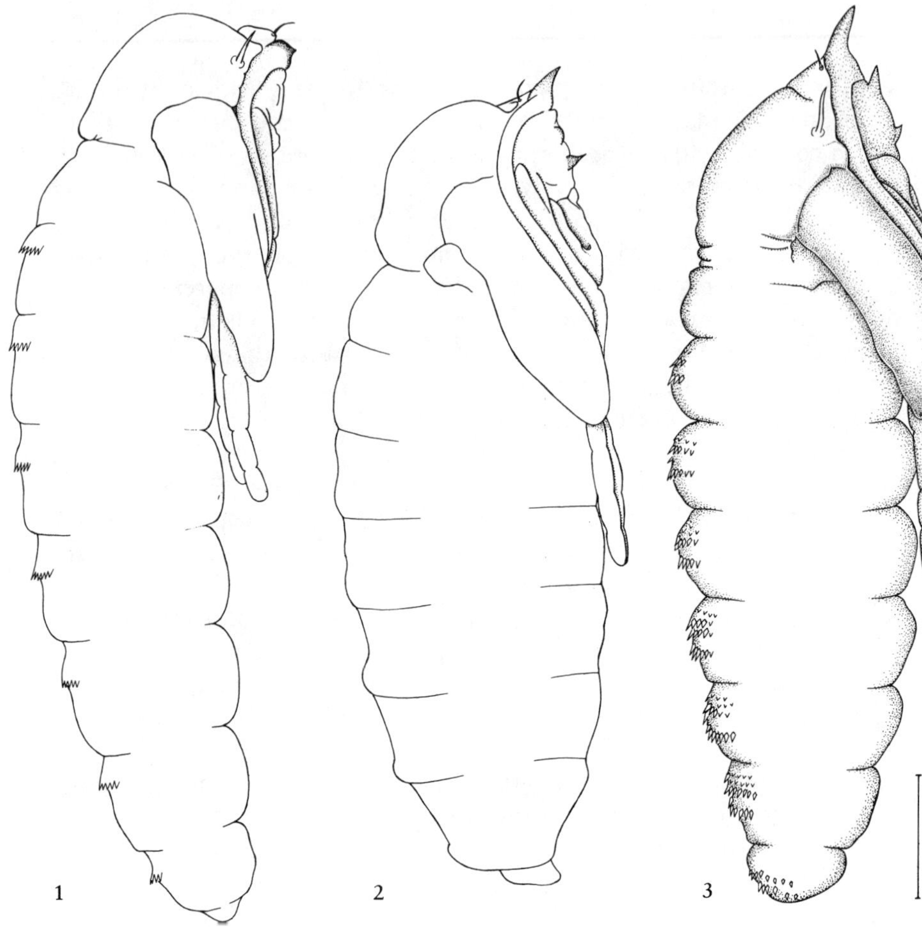

Figures 1–3. Pupae, lateral view. **1,** *Contarinia* sp. **2,** *Rhopalomyia pomum.* **3,** *Asphondylia atriplicis.* Scale line = 0.5 mm.

Adult

Cecidomyiid adults are delicate, fragile flies with reduced wing venation. They are nonetheless structurally diverse, having acquired many adaptations for different ways of life. Examples of representative genera are illustrated in figures 8–17. Adaptations for faster flight may include shorter antennae and a reinforced leading wing edge from a shortened R^5 vein, the major vein behind the anterior border of the wing (figs. 12–13). Secondarily shortened legs (figs. 12–13, compare with fig. 16) may allow adults to scuttle about more rapidly. To identify adults, see the generic

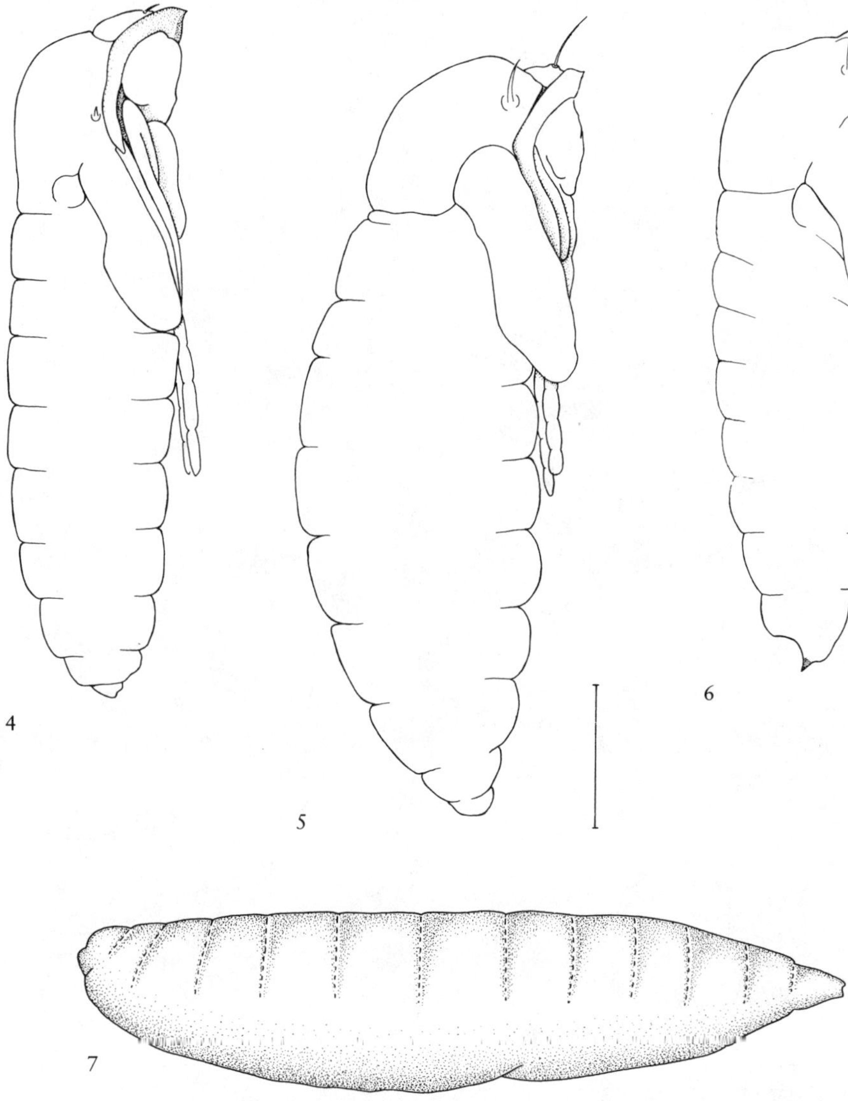

Figures 4–7. Pupae and puparium, lateral view. **4,** *Neolasioptera* sp. **5,** *Dasineura gleditchiae*. **6,** *Monarthropalpus flavus*. **7, Puparium of** *Mayetiola destructor*. Scale line = 0.5 mm.

key in Gagné (1981a). References given under the various groups in chapter 3 should also be helpful in identification.

The adult head is composed mainly of the eyes, which are joined at the top of the head. Some reduction in eye size and number of facets occurs in certain groups, particularly the Oligotrophini, in which the lateral eye

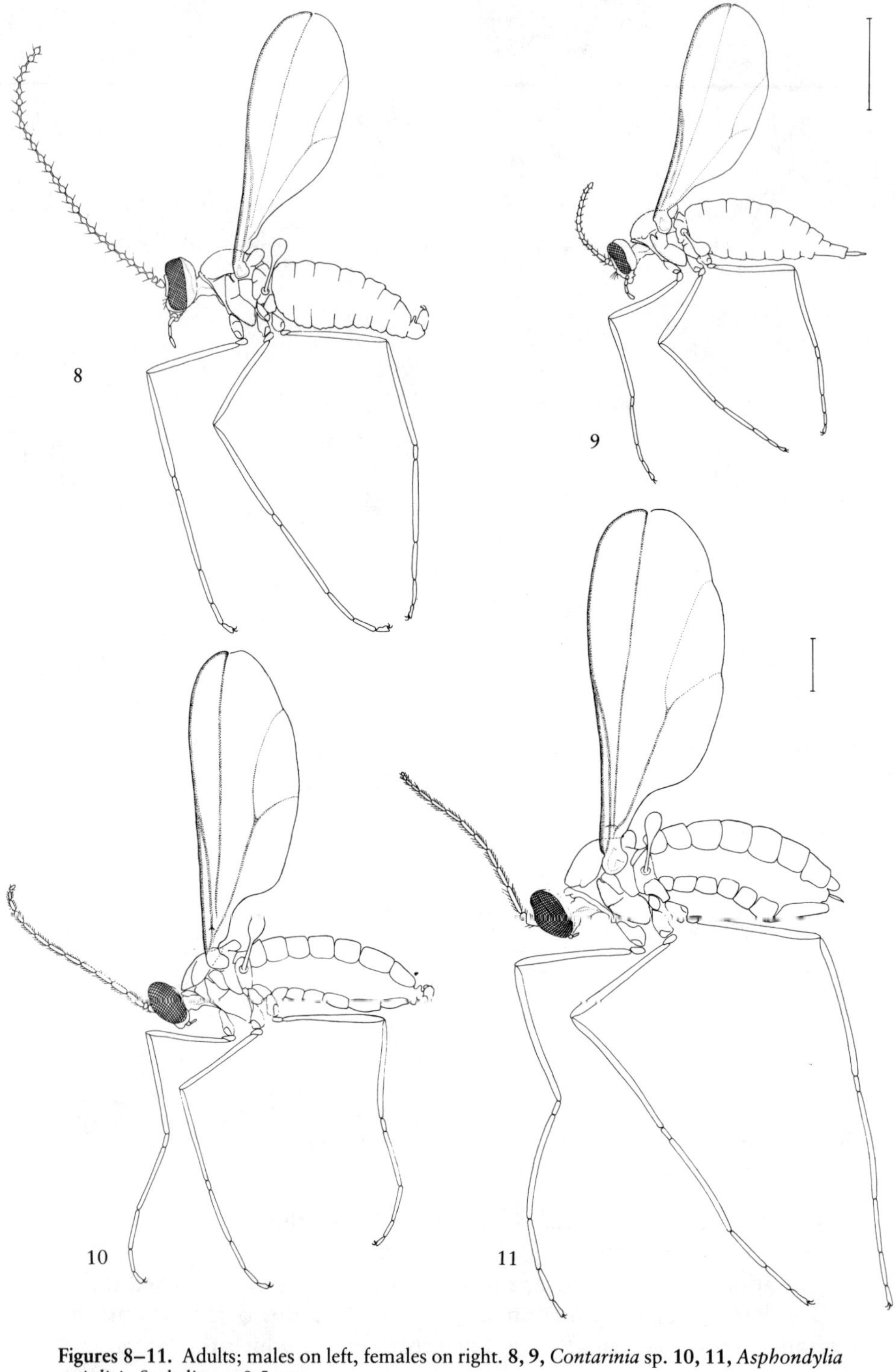

Figures 8–11. Adults; males on left, females on right. **8, 9,** *Contarinia* sp. **10, 11,** *Asphondylia atriplicis*. Scale line = 0.5 mm.

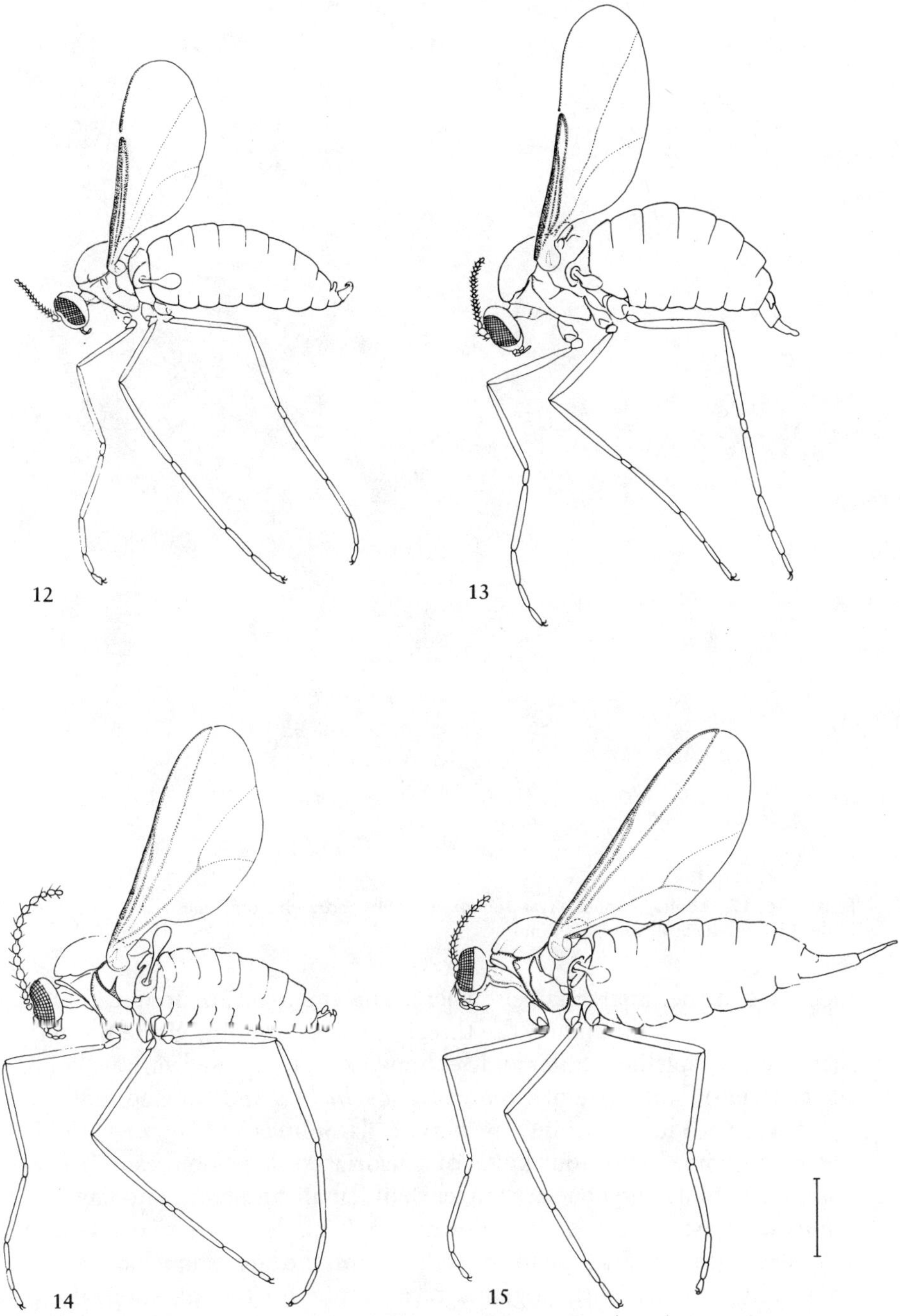

Figures 12–15. Adults; males on left, females on right. **12, 13,** *Neolasioptera lathami.* **14, 15,** *Dasineura gleditchiae.* Scale line = 0.5 mm.

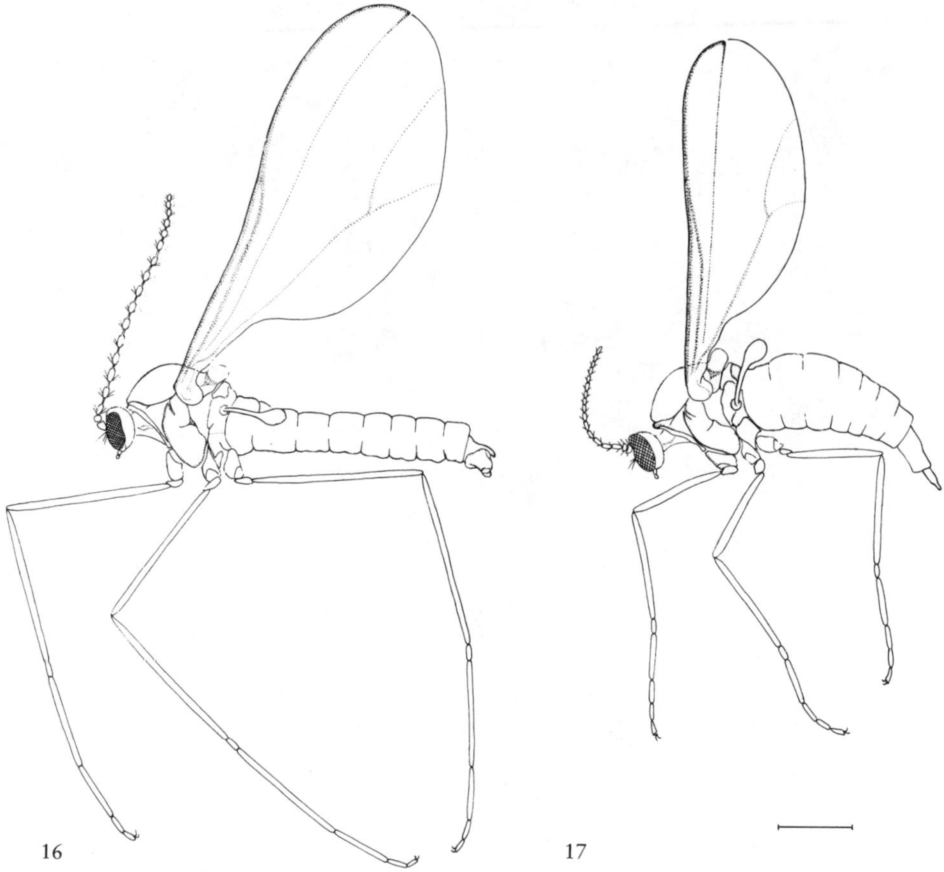

16 17

Figures 16, 17. Adults, *Rhopalomyia pomum* (partially redrawn from Jones et al. 1983). 16, Male. 17, Female. Scale line = 0.5 mm.

facets are farther apart than elsewhere. The antennae are unique among Diptera for their circumfila, the threadlike sensoria that girdle the segments of the flagellum. Slifer and Sekhon (1971) illustrated various kinds of sensoria on antennae of *Contarinia sorghicola*, and showed that circumfila are produced by bifurcate nerve cells. Solinas and Nuzzaci (1987) illustrate in fine detail four kinds of sensoria on flagellomeres of a *Mycodiplosis*. Male antennae are longer than female antennae, and flagellomeres of most male Cecidomyiidi are each made up of two separate nodes. Mouthparts include the dorsal labrum, the lateral labella, and the ventral hypoproct, which together form a tube for the ingestion of liquids, and the maxillary palpi. Some species have elongate mouthparts, corresponding with a flattened head, long neck, or other body modifications, and adapted for specialized feeding (Collin 1922, Gagné and Boldt, in press, and Gagné and Byers 1985). In others, the mouthparts are greatly reduced and are probably nonfunctional (Jones et al. 1983). Palpi are primitively four-segmented but become reduced to as few as one segment, usually in conjunction with general reduction of the remaining mouthparts.

The already simple wing with two or three major veins has undergone some modification in the family, chiefly in the length of the R^5 vein. In some Oligotrophini, the length of the R^5 vein relative to total wing length is positively correlated with body size (Sylvén and Carlbäcker 1981).

The empodium, a padlike appendage between the tarsal claws, is usually about as long as the claws. In some species, however, the empodium is always longer than the claws. Among the species with a longer empodium are most of those that live on conifers, even though the cecidomyiids of conifers are from diverse lineages. Some sexual dimorphism occurs in leg length; for example, legs of some male *Rhopalomyia* are longer than those of females (figs. 16–17).

The female postabdomen is very adaptable and has evidently become elongate many times. The progression from short to long can be seen in several monophyletic species groups and genera. The two cerci are fused into one lobe in most Lasiopteridi, and this modification has occurred separately in *Taxodiomyia* in the Cecidomyiidi. In most gall midges, the ovipositor is long-attenuate and suitable for placing eggs in crevices between bud scales or flower parts. Some ovipositors have become modified for piercing plant tissue: the ovipositor is variously shaped like a needle (*Asphondylia*), a saw (*Lygocecis*), a blade (*Procystiphora*), or a hook (*Cystiphora*, *Monarthropalpus*).

3 Classification

The first cecidomyiid described from North America was the Hessian fly. Thomas Say, who named the species in 1817, argued that it was a native American insect. But as Fitch (1846) subsequently showed, the fly had been a pest of wheat in Europe long before it was first noticed in 1778 on Long Island, New York, from where it quickly spread to all wheat-growing regions of the country.

By 1900, fewer than 150 North American gall midges had been described, but almost all of them were associated with particular damage or galls. The most noteworthy work was done by Fitch (1845, 1859, 1861), Osten Sacken (1862, 1863, 1869, 1870, 1871), and Walsh (1864), each providing careful and fine written descriptions. They occasionally added a few illustrations to their descriptions.

The generic classification remained sketchy, and most species were placed in *Cecidomyia*, then a catch-all genus. Known gall midges were not yet too numerous to manage that way because one could still identify most known species by their hosts and damage. Until 1907, the galls were seemingly as important for recognition of a species as were the anatomical characters of the gall midges themselves. Descriptions of adults and even of larvae were made, but few species in either North America or Europe were described from adults caught in flight. A fine example of this kind of work was a paper by Beutenmüller that appeared in 1907. In it were described 20 new species, each reared from a particular kind of gall on a known host and accompanied by a fine illustration of the gall and, in most cases, of a larva or part of a larva.

Appearing in that same year was a paper by Felt (1907a) that changed the development of cecidomyiid taxonomy in North America. In it, Felt perfunctorily described 179 species of gall midges, of which only 15 were reared from known hosts. The remainder had been caught in flight or at rest on plants.

Beutenmüller (1907) criticized Felt for describing species with no definite host associations and for basing names of 96 of the 179 species on plants from which the specimens were swept or in whose vicinity they were found. Felt (1907c) responded that the published descriptions were preliminary but based on more detailed knowledge than appeared in the descriptions. He wrote that some of the species were probably not plant-feeders, and indeed many of them belonged to the Lestremiinae and Porricondylinae, which feed on fungi, not plants. Felt noted also that other groups were studied on the basis of adults only. It is true that one could not forever hold the study of cecidomyiids as hostage to their hosts. Felt reasonably expected that a natural classification based on adults would encompass and anticipate all other information about the group. That may be true for some groups of insects, but not for all gall midges.

The genus *Calamomyia* illustrates the hindrance that arises from describing isolated adults. Males can readily be placed to genus but females only doubtfully so, and species are difficult to separate on the basis of one sex only (Gagné 1981b). For an adequate description, one should have larvae as well as associated males and females, in which case the host will also be known. The genus now contains 18 described species, 11 of which are each known from a single specimen caught in flight and may never be assigned with certainty to any reared species. No benefit accrues to having those names available; yet they will forever have to be carried in catalogs and worried over every time one rears a *Calamomyia* from a new host.

Felt did not respond to Beutenmüller's criticism about naming species for plants on or near which the flies were caught, but that practice gave us specific names such as *liriodendri* ("of yellow poplar") and *aceris* ("of maple") for species that belong, respectively, to genera restricted to grasses and oaks.

In time, Felt developed his preliminary descriptions only slightly more fully, in a series of work on American gall midges (Felt 1913, 1915a–b, 1916, 1918b–c, 1921, 1925), but he described additional new species in a continuous stream of short taxonomic papers between 1908 and 1939. He named 965 species from North America alone. Many of them are now known to be synonyms: 11 species of Lestremiinae were each described under four or more names; *Neolasioptera vitinea* (Felt) was described under three names, each time in a different genus. The task was clearly too large for one person with many other interests.

Chief among the problems affecting Felt's study of the gall midges was the inadequate classification that he used. The generic and suprageneric classification of the plant-feeding gall midges used even now stems in large part from the work of two Europeans, J.-J. Kieffer and E. H. Rübsaamen, who began publishing on gall midges in the late nineteenth century, and whose work culminated in Kieffer (1913) and Rübsaamen and Hedicke (1926–1939). But that classification was developed for the fauna of the Palearctic Region. Felt adapted it to the North American fauna, but groups such as the Alycaulini that are restricted to the Americas did not fit into it.

Today, several catch-all genera are still used to hold unrelated species, but these are continually being eroded. The concordance between the Eurasian and North American classifications continues to improve, and it is now clear that the North American fauna is made up of not only Holarctic but indigenous and Neotropical elements. Some large tribes cannot be divided with confidence, and they may never be divided to everyone's satisfaction or even in a truly natural way. But, clearly, knowledge of larvae and their hosts is necessary to classify adequately what is probably the largest family of flies.

Taxonomy

The Genera

Genera are subjective categories. One should not look upon any two or more as representing similar or objective criteria. For example, all genera are not equivalent in age, particularly if one compares a genus of Lestremiinae with one of the Cecidomyiinae, a much younger group. It would not be practical to base them on age even when this can be inferred. The result would be a superfluity of generic and higher taxa names for old, species-poor groups of no great interest, but a dearth of such names for more recent, species-rich, economically important groups.

Genera should be monophyletic, but that condition is often difficult to determine in the gall midges, in which similar adaptive strategies have developed many different times. Some of the most distinctive and taxonomically helpful anatomical characters, such as the loss of particular setae or the presence of an elongate ovipositor, have apparently developed many times. The larval niche, for example, correlates well with the loss of certain larval papillae, reduction of the spatula, and changes in body form (Sylvén 1975).

The truly monophyletic genera are mostly made up of species that fill a certain niche on a genus of plants. But such groups cannot always be related to other genera within their tribe. *Caryomyia*, with its many species forming complex leaf galls on hickory, certainly belongs in the tribe Cecidomyiini, but its closest relative is not apparent. Relationships even among species of *Caryomyia* are obscure. When the geographic range of hickories was severely restricted during successive glacial periods, many species of *Caryomyia* must certainly have become extinct. After having been isolated in the various refugia, those species that remained or their resulting daughter species presumably migrated north with the hosts during the interglacial periods. The species that are extant now are presumably of mixed age and provenance, which complicates attempts to make sense of the group, using the few anatomical characters available for discrimination.

Several genera, such as *Dasineura*, are evidently polyphyletic and cannot now be divided with any confidence. While monophyletic groups within those genera could be treated as separate taxa, the truth is that catch-all genera serve a practical purpose. If a new species is discovered that cannot be placed in any particular available genus, a broadly defined, paraphyletic genus can be a handy, temporary place for it. The advantage is that the name is available to report on the species' biology or illustrate some anatomical structure.

Species and Some Problems with Their Names

Several taxonomic matters should be mentioned here, some peculiar to gall midges. The names of many species are based only on galls. For example, Osten Sacken (1875) described particular leaf galls on *Urtica* as *Cecidomyia urnicola* (now in *Dasineura* in the broad sense). According to present rules of nomenclature, such a name, as long as it was proposed before 1931, is available because it is based on the work of an animal. One might counter that a gall is the work of a plant, but it is at least a manifestation of the interaction between two organisms. Even before 1931, not everyone agreed that such a practice was proper. In fact, some did not agree that a name based on the larval stage was available because that was not the "perfect form" of the gall midge (Gagné 1972). Early workers sometimes renamed species based on galls or galls and larvae, but the earlier name must now be used.

Species named without a description or other indication are not available. One such name is *Rabdophaga coloradensis*, which was used for the first time in Felt (1940) but without any description of the insect. Although its damage was described, the name is still not available for use because the book appeared after 1930.

Some specific names were coined as two separate words, for example Walsh's (1864) *Cecidomyia salicis batatas* for the gall midge that forms the potatolike stem gall on willow. But that name was stillborn, unavailable because a specific name has to be a single word based on a name of a single entity. That name became available when Osten Sacken (1878) listed it in a catalog as one word, *salicisbatatas*. Osten Sacken automatically became the author because he was the first to use the name in that form. *Cecidomyia bigeloviae-strobiloides* Townsend was available because the words were united by a hyphen. But because present rules require that the word be spelled without a hyphen, the name is written as *bigeloviaestrobiloides* and is considered available from the original description. The name proposed as *Cecidomyia pini inopis* Osten Sacken (1862) is considered valid with the original author and date; although the name was originally described as two words, it referred to a single entity, i.e., the plant *Pinus inopis*.

The Family and Its Subdivisions

The family Cecidomyiidae belongs to the order Diptera, which contains the true flies. Within that order, the family is classified within the suborder Nematocera and, further, in the infraorder Bibionomorpha, the land midges. Within the latter group, the closest apparent relatives of cecidomyiids are the fungus gnats, or Mycetophiloidea (Hennig 1973). The oldest known fossil Cecidomyiidae are from the Cretaceous Period, but none of them belongs to a plant-feeding group (Gagné 1977). The family evidently diversified with the flowering plants during the Cenozoic Era, so that, even though the family has ancient roots, most of its members are very young (Gagné 1973a). As a group with recent explosive evolution, the Cecidomyiidae are analagous to the acalyptrate Diptera, which also diversified greatly during the Cenozoic. Approximately 4,460 valid species are known, a small fraction of the number that must exist.

The Cecidomyiidae are usually divided into three subfamilies (table 1), although these are not equivalent in age. The subfamily Lestremiinae is the most primitive (fig. 18); none of this group is associated with living plants. All species feed on fungus in decaying organic matter, the primitive feeding habit of the family. Genera are mostly species-poor and wide-ranging. Many species also are widespread, an indication of great age. Of about 370 species known, 91 occur in North America. The Lestremiinae are paraphyletic to the remainder of Cecidomyiidae. This means that, while the remaining cecidomyiids are monophyletic and presumably derived from the Lestremiinae, it is not possible now to hypothesize with confidence which lestremiine is ancestral. This subfamily will not be treated further here. For a review and general information about the subfamily, see Mamaev (1968); for classification, see Kleesattel (1979); for a key to Nearctic genera, see Gagné (1981a).

The two remaining subfamilies are the Porricondylinae and the Ceci-

Table 1. Cecidomyiid genera mentioned in this book, listed according to family group categories. Subfamilies end in "inae," supertribes in "idi," tribes in "ini," and subtribes in "ina." All currently used suprageneric divisions of the subfamily Cecidomyiinae are listed.

- Lestremiinae
- Porricondylinae
  - Asynaptini
    - *Asynapta*
    - *Camptomyia*
- Cecidomyiinae
  - Cecidomyiidi
    - Aphidoletini
      - *Aphidoletes*
      - *Monobremia*
    - Asphondyliini
      - Asphondyliina
        - *Asphondylia*
      - Polystephina
        - *Kochiomyia*
        - *Polystepha*
      - Schizomyiina
        - *Bruggmannia*
        - *Schizomyia*
        - *Stephomyia*
    - Cecidomyiini
      - *Apagodiplosis*
      - *Blaesodiplosis*
      - *Caryomyia*
      - *Cecidomyia*
      - *Contarinia*
      - *Halodiplosis*
      - *Harmandia*
      - *Macrodiplosis*
      - *Monarthropalpus*
      - *Paradiplosis*
      - *Pinyonia*
      - *Planetella*
      - *Plemeliella*
      - *Prodiplosis*
      - *Putoniella*
      - *Resseliella*
      - *Sequoiomyia*
      - *Taxodiomyia*
      - *Thecodiplosis*
      - *Zeuxidiplosis*
    - Clinodiplosini
      - *Ametrodiplosis*
      - *Clinodiplosis*
      - *Hyperdiplosis*
      - *Parallelodiplosis*
      - *Sitodiplosis*
    - Lestodiplosini
      - *Cartodiplosis*
      - *Feltiella*
      - *Lestodiplosis*
    - Mycodiplosini
      - *Mycodiplosis*
    - Other Cecidomyiidi
      - *Ctenodactylomyia*
      - *Karshomyia*
      - *Lobodiplosis*
      - *Obolodiplosis*
      - *Olpodiplosis*
      - *Pilodiplosis*
      - *Youngomyia*
  - Lasiopteridi
    - Alycaulini
      - *Asteromyia*
      - *Astictoneura*
      - *Calamomyia*
      - *Chilophaga*
      - *Edestochilus*
      - *Edestosperma*
      - *Meunieriella*
      - *Neolasioptera*
      - *Protaplonyx*
    - Brachineurini
    - Lasiopterini
      - *Hybolasioptera*
      - *Isolasioptera*
      - *Lasioptera*
      - *Ozirhincus*
    - Ledomyiini
      - *Ledomyia*
    - Oligotrophini
      - *Acericecis*
      - *Camptoneuromyia*
      - *Celticecis*
      - *Craneiobia*
      - *Cystiphora*
      - *Dasineura*
      - *Ficiomyia*
      - *Iteomyia*
      - *Janetiella*
      - *Kaltenbachiola*
      - *Lygocecis*
      - *Mayetiola*
      - *Oligotrophus*
      - *Procystiphora*
      - *Rabdophaga*
      - *Rhopalomyia*
      - *Sackenomyia*
      - *Semudobia*
      - *Walshomyia*
    - Rhizomyiini
    - Trotteriini
      - *Trotteria*
  - Stomatosematidi
    - *Didactylomyia*
    - *Stomatosema*

domyiinae. These are related on the basis of several shared, derived characters, most conspicuous of which is the very short, first tarsal segment (see figs. 8–17). The connection between the short first and the longer second segment is easily broken. This allows long-legged gall midges to detach themselves readily when a leg is caught on a sticky substrate. That modification may have evolved as an adaptation for passive roosting on spider webs, a habit of Porricondylinae and many nonphytophagous Cecidomyiinae. Cecidomyiids that one finds on spider webs are relatively long-lived, do not emerge from the pupal stage synchronously, and have many successive generations over a season. Some

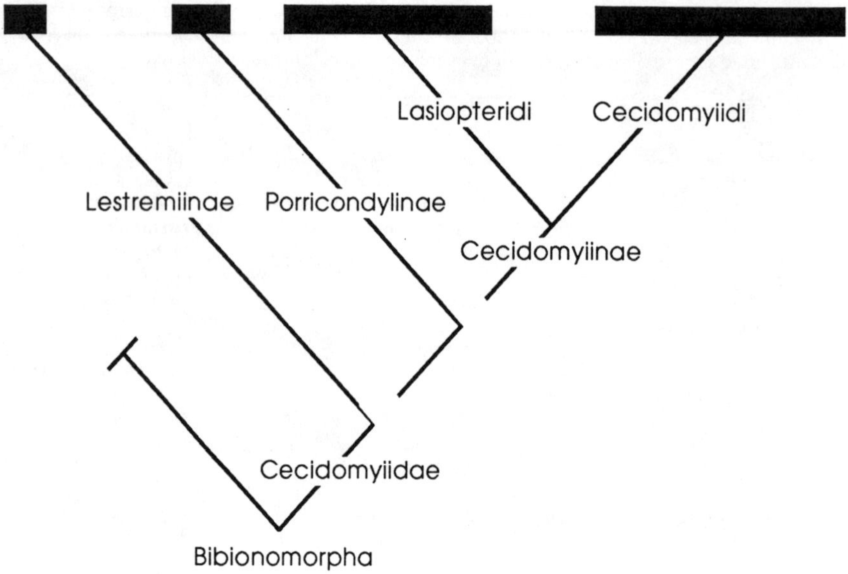

Figure 18. Simplified phylogenetic scheme of the Cecidomyiidae. Lines between subfamilies are broken to show that each of the two basic subfamilies is paraphyletic to the remaining cecidomyiids. The boxes at the top of the chart are proportionate to the number of described species in the various subfamilies or supertribes.

species roost at regular times (Kaufman 1973, Sivinski and Stowe 1981, Young 1985). Adults of at least *Didactylomyia longimana* (Felt) are no longer merely passive inhabitants of the webs; they evidently share the spider's prey (Sivinski and Stowe 1981). Passive roosting is a habit apparently no longer needed by the shorter lived plant-feeders, although they still have the short first tarsal segment.

As the Lestremiinae are paraphyletic to the remainder of the family, so the Porricondylinae are to the Cecidomyiinae (fig. 18). Most if not all Porricondylinae share with the Lestremiinae the primitive fungus-feeding habit. The genera are distinct, species-poor, and widespread, indicating a long evolutionary history for this subfamily also. Of approximately 475 described species, 83 are known from North America. Only two porricondyline genera, *Asynapta* and *Camptomyia*, are treated further here. Each has some species regularly found associated with living plant tissue, although they may be only secondary feeders on plants. For further information on Porricondylinae in general, see Mamaev (1968), Panelius (1965), and Wyatt (1967); for a key to Nearctic genera, see Gagné (1981a).

Cecidomyiinae make up the youngest subfamily of gall midges. It is tremendously diverse and rich in species. Of about 3850 described species, 968 are known from North America. Many more remain undescribed. The subfamily contains some primitive fungus-feeders, but mainly plant-feeders and predators. The large number of plant-feeders is due to the fact that most of these cecidomyiids are narrowly adapted and synchronized to their hosts. Some of the shared, derived characters of the Cecidomyiinae are the reduction of the R^5 vein, the reduced number of 12 or 13 flagellomeres, the loss of a ventral bridge between the gonocoxites, and the loss, on the larva, of one pair of dorsal papillae on the eighth abdominal segment. Two monophyletic supertribes, the Lasiopteridi and the Cecidomyiidi, make up most of the Cecidomyiinae (fig. 18). Two genera that remain cannot easily be placed in either of those supertribes and have been separated in another supertribe, the Stomatosematidi (Gagné 1975b), but these are not plant-feeders and are not considered further here.

The Lasiopteridi (fig. 18) comprise approximately 1,700 described species, 475 from North America. A few tribes, the Brachineurini, Ledomyiini, and Rhizomyiini, are fungus-feeding and species-poor. Three other tribes—the Alycaulini, Lasiopterini, and Oligotrophini—are made up almost entirely of plant-feeders. Some Oligotrophini produce structurally complex galls, some produce simple leaf or bud galls, still others live freely in flower heads. Alycaulini and Lasiopterini live mainly in stem swellings. Trotteriini, another tribe, are inquilines in galls made by *Asphondylia* spp.

The Cecidomyiidi (fig. 18) are a larger and more diverse group than the Lasiopteridi. Of about 2,100 described species, 555 are known from North America. Several tribes comprise many kinds of general and specific fungus-feeders or species that are predaceous on insects and mites. Plant-feeding itself probably developed several times in this supertribe. Most Clinodiplosini in the Holarctic Region are polyphagous fungus-feeders, but some appear to be monophagous fungus-feeders, and a few feed in and kill plant buds. In the Neotropical Region, however, many Clinodiplosini form complex galls and are host specific. The other plant-feeding Cecidomyiidi are extremely diverse and rich in species and occur on most groups of gymnosperms and angiosperms. Some live freely in flower heads and others form simple and complex galls.

Synopsis of Genera Found on Plants

In the synopsis that follows, the genera within a tribe are listed in alphabetical order. To find a particular genus it is best to use the index. There, the italicized number will refer to the page in this synopsis.

Information for most genera includes the number of described species, biological and economic importance, distribution, relationships, the more distinctive recognition characters, and the best sources for further information.

For further study of many of these genera and higher categories, one will have to depend on available catalogs and on citations listed in them. For a list of taxa of the Nearctic Region, see Gagné (in prep.); for the Palearctic, Skuhravá (1986); the Neotropical, Gagné (1968b); the Afrotropical, Harris (1980); the Oriental, Gagné (1973d); the Australasian and Oceanian, Gagné (in press). Other general works not specifically mentioned farther on in this chapter, except where pertinent under some tribes, include: Gagné (1981a; q.v. also for further discussion of anatomical characters), Harris (1968), Kieffer (1913), Mamaev (1968), Mamaev and Krivosheina (1965), Möhn (1955), Rübsaamen and Hedicke (1926–1939), and Yukawa (1971).

Subfamily Porricondylinae

Most species of this subfamily feed on fungi growing in decaying organic matter. Only *Asynapta* and *Camptomyia* include species regularly associated with plant tissue. These two genera belong to the same tribe, the Asynaptini. Refs.: Mamaev (1968), Panelius (1965).

Tribe Asynaptini

Asynapta. This cosmopolitan genus presently includes 33 described species, 11 from North America. Several live among the scales of conifer cones in Europe and North America, and one lives in prickly pear cactus, *Opuntia* spp., in the American Southwest. It is possible that these attack plant tissue after it has already been damaged. Larvae are white, long and slender, and lack a spatula (fig. 27). The end of the adult abdomen is characteristically held in a dorsally recurved position. The cubital vein is unforked. Refs.: Panelius (1965), Parnell (1971).

Camptomyia. This genus is also cosmopolitan and includes 48 described species, 6 of them in North America. Many live under dead bark of various trees, but a few live among scales of conifer cones. Larvae are white or yellow, and usually have a narrow, short spatula as well as a pair of hooklike lobes on the terminal segment (figs. 40, 41). Adults superficially resemble those of *Asynapta* except that the cubital vein is forked. Refs.: Panelius (1965), Parnell (1971).

Subfamily Cecidomyiinae: Supertribe Lasiopteridi

The more primitive forms of this group are fungus-feeders, although they account for only a small proportion of the supertribe. Most Lasiopteridi have an irregular number of flagellomeres, seldom numbering exactly 12. In addition, male flagellomeres are primitively made up of a node and a distal neck that is longer than on female antennae. The cerci of the ovipositor are most often fused into one caudal lobe. The male aedeagus is supported on each side by gonocoxal lobes. Most ventral papillae of the larvae are setose. Ref.: Sylvén (1975).

Tribe Alycaulini

Restricted to the western hemisphere, this tribe contains 102 described species from North America and approximately the same number from Central and South America (Gagné 1968b, Wünsch 1979). These gall midges are found mainly in stem swellings, although some live in leaf petiole and vein swellings, even blister galls on leaves, and a few in the flower receptacles and achenes of Asteraceae. Larvae pupate in the galls, and most species have a fungal symbiont, although some species and entire genera, e.g., *Chilophaga*, apparently do not. Adults (figs. 12, 13) are characterized by short antennae, compact bodies, a very short R^5 vein, and a covering of scales that usually forms striking light and dark patterns. Larvae (figs. 29, 59–64, 66–67) have reduced lateral papillae and usually a strong spatula, serviceable for cutting a tunnel through woody stem swellings. In all these characters the tribe closely resembles the Lasiopterini, but whether the characters are convergent cannot be determined at this time. The main difference between the two tribes lies in the structure of the ovipositor and sclerotization of the male abdominal tergite 7. Both types of ovipositor may have been derived separately from a simple one such as is found in *Dasineura*. Refs.: Gagné (1969c, 1981a), Möhn (1964a, b, 1975), and Wünsch (1979).

Asteromyia. Eight valid species of this North American genus are known. All form galls on plants of the tribe Astereae (Asteraceae). Except for *A. tumifica* (Beutenmüller), which occurs at the base of goldenrod stems, all are responsible for blister galls on leaves and green stems. Most have a fungal symbiont responsible for forming a black, hard cast around full-grown larvae. This genus is characterized by short palpi and antennae. The ovipositor is also short for the tribe. Larvae have a reduced number of papillae, and the spatula is short or lost (figs. 60, 61). Refs.: Gagné (1968a, 1969a), Weis (1982c).

Astictoneura. This genus contains one North American species that causes an imbricated, broomlike stem gall on a grass (*Muhlenbergia*). The gall is similar to those made on other grasses by *Chilophaga* spp. This is the only alycauline without light-colored scales at the juncture of veins C and R^5. Larvae have no spatula and, unlike larvae of *Chilophaga*, have an unmodified terminal segment. Ref.: Gagné (1969c).

Calamomyia. This North American genus contains 18 described species. All probably live in stems of grasses, but only six have been definitely associated with particular grasses. Most others were caught in flight and are known from a single specimen. Two distinct species live on separate species of *Spartina*, so the actual number of species must be large. Hosts are distributed among five tribes of Poaceae. Larvae may be single or gregarious. When full-grown, they form silken cocoons in the stems, where they pupate. Adults of some species have been reared in fall, others in spring. Infested grasses show no external sign of damage unless their fungal associate shows through stem tissue as a black discoloration. Infested stems break easily in winter. Adults have black and white bands of scales on the tarsi, and male abdominal tergites have peculiar ventrolateral extensions. Larvae have a tridentate spatula and an unmodified terminal segment (figs. 66, 67). Refs.: Gagné (1969c, 1981b).

Chilophaga. This genus contains four described North American species, three that form characteristic broom galls on grasses and one known only from a specimen caught in flight. Larval feeding causes a foreshortening of the stem that results in a broomlike gall. Larvae live at the base of the leaf sheaths and pupate there. Adults have banded legs. The spatula of the larvae is either short or absent, and the terminal segment has two caudal projections (fig. 59). Refs.: Gagné (1969c), Gagné and Stegmaier (1971).

Edestochilus. The one known species in this genus causes an onionlike, basal swelling on stems of a grass. Adult palpi are reduced to three segments. Males have slightly sclerotized mediobasal gonocoxal lobes. Larvae lack a spatula and have two characteristic caudal appendages and a blunt projection between them (fig. 62). Ref.: Gagné (1969c).

Edestosperma. This genus contains only one known species, which lives in seeds of a grass. The ovipositor is bilaterally flattened and rigid, and the cercus is slightly sawtoothed ventrally. The larva has a clove-shaped spatula and an unmodified terminal segment. Ref.: Gagné (1969c).

Meunieriella. This genus is rich in species in Central and South America, but only one species, *M. aquilonia* Gagné, has been described from North America. That species is evidently the only one responsible for its own gall; the others are associated with other cecidomyiids and may be inquilines. Males have peculiar terminalia with extremely narrow and elongate gonopods. Refs.: Gagné and Valley (1984), Möhn (1975), and Wünsch (1979).

Neolasioptera. This is a catch-all genus containing 64 described North American species. It includes mostly stem gall makers on many families of plants, but also includes species that attack achenes of Asteraceae. At least one, an undetermined species reared from galls of *Lasioptera ephedricola* Cockerell on *Ephedra*, appears to be an inquiline. At present nothing can be said with great confidence about relationships within the group. A pupa is illustrated in figure 4 and adults in figures 12 and 13. Larvae in woody stems usually have especially strong spatulas, which are used in cutting an exit channel (figs. 29, 63, 64, also as in 65). Refs.: Gagné (1969c), Möhn (1964a, b, 1975), Wünsch (1979).

Protaplonyx. The only known species forms blister galls on leaves and green stems of a chenopod in the Great Basin. The ovipositor is modified for piercing plant tissue. Ref.: Gagné (1969c).

Tribe Lasiopterini

This tribe is represented in North America by 37 described species, but is much more numerous in the Old World, with 220 known species. Many of them attack desert Chenopodiaceae in Asia. The tribe is not known from South America, but *Isolasioptera* occurs as far south as El Salvador. Larvae pupate in the galls, although a species in Japan is an exception (Yukawa 1983), and most species have a fungal symbiont. Some species are inquilines, and it appears that a higher proportion of New World than of Old World species are inquilines. This tribe has distinguishing characters similar to the Alycaulini and differs only in characters of the male and female postabdomens. Ref.: Gagné (1969c).

Hybolasioptera. This genus contains one European and one American species. Both live in stems of rye native to their respective continents. The ovipositor is partially sclerotized and strongly flattened bilaterally; its cerci are completely sclerotized. Larvae lack a spatula. Refs.: Gagné (1969c), Möhn (1966–1971).

Isolasioptera. This genus is known from one species in swollen veins of *Ageratina* (Asteraceae: Eupatoriae) from Mexico to El Salvador. It is the southernmost genus of Lasiopterini in the Western Hemisphere, and has some notable, primitive character states, namely the longest R^5 vein of this tribe and an entirely sclerotized male seventh abdominal tergite. Ref.: Möhn (1964b).

Lasioptera. This is a diverse genus of 34 described North American species. Most are found in stems, where some are inquilines. A few species live in leaf spots or fruit. No American species occurs on grasses, although many Old World species do. The 88 described Old World *Lasioptera* spp. are mostly all gall makers. As with *Neolasioptera* in the tribe Alycaulini, *Lasioptera* currently serves as the catch-all genus of Lasiopterini. Some larvae are similar (fig. 65) to those of *Neolasioptera.* Refs.: Gagné (1969c), Möhn (1966–1971).

Ozirhincus. An immigrant that lives in achenes of the introduced weed *Achillea millefolium* (Asteraceae) is the only representative of this genus in North America. It and 12 other species are known from Europe, all from achenes of Anthemideae (Asteraceae). The distinctive long mouthparts and associated adaptations of the head and thorax evidently allow the adults to feed in flowers of their host. Refs.: Gagné (1969c), Möhn (1966–1971).

Tribe Ledomyiini

This tribe contains 15 North American species, which for convenience and for the time being are placed in *Ledomyia* (Gagné 1975b). Some are known to be mycophagous (Larew et al. 1987). Those that have been reared from plants are recorded in chapter 8, although they may have been only secondary feeders in galls formed by other cecidomyiids. Adults have a regular number of flagellomeres in each species, from 8 to 12; female cerci are not fused; and larvae bear 10 dorsal papillae rather than the six found in all other Lasiopteridi. Ref.: Gagné (1985).

Tribe Oligotrophini

Diverse genera are placed under this tribe only because they lack characters to sort them convincingly into definite subgroups. Contained here are genera with primitively divided cerci and others with fused cerci; the primitive number of 12 flagellomeres or, as is usual, an irregular number; toothed or simple claws and both toothed and simple claws.

Toothed claws are presumably ancestral to simple claws, but it is also apparent that tarsal teeth have been lost many times and not necessarily in association with other character changes. Representative adults are illustrated in figures 14–17 and pupae in figures 2 and 5.

Acericecis. Only one North American and one European species are known for this genus. Both occur on maple, but the American *A. ocellaris* lives in the open on the underside of leaves, while the European *A. vitrina* (Skuhravá and Skuhravý) lives inside leaf blister galls. Pupation occurs in the ground. Adult antennae have a regular number of 12 flagellomeres, the R^5 vein is long and the costa unbroken at its juncture with R^5, tarsal claws are simple, and the ovipositor is short with separate cerci. Larvae are ellipsoidal, have very short setae, a reduced number of papillae, and no spatula (figs. 22, 32). Refs.: Gagné (1983b), Skuhravá and Skuhravý (1986).

Camptoneuromyia. Six Nearctic and nine Neotropical species are known. The Palearctic *C. petioli* Mamaeva is evidently a species of *Ledomyia* in the strict sense, leaving *Camptoneuromyia* a strictly New World genus. All camptoneuromyias may be inquilines in galls made by other cecidomyiids. Larvae pupate in or near the galls. An undetermined species has been observed feeding on fungus immediately below the epidermis of *Asteromyia carbonifera* galls on goldenrod (Weis 1982b). Males have short flagellomeres with no necks and a short R^5 vein, conditions that doubtless occurred many times in the Lasiopteridi. The ovipositor resembles that found in *Dasineura.* Larvae have a full complement of papillae and resemble *Dasineura* larvae (figs. 34–36). Refs.: Gagné (1969c), Möhn (1975).

Celticecis. This genus is known from North America, Japan, and Turkey (pers. observ.) and is restricted to hackberries, *Celtis* spp. All species form complex, dehiscent leaf and stem galls. Pupation occurs in the galls. Of the 10 described American species, five are known only from the galls. One kind of gall may occur on more than one species of hackberry. Tarsal claws are toothed, the R^5 vein is relatively long, reaching to the apex of the wing, and the female cerci are fused. Larvae have a short, broad spatula and reduced papillae, only four terminal papillae present (figs. 37, 38). Refs.: Gagné (1983a), Yukawa and Tsuda (1987).

Craneiobia. One Nearctic and one Palearctic species are placed here. Both form tubelike galls on leaves of *Cornus.* Adults resemble some *Dasineura* with their toothed claws, slightly shortened R^5 vein, and long ovipositor. The larvae, however, have a pebbled skin and have lost some lateral papillae and the setae from the ventral papillae of the eighth abdominal segment (figs. 51, 52).

Cystiphora. This is a Holarctic genus with one species native to North America and six to Europe. Two European species have been introduced to western North America for the control of noxious range weeds. All species live in blister galls in stems and leaves of the asteraceous tribe Cichoriaeae. Larvae pupate in the galls. The female eighth segment is bulbous, and the fairly short ovipositor is telescoped and tapered abruptly to a point suitable for inserting eggs directly into plant tissue (DeClerck and Steeves 1988). Larvae have a clove-shaped spatula but a reduced number of papillae. Refs.: Gagné (1975d), Harris (1975).

Dasineura. Conveniently, no one now knows what the type species of *Dasineura*, the largest genus of Cecidomyiidae, looks like, so the genus can continue to be a catch-all category. The type species was caught in flight and is presumed lost. It is possible that it does not fit the general concept of *Dasineura.* In this book the genus is used for simple and complex gall makers, but mainly for leaf rollers or those species that live freely in flowers or flower heads. The genus now contains 110 Nearctic species. *Dasineura* spp. are generally defined as Oligotrophini with usually a short R^5 vein, a long, soft ovipositor with fused cerci, and toothed claws. Most otherwise have no characters conspicuous enough to distinguish them readily. Exceptions include *D. rhois* (Coquillett), which, with its divided, free female cerci and short ovipositor, will eventually need to have a genus erected for it. *Dasineura* does contain apparently monophyletic groups. One such lives in leaf rolls of Fabaceae. Some *Dasineura* spp. e.g., *D. balsamicola*, are inquilines and require a gall maker from another genus to initiate a gall on a particular host. Several other species, two on grasses and two on clovers, are European immigrants. Representative larvae are illustrated in figures 19, 34–36, 55, 56, a pupa in figure 5, and adults in figures 14 and 15. Ref.: Sylvén (1975).

Ficiomyia. Known from only two species from figs in southern Florida, this genus is quite different from other North American Lasiopteridi. The antennae have more than 30 flagellomeres, eyes are very large, mouthparts are reduced, gonocoxites are lobed, and the body including the wings is patterned with dense, alternately light and dark patches of scales.

Iteomyia. This Holarctic genus contains several species that form spherical or conical leaf galls on *Salix* spp. Two species have been described from North America, two from Europe. Females have a long, telescoped ovipositor with a short, fused cercus, but the eighth tergite is not longitudinally divided and is barely longer than the seventh. The male genitalia resemble those of *Oligotrophus* in that the aedeagus is recurved and much longer than the claspettes. The tarsal claws are untoothed. Larval papillae are short-setose and the spatula is lost.

Janetiella. This is a catch-all Holarctic genus with 14 Nearctic species. It is so convenient to have such a category for *Dasineura*-like species with simple claws, that until genera are described for the various species groups now contained here, one ignores the fact that none of the included Nearctic species has any apparent relationship to the type species, the European *J. thymi* (Kieffer). Most species now placed here fit into monophyletic groups for which new genera will eventually be described. Among these groups is a series of species that cause needle galls on *Pinus edulis* and *Pinus monophylla* in western North America, a series in conifer cones, and a series of species that cause various swellings and leaf galls on grape. In fact, species in this last group, although their tarsal claws are simple, are more closely related to other species presently listed under *Dasineura*, placed there because they have toothed claws. The claws being simple or toothed can be seen to be a superficial character in these species on grape, all of which have a characteristic, dorsally recurved ovipositor. The group on grape, at least, is Holarctic, including as it does the European *J. oenophila* (Haimhoffen). Most American species placed in *Janetiella* live in galls, so many have reduced larval papillae and spatulas (figs. 48–50, 53, 54).

Kaltenbachiola. This genus contains two North American and one European species, all from spruce cones. Larvae have reduced lateral papillae and no spatula. Females have long, narrow sclerites on the seventh abdominal segment and an elongate, divided eighth tergite. As in *Dasineura* in the broad sense, the R^5 vein is short, the tarsal claws are toothed, and the fused female cercus is long.

Lygocecis. This Holarctic genus comprises at least four American species, two of them undescribed, and one undescribed Japanese species, all from woody stem galls of willows. Larvae pupate in the galls. These species have a strongly sclerotized, bilaterally flattened, sawlike ovipositor, presumably used for inserting eggs into willow shoots. The postabdomens of both sexes are highly modified. Male genitalia have very long, glabrous claspettes, long gonocoxites, and short gonostyli. Tarsal claws are untoothed. The spatula is very large, strongly sclerotized, and has many teeth (fig. 44). Larval papillae are reduced in number.

Mayetiola. Strictly defined, this is a chiefly Palearctic genus with 27 European species, of which one, the Hessian fly (*M. destructor*), was introduced into North America. The only native North American species is *M. ammophilae* Gagné. All species live in stems of grasses, specifically the Festucoideae (our Festuceae, Hordeae, Avenae, Agrostideae). Unlike most cecidomyiids, the third larval instar does not feed. It develops inside the skin of the second larval instar, and the pupa later develops within the same skin. The puparia resemble seeds in shape but are dark brown (fig. 7). Adults are fairly large and have a long R^5 vein, claws with short teeth, and a fairly short ovipositor with fused cerci. Larvae have a slightly reduced spatula but a full complement of papillae. Ref.: Ertel (1975).

"Mayetiola." Species combined with the name *Mayetiola* in quotation marks have no genera to accommodate them now. These species have no particularly close connection to *Mayetiola* on grasses or to one another. The species placed here usually have simple or toothed claws, but always a long R^5 vein that reaches the wing apex. Other than that they are quite variable and live on very different hosts.

Oligotrophus. Three species, all gall formers on juniper, have been described from the Nearctic Region. *Oligotrophus betheli* Felt is a serious pest of ornamental junipers. The genus is represented in Europe and Japan by five species on junipers. Other species from various hosts have been described in *Oligotrophus*, but the genus is now considered to include only species that attack junipers. In addition, the genus probably includes the European *Schmidtiella gemmarum* Rübsaamen, the three Palearctic species of *Arceuthomyia*, and the Japanese *Rhopalomyia uetsukii* Inouye, which form branch tip or fruit galls on junipers. All of these taxa were differentiated from *Oligotrophus* chiefly by the number of palpal segments, an unimportant criterion. Houard (1908) illustrated many European juniper galls presumably formed by *Oligotrophus* spp., and White (1950) illustrated an assortment of galls from Texas. The genus evidently has some affinities to *Walshomyia.* The two genera appear to be an early offshoot from the rest of the tribe, before the ovipositor became elongate and the cerci became joined into one lobe. Their palpi have a reduced number of one to three segments (with spiniform setae in *Oligotrophus*), the tarsal empodia are conspicuously longer than the tarsal claws, and the male terminalia are exceptionally large with a dorsally recurved aedeagus that is usually much longer than the attendant claspettes. Larval papillae are reduced in number and the spatula is absent. Each of the derived characters except the peculiar shape of the male genitalia is shared, probably independently so, with various other Oligotrophini.

Procystiphora. This genus is known from two North American and one European species, all associated with stems of *Juncus* (Juncaceae). Females have a bilaterally flattened, strongly sclerotized ovipositor suitable for inserting eggs in plant tissue. The eighth tergite is longitudinally divided. Tarsal claws are toothed and shorter than the empodia. Larvae have a full complement of papillae, but the terminal segment is modified into two lobes, each bearing four short terminal papillae (fig. 39). The spatula is reduced in size. Refs.: Gagné (1975d), Meyer (1984).

Rabdophaga. This is a Holarctic genus of 40 North American and about 35 European species that attack buds and stems of willows. One exception is *R. cephalanthi* Felt on *Cephalanthus*, which is retained in this genus because it appears to fit. *Rabdophaga* spp. constitute the preponderant group of cecidomyiids on willows, on which they cause a great variety of galls, including woody stem swellings that on other plants in North America are most often the domain of *Neolasioptera*. Pupation takes place in the galls. Only a few Nearctic species have been studied in detail, although the European species have received a lot of attention. Body size and number of flagellomeres can vary greatly in *R. rigidae* (Osten Sacken) and are dependent in part on the species or vigor of the host (Smith et al. 1983). Inasmuch as most American species were differentiated on the basis of the number of flagellomeres, many are probably synonyms. It is possible that *Rabdophaga* as presently defined is polyphyletic. The bud gall makers, e.g., *R. strobiloides* (Osten Sacken) and *R. rigidae*, are distinct from some of the stem gall makers taken as a group. The bud gall makers are generally larger, have a short ovipositor with a rounded cercus, and a short larval spatula (fig. 42). Tarsal claws may be toothed or not, depending on the species. Some of the stem gall makers are small and have a long ovipositor with an attenuate cercus and a full-shafted spatula (fig. 43). Refs.: Nijveldt and Yukawa (1982), Stelter (1982), Walsh (1864).

Rhopalomyia. This genus has 126 described species distributed over the Holarctic Region and southern South America. Eighty-six are known from the Nearctic Region. Almost all occur on Asteraceae, where they form complex galls. Those that form galls on other plant families are placed here on the basis of some resemblance to those on Asteraceae, but need to be restudied. Larvae of many species have lost the spatula and many papillae. Adults have reduced palpi and some dimorphism, male legs and wings often being more slender and longer than those of females. The more than 60 Holarctic species form a monophyletic assemblage that has been segregated into the subgenus *Diarthronomyia* (Gagné 1975c). Those species are restricted to *Artemisia* and *Chrysanthemum* (Asteraceae) and differ from the remainder of the genus by having only caudal setae on the adult abdominal terga. A pupa is illustrated in figure 2 and adults in figures 16 and 17. Refs.: Gagné (1975c), Jones et al. (1983).

Sackenomyia. This is a Holarctic genus known from three Nearctic and two Palearctic species that form blister galls on leaves of *Viburnum.* Larvae are very active when disturbed. The ovipositor is short but dorsally recurved and bilaterally flattened for inserting eggs into plant tissue. Larvae have a spatula but a reduced number of papillae (figs. 45, 46). Ref.: Gagné (1975d).

Semudobia. This genus contains five species, all occurring in North America. Two of them are exotic species that were introduced from Europe in seed of ornamental birches. All live in birch seeds or bracts. Larvae have a diminutive spatula and reduced papillae (figs. 57, 58). Adults have three-segmented palpi and toothed tarsal claws. The female eighth tergite is undivided longitudinally, and the fused cercus is fairly short. Refs.: Roskam (1977), Roskam and Van Uffelen (1981).

Walshomyia. This genus, known from five described western Nearctic species and found in bud and fruit galls of *Juniperus*, *Cupressus*, and *Libocedrus* (Cupressaceae), has much in common with *Oligotrophus.* Abdominal sclerites of *Walshomyia* spp. are extensive and strongly sclerotized. Palpi, reduced to one or two segments, do not have the spiniform setae found in *Oligotrophus.* See comments under *Oligotrophus* for generalities concerning *Walshomyia.* Ref.: Gagné (1969a).

Tribe Trotteriini

This tribe includes only *Trotteria*, a genus of almost cosmopolitan distribution. Two species are known from North America, five from Central and South America, fourteen from Europe, and one from Africa. The genus is peculiar, and its affinities within the Lasiopteridi are unclear. Larvae are inquilinous in galls formed by Asphondyliini. They will cause the death of the gall-making larva if the latter is not far enough developed when the *Trotteria* begins feeding. The one to several larvae in each gall pupate in the galls. *T. caryae* (Felt) from North America was reared from galls of *Asphondylia monacha* Osten Sacken and *Asphondylia thalictri* Felt. Adults are covered with thick, black and white, sometimes iridescent scales. Male and female flagellomeres are similar, shorter than wide and without necks, and vein R^5 is fairly short and bowed. The head is modified, with the eyes placed far ventrad. Legs bear spinous setae, and the hind coxa and femur are enlarged. The ovipositor is elongate; in most species the fused cercus is attenuate and needlelike, but in *T. karnerensis* Felt the ovipositor resembles that of a *Dasineura.* Larvae have robust mouthparts and a full complement of papillae. Adults resemble in gross aspect Lasiopterini and Alycaulini because of the short flagellomeres, without necks in males, the short R^5 vein, and the scale covering, all characters that could be separately derived. The ovipositor of *Trotteria* has none of the unique modifications of either of those two tribes. Refs.: Gagné (1969c, 1984), Möhn (1966–1971, 1975), Parnell (1964).

Subfamily Cecidomyiinae: Supertribe Cecidomyiidi

This group contains many fungus-feeders and predators not referred to in this book because they are not associated with plants. The plant-feeding Cecidomyiidi possibly developed that habit separately. Members of this tribe typically have 12 flagellomeres. Male flagellomeres are binodal (except in most Asphondyliini) and their circumfila are looped; flagellomeres of the female are cylindrical. Most adults have a long, narrow lobe at the top of the head behind the eyes, which bears a pair of large proclinate setae. Female cerci are free, although they are sometimes closely appressed or so reduced as to seem fused. The aedeagus is free and appears to be more sclerotized than in the Lasiopteridi. Ventral papillae of the larvae are generally asetose.

Tribe Aphidoletini

This tribe contains two genera, *Aphidoletes* and *Monobremia*, united not so much on anatomical similarities as because they both include predators of aphids. Their relationship with other Cecidomyiidi is unclear; they are quite unlike other predaceous genera of the supertribe. Ref.: Harris (1966).

Aphidoletes. This genus contains three known species, all Holarctic and all predators of aphids. *A. aphidimyza* is sometimes found in galls formed by aphids. Males are characterized by some extremely long loops of the antennal circumfila. The larva (figs. 23, 125) has fairly long antennae and prominent, rounded ventral lobes. Ref.: Harris (1966).

Tribe Asphondyliini

This tribe is rich in species, all of them gall makers. It is represented on all continents but is especially diverse in the Neotropical Region. The North American species are distributed among six genera in three subtribes. The character that best distinguishes the tribe is the presence in the female of an enlarged seventh abdominal sternite (fig. 11). In addition, most species have parallel-sided flagellomeres (figs. 10, 11) with anastomosing circumfila. Anatomical modifications peculiar to the tribe include reduction of the terminal segment of the larva, "tailed" larvae (fig. 134), secondary development of setae on some papillae, loss of a pupal cocoon, and modification of the ovipositor for piercing plant tissue. This tribe appears related to the Cecidomyiini because both tribes have a pair of stubby, usually dorsally recurved, papillae on the terminal larval segment, although in Asphondyliini they are often atrophied through simplification. Adult antennae are greatly modified in most Asphondyliini, but the presence of binodal, trifilar flagellomeres in some unite this tribe with the Cecidomyiidi. Refs.: Möhn (1959, 1960, 1961, 1962, 1963).

Subtribe Asphondyliina

This subtribe contains 13 genera, but most are known from one or a few species. Eleven genera are restricted to the Neotropics and one other is known only from North Africa. Ref.: Möhn (1961).

Asphondylia. This is a very large genus with 247 described species, 60 of them from North America. In addition, the damage of 43 presumably undescribed species are treated in this book. *Asphondylia* occurs on all continents and includes many agricultural pests; its species are most often responsible for bud, flower, and fruit galls, and the galls are always accompanied by a fungal symbiont. Pupation occurs in the gall, but only rarely does a species diapause in that stage. Diapause may be passed in the first, second, or third larval instar, but appears to be passed in the first for most species. Full-grown larvae have a diminutive terminal segment and usually a strong, quadridentate spatula (figs. 28, 128, 129). Pupae are pigmented and have various head and abdominal structures that aid in cutting through gall tissue (fig. 3). Females have a pair of dorsal, cercilike lobes at the end of the eighth segment and an ovipositor shaped like a long, fine needle (fig. 11). Males have a bidentate gonostylus (fig. 10). Refs.: Möhn (1961), Richter-Vollert (1964).

Subtribe Polystephina

This subtribe contains two genera, *Polystepha* from oaks in Europe and North America and *Kochiomyia* from Old World Chenopodiaceae. Ref.: Möhn (1961).

Polystepha. This genus is known from 18 nominal species from North America east of the Great Plains, two from El Salvador, and three from Europe. All cause blister or pill-like leaf galls on oaks, and all are univoltine. North American hosts are apparently exclusively red and black oaks; European hosts are white oaks, which are the only oaks that naturally occur there. Larvae are orange or white, have a spatula, two short, cone-shaped terminal papillae, and a rugose integument (figs. 135, 136). Larvae drop to the ground with the leaves; at least some escape from the well-rotted galls in spring and only then form a cocoon in which they pupate. Females have a short but sclerotized and roughened ovipositor, probably used for inserting eggs into plant tissue. Of the 17 species known from North America, only the nine that have been associated with particular galls are keyed in chapter 8. Two others, *P. canadensis* (Felt) and *P. caryae* (Felt), were reared from oak leaves, but their galls were not described. The remaining six species are each known from one specimen caught in flight, some over plants reflected in their specific names. These are: *P. aceris* (Felt), *P. connecta* (Felt), *P. cornifolia* (Felt), *P. multifila* (Felt), *P. rhoina* (Felt), and *P. transversa* (Felt). Ref.: Möhn (1961).

Subtribe Schizomyiina

Three of the 12 genera of this tribe occur in the Nearctic Region. *Schizomyia* has a worldwide distribution, but the other two North American genera are restricted in our range to southern Florida and are otherwise Neotropical. Most of the remaining nine genera are Neotropical. Ref.: Möhn (1961).

Bruggmannia. This genus is Neotropical. Only two of the six known species enter our range, on *Pisonia* (Nyctaginaceae) in southern Florida and Mexico. *Bruggmannia* spp. are responsible for woody, blister leaf galls on plants of various families. Pupation occurs in the galls. Our species have weakly nodular flagellomeres with looped circumfila, short, three-segmented palpi, and an ovipositor shorter than the seventh sternite. Larvae are notable for the long extension of the terminal segment (fig. 134). Ref.: Möhn (1961).

Schizomyia. This genus is worldwide in distribution; 42 species have been described, 16 from a wide variety of plants from North America. Most form bud galls, but some cause complex leaf galls. The larvae of most species drop to the soil to pupate. Adults superfically resemble asphondylias but *Schizomyia* spp. lack the pair of cercilike lobes at the end of the female eighth abdominal segment, the ovipositor is longer and finer, the gonostylus has a long, distal crown on the apex, and the gonocoxite is lobed apically. Larvae are very active, usually bright red or orange. The spatula is clove-shaped or reduced, and the eighth abdominal tergum is extended into a tail (figs. 130–133). Ref.: Möhn (1961).

Stephomyia. This Neotropical genus has at least three species in southern Florida and one other from Brazil, all from leaf galls on *Eugenia* (Myrtaceae). Pupation occurs in the gall. Adult palpi are one-segmented. The ovipositor is shortest and has the largest cerci of any Asphondyliini. Larvae have a simple, evenly rounded terminal segment and a long, clove-shaped spatula. Ref.: Möhn (1961).

Tribe Cecidomyiini

This tribe is diverse and is made up entirely of plant-feeders. Included genera have one pair of larval terminal papillae that are stubby and recurved dorsally, and the remainder usually have short setae of equal length. Males of most of these genera have two instead of three circumfila on each flagellomere, possibly because searching for females is reduced in short-lived cecidomyiids that emerge in large numbers synchronously. Refs.: Gagné (1973c), Harris (1966).

Apagodiplosis. This genus is known from a single Nearctic species, *A. papyriferae* (Gagné), which lives in birch buds. Male flagellomeres are cylindrical but trifilar. The ovipositor is short and distinctive for its sclerotized, narrow, dorsal sclerite. The spatula is short with large cerci, but weakly sclerotized; larval papillae are diminutive (fig. 114). Ref.: Gagné (1967).

Blaesodiplosis. This Nearctic genus is restricted to leaf galls on hawthorn, rose, and shadbush (Rosaceae). Four species have been described. Similar leaf galls on ninebark (also Rosaceae) are caused by a group of species in *Contarinia.* The relationship of *Blaesodiplosis* to other genera is unclear. Adult mouthparts are reduced. Females have three circumfila ringing each flagellomere, a rare feature in females. Terminal papillae of the larvae are uniformly setose (figs. 109, 110). Possibly the stubby papillae are secondarily reduced.

Caryomyia. This Nearctic genus is restricted to complex leaf galls on *Carya* spp. Only 14 species have been named, but more than 40 kinds of galls are known (see Wells 1916), each presumably made by a different caryomyia. Several kinds of galls may be found on one leaflet, and many can be found on more than one species of hickory. Except for the polythalamous gall of *C. cynipsea* (Osten Sacken), galls are monothalamous. Larvae are ellipsoid and fairly flat. The spatula is variable, with the anterior end either very wide or considerably narrowed (figs. 91–93). Pupation occurs in the galls. Adults are brown and robust with wide wings. The ovipositor is short and has a rigid dorsal surface.

Cecidomyia. In the strict sense this genus contains 9 described Nearctic species and seven Palearctic species. All feed on resin of conifers, mainly pine. Most species do not appear to be host specific, as six were found in a small grove of *Pinus virginiana* in Maryland. Larvae usually live fully immersed in the resin except for their enlarged, backward-thrust hind pair of spiracles that protrude through the medium. The species of one section of the genus are gregarious and, when fully fed, pupate in silken tunnels in the resin mass; species of the other section are mainly solitary, and their full-grown larvae crawl away from the feeding site to pupate on a needle or branch in a white cocoon. Although the larval terminal papillae are modified and the spatula is usually reduced or lost, the stubby pair of terminal papillae that relates this genus to other Cecidomyiini is usually present. Representative larvae are illustrated in figures 31, 86–88. Refs.: Gagné (1978b), Nijveldt (1987).

"Cecidomyia." In the past many species were placed in *Cecidomyia* in the broad sense. It served as an omnibus category for many unrelated species that did not fit in particular genera or that were known only from immature stages or plant damage. Some of these are no better known now than they were when first described; here, these species are still combined with that generic name, but it is given in quotation marks. An example of a larva of one of these species, *"Cecidomyia" glutinosa* Osten Sacken, is illustrated in figures 105, 106.

Contarinia. This genus serves as a catch-all category for Cecidomyiidi. Under this genus are placed 60 Nearctic species. Many live freely in flowers or flower heads or in leaf folds. Some of these have several generations per year, and most leave the host to pupate in the ground. Other species apparently form natural groups, as do the several species in leaf galls of ninebark (Rosaceae) and others on grasses (see *Contarinia sorghicola* species group, below), and will eventually be segregated as genera. *Contarinia* includes species with elongate, tapered ovipositors and bifilar male flagellomeres, although some species with short ovipositors or trifilar male flagellomeres have been placed here in recent years. Representative larvae are illustrated in figures 20, 26, 68, 69, 71–77; a pupa in figure 1; and adults in figures 8, 9. Refs. Gagné (1973c), Harris (1966).

Contarinia sorghicola species group. More than 30 species in this group have been described, and the name *Stenodiplosis*, now a subjective synonym of *Contarinia*, would be available. All live in grass spikelets and are closely host specific. The group is distributed throughout the world. Five described species are known from North America, two of them introduced. Several other, undescribed species are also known. The most notorious species is the sorghum midge, first brought to this country from Africa with Johnson grass and later also with sorghum. *Contarinia bromicola* (Marikovskij and Agafonova), a pest of bromegrass, is another important but more recently introduced species. The biology of all species is generally similar. Females lay eggs in developing spikelets. Upon hatching, larvae crawl into and settle in the florets, where their feeding aborts the ovaries. These species can evidently travel easily with commerce: the sorghum midge is found throughout the world wherever sorghum is grown. Most species have several generations each year as long as flowers are produced. Pupation occurs in the florets. Larvae are white, have lost the spatula, and the papillae are greatly reduced (figs. 21, 70). Adults lack the group of lateral setae on the abdominal tergites. Females have a very long, fine ovipositor. Refs.: Harris (1970, 1979).

Halodiplosis. This genus contains one Nearctic and 37 Palearctic described species. All form bud galls on Chenopodiaceae; our species lives in the West on *Sarcobatus*. Males have trifilar flagellomeres, and females have the distal half of the ovipositor thickly covered with long, soft, pliable hairs.

Harmandia. Species of this Holarctic genus form convex or spherical leaf and petiole galls on *Populus* spp. Only two of the six described North American species have been reared from particular galls; the remainder were caught in flight. Five species are known from Europe. Larvae are large, up to 6 mm long. The spatula varies greatly in shape among species. Terminal papillae are on raised bumps, but their setae are short (fig. 85).

Macrodiplosis. This genus is Holarctic, its species restricted to oaks, on which they form leaf folds in spring. Nine North American species are known, five from particular oak leaf galls, the remaining five from specimens caught in flight. Larvae are white, large, up to 6 mm long, and their terminal papillae are situated on raised bumps (figs. 111, 112). Full-grown larvae drop to the soil. All species are univoltine.

Monarthropalpus. The one known species, *M. flavus*, is a Palearctic immigrant, unique for its short, hooklike ovipositor used for inserting its eggs in plant tissue. The larva has a long, tapered terminal segment with four papillae (figs. 103, 104). A pupa is illustrated in figure 6.

Paradiplosis. Three or four species are known in this genus, two from North America, one from Europe, and possibly one from Japan. All, except the Nearctic *P. obesa* (Felt), whose host is not precisely known, live in needles of fir. Males have trifilar flagellomeres, wide, short gonostyli, and a short, wide aedeagus. Females have a short ovipositor and large cerci. Larvae (fig. 83) essentially do not differ from many *Contarinia* spp. Ref.: Osgood and Gagné (1978).

Pinyonia. This genus is known from one species, *P. edulicola* Gagné, found in needle galls of pinyon pine. Adult palpi have a reduced number of segments. The male has bifilar flagellomeres and a long aedeagus relative to the hypoproct. Larvae have tiny papillae and no spatula (fig. 94). Ref.: Gagné (1970).

Planetella. This is a large Holarctic genus that lives exclusively on *Carex* and *Cyperus* (Cyperaceae). Of the 22 described Nearctic species, only one has been reared, although other, unnamed American species have been reared. Twenty-six Palearctic species are known, most of them taken from *Carex*. Adults are striking for their large body size, sometimes more than 10 mm long, and the fact that the scutum is produced anteriorly over the head. Antennae are remarkable in that they have 13 or more flagellomeres, unlike the remainder of Cecidomyiidi, which have only 12. Larvae (figs. 99, 100) have 10 dorsal papillae, four more than other Cecidomyiidi. Ref.: Gagné (1973c), Rübsaamen and Hedicke (1926–1939).

Plemeliella. An undescribed species from spruce seeds is tentatively placed in this genus. The genus is based on the Palearctic *P. abietina* Seitner, also from spruce seeds. *Plemeliella* larvae have a shorter and narrower spatula and shorter papillar setae than *Contarinia* spp. The gonostylus is characteristically short-ovoid. Ref.: Seitner (1908).

Prodiplosis. This genus is known from the Holarctic and Neotropical Regions. Nine species occur in the Nearctic Region, many with unusually wide host ranges. Some species live on more than one plant family. Females are similar to those of some *Contarinia* spp., but males, unlike *Contarinia*, have a long, dorsally recurved aedeagus and dorsoventrally articulated instead of mesolaterally recurved gonostyli. Larvae resemble those of some *Contarinia* spp. (see fig. 20) Ref.: Gagné (1986b).

Putoniella. This genus is known from two species, one from Europe, the other, apparently undescribed, from eastern North America. Both curl leaves of plum. Male antennae have short circumfilar loops and an aedeagus that is much longer than the hypoproct. The ovipositor is short and not protrusible and has large cerci. The larva (figs. 78, 79) has three pairs of stubby terminal papillae, one pair setiform. Ref.: Harris (1966).

Resseliella. Thirty-nine species of this genus, a remarkably homogeneous group, have been described from all over the world, 16 of them North American. Larval niches are quite varied, including in and among flowers, beneath bark, in stems, among scales of pine cones, and in simple galls. Some species may be inquilinous. Among Nearctic taxa, seven are known from single specimens caught in flight. Tarsal claws are unlike those of other Cecidomyiini because they are angled at the basal third and are sometimes toothed. The ovipositor is long and telescoped, but the cerci are not greatly reduced in size as they are in *Contarinia.* Larvae are robust and have two recurved lobes on the terminal segment (fig. 84). A Japanese species that lives in resin under bark of *Cryptomeria* (Taxodiaceae) is remarkable for enlarged hind papillae that are situated at the rear of the body, somewhat as in *Cecidomyia* (Sanui and Yukawa 1985).

Sequoiomyia. This genus is known from two species, one in seeds of bald cypress in North America, the other in seeds of a Miocene fossil of *Sequoia* in Germany. *Sequoiomyia* is similar in many respects to *Taxodiomyia,* also from bald cypress, except that the ovipositor of *Sequoiomyia* is short and the cerci separate. Larvae have a short, wide spatula and only two pairs of terminal papillae, one with short setae, possibly homologous to the stubby setae of *Contarinia,* the other with long setae. Ref.: Gagné (1969b).

Taxodiomyia. This genus is known from three described North American species, all of which form complex galls on branchlets and needles of bald cypress. Several additional kinds of galls are known. Pupation takes place in the galls. The ovipositor of *Taxodiomyia* is long, and the cerci are united into one lobe, a development parallel to that seen in most Lasiopteridi. Larvae (figs. 107, 108) have only two pairs of terminal papillae, one of them stubby. Ref.: Gagné (1969b).

Thecodiplosis. Four Nearctic and two Palearctic species are placed here. All live at the base of pine needles, where they usually cause a swelling and prevent the needles from growing to their normal length. The costal vein is not broken at its juncture with R^5, and the empodia are longer than the claws. This latter character is shared with most cecidomyiids that live on conifers. Males have bifilar flagellomeres, and females have an elongate ovipositor. The two stubby terminal papillae of the larvae are close together (figs. 80, 81). Ref.: Soné (1985).

Zeuxidiplosis. This genus is known from one Palearctic species, *Z. giardi* (Kieffer), introduced into North America and elsewhere for control of St. Johns-wort. Adults generally resemble many other *Contarinia* except that the palpus has only three segments, and the third is inserted basad of the apex of the second. Larvae are distinct for their dorsal, pleural, and terminal papillae situated on lobes (fig. 82). Ref.: Harris (1966).

Tribe Clinodiplosini

Most species are mycophagous and presumably generalists, but a few North American species in the genera treated here are associated with particular galls and appear to be gall makers. Still others are inquilines. In the Neotropics many Clinodiplosini are gall makers, and the tribe is much more diverse there than in the Holarctic. Ref.: Harris (1966).

Ametrodiplosis. One Nearctic and 12 Palearctic species are known for this genus. Our species was reared from seeds of a sedge. Males of *Ametrodiplosis* have the base of the costal vein strongly convex and short antennal circumfilar loops.

Clinodiplosis. This is a very large, cosmopolitan genus, the catch-all category for the Clinodiplosini. Forty-eight Nearctic species are now assigned here. Most are mycophagous; many are secondary feeders in galls, but some are evidently plant-feeders, e.g., *C. rhododendri* (Felt) in rhododendron leaf buds. They do not form complex galls in the Holarctic Region. Some species are regularly taken from grass heads. If these species are secondary feeders, they still show some specificity. East and West Coast rosette galls on spirea caused by a *Contarinia* sp. have different species of *Clinodiplosis* among the outer leaves. Males of *Clinodiplosis* typically have a large, angled mesobasal lobe on the gonocoxites and a long, parallel-sided hypoproct. The ovipositor is protrusible. The larval terminal papillae are composed of two pairs with short, stubby setae, one pair with short, thin setae, and one with long, thin setae (figs. 24, 115, 116). Any *Clinodiplosis* sp. in the keys to damage that is known only from the larval stage is only tentatively placed in this genus. Ref.: Skuhravá (1973).

Hyperdiplosis. Species from the Nearctic, Neotropical, and Oriental regions have been referred here, but even the five Nearctic species are not necessarily monophyletic. The deeply divided hypoproct with splayed lobes may have developed separately many times. No plant-feeding species is now referred to this genus. *Hyperdiplosis recurvata* (Felt) has been reared from pine cones, but also from decaying nectarines. Larvae are similar to those of *Clinodiplosis.*

Parallelodiplosis. This genus also is a cosmopolitan group that is not necessarily monophyletic. Most of the 16 Nearctic species referred here have been reared from secondary plant damage, but some are regularly found in galls made by other cecidomyiids or other gall makers. One is apparently responsible for a leaf spot gall on dogwood. At least the Nearctic species have in common the laterally divided second through sixth adult abdominal tergites. Parts of a larva are illustrated in figures 117 and 118. Refs.: Gagné (1973c), Harris (1966).

Sitodiplosis. This genus is known from three European species, one of which is evidently an immigrant in North America. In Europe it is considered a pest of wheat and other grains, but in North America it is associated with glume blotch, a fungal disease of wheat (Wellso and Freed 1982). Larvae (fig. 119) have a pebbled integument and crawl about on the surface of the developing grain. Adults are almost unique in the tribe Clinodiplosini for their bifilar flagellomeres.

Tribe Lestodiplosini

This tribe contains predators of insects and mites. Some prey on gall midges or other gall makers. When they are reared from leafy galls or flower heads that may harbor more than one insect, it is not always possible to know which the larvae were attacking. Larvae have long antennae and ventral pseudopods. Refs.: Baylac (1987), Harris (1966), Nijveldt (1969).

Cartodiplosis. This genus contains one species, *C. nyssaecola* (Beutenmüller). It evidently preys on a phylloxerid (Homoptera), which forms galls on sour gum (fig. 256a). The galls were once attributed to the predator. Terminal segments of a larva are shown in figure 127.

Feltiella. This genus has an almost worldwide distribution and contains predators of mites, some of which can be found in mite galls. Fourteen species are known from North America. Larvae (fig. 126) are generally similar to some *Lestodiplosis*, but adults have differently shaped tarsal claws.

Lestodiplosis. This cosmopolitan genus of more than 175 described species, 48 of them Nearctic, is a catch-all genus. Some species are mite predators, others insect predators. Larvae are common inhabitants of flower heads and leafy cecidomyiid galls. Adults often have spotted wings and, less commonly, striped legs. Larvae are illustrated in figures 30, 33, and 123.

Tribe Mycodiplosini

Mycodiplosis. This genus is cosmopolitan and is represented in North America by seven species. Larvae of most species feed on rust fungi, which occasionally grow on leaves that are part of leafy galls. Larvae (figs. 25, 121) have a robust head with longer antennae than plant-feeding gall midges. Ref.: Holz (1970).

Other Cecidomyiidi

Under this category are placed an assortment of genera that are not necessarily related to one another and cannot be placed in previously designated tribes.

Ctenodactylomyia. This genus is known from one species that forms hard blisters on leaves of *Coccoloba* in Florida and the West Indies. Its affinities are not certain. The larval terminal papillae are reduced, but one pair is stubby (figs. 96, 97). Male and female flagellomeres are cylindrical and otherwise similar. Male genitalia have a mesobasal gonocoxal lobe and very long gonostyli. The female abdomen is not protrusible. Ref.: Gagné (1973c).

Karshomyia. This is a Holarctic genus with many described and undescribed species, 9 described from the Nearctic Region. Only one, *Karshomyia caulicola* (Coquillett), has been associated with plant damage, but its larvae may have been feeding on fungi growing on decaying tissue. Adult abdominal terga are similar to those of *Lobodiplosis* in that they are transversely divided. Male genitalia have bizarre mesal lobes on the gonocoxite. Terminal larval segments are illustrated in figure 124. Ref.: Gagné (1973b).

Lobodiplosis. As with *Karshomyia*, *Lobodiplosis* has numerous species, many of them still undescribed. It is also Holarctic and is related to *Karshomyia*. Six Nearctic and one Palearctic species have been described. All are presumably fungus-feeding, but *L. triangularis* Felt was reared once from pine cones, where it may have been feeding on fungus. Adults resemble *Karshomyia* except that male genitalia have a characteristic apical, setose lobe on the gonocoxites. Ref.: Gagné (1973b).

Obolodiplosis. This genus is known from one eastern Nearctic species that causes a marginal leaf roll of black locust. Male genitalia have long, narrow gonostyli and a hypoproct that is modified into two lobes that flank the aedeagus. The ovipositor is not protrusible, and the cerci are large. The terminal segments of a larva are illustrated in figure 95.

Olpodiplosis. The one known species in this genus causes a complex, flask-shaped gall on *Helianthus*. The male genitalia are generally similar to those of Cecidomyiini, but the adult head and female abdomen show some resemblances to the Asphondyliini. It is possible that these similarities are separately derived. The larvae have elongate papillar setae (fig. 113), unusual in a species that forms a complex gall. Ref.: Gagné (1973c).

Pilodiplosis. This genus is known from one Nearctic species that makes woody, convex swellings on leaves of sunflowers. Male genitalia are unique for their pointed gonostylar tooth, the large, wide, hypoproctal lobes, and broad, short aedeagus. Female antennae have ramifying circumfila and a short but protrusible ovipositor with large cerci. The spatula is broad and blunt anteriorly; the terminal larval segment has only six papillae, but all are setiform (figs. 101, 102).

Youngomyia. This genus is known from one species that is an inquiline or predator in galls made by other gall midges, especially *Schizomyia* spp., on a variety of plant hosts. The male genitalia are unique for the spiny hypoproct and long, bowed gonostyli. Females have large cerci covered by blunt-tipped setae. Larvae have eight terminal short-setiform papillae (fig. 122).

4 Generic Key to Larvae of North American Cecidomyiidae Associated with Plants

This key is for full-grown or third-instar larvae only. These larvae are far better known and longer lived than the earlier larval instars, and they usually offer additional characters in the distinctive spatula. Larvae of all known North American genera of Cecidomyiidae associated with plants are treated here, including predaceous and fungus-feeding forms likely to be found in association with plant damage. The key should be especially helpful for identifying larvae when two or more genera co-occur in particular plant damage or galls. As examples, galls on balsam fir needles made by *Paradiplosis tumifex* often also contain *Dasineura balsamicola*, and flower heads of *Baccharis* contain four species of gall midges that belong to four different genera. The keys by Möhn (1955) and Mamaev and Krivosheina (1965) for Palearctic genera complement this one. The study by Sylvén (1975) on Palearctic Lasiopteridi will also be helpful.

This key is a practical one, making crucial use of the host plant group or site of infestation. A key that exactly reflects gall midge classification is not feasible because many distantly related genera have independently undergone similar losses and other modifications. Some genera, e.g., *Contarinia* and *Prodiplosis*, are so similar that they are not distinguishable in the larval stage. Nevertheless such genera will still be identifiable by their host association. Larvae of *Ficiomyia* and *Halodiplosis* are unknown, so these genera are not in the key. A few undescribed taxa that may represent new genera are included.

1 Antennae long, at least 4, usually 5–6 times as long as basal diameter (figs. 23, 25, 30, 33); predaceous or feeding on rust fungi 2
Antennae short, not longer than 3 times, usually about 2 times as long as basal diameter (figs. 19, 20, 35); plant- or fungus-feeding 4

2(1) Head directed anteroventrally, robust; terminal segment with 4 stubby papillae of equal size and 4 setiform papillae of equal length (figs. 25, 121); usually associated with rust fungi . . . . *Mycodiplosis*
Head directed anteriorly, not as robust but with long posterior tentoria (fig. 33); terminal segments with setiform papillae only (figs. 123, 126, 127); predaceous on other arthropods. 3

3(2) Ventral pseudopods convex (fig. 23); anus ventral; predaceous on aphidoids . . . . *Aphidoletes*
Ventral pseudopods tubelike (fig. 30); anus dorsal (figs. 123, 126, 127); predaceous on various arthropods . . . *Cartodiplosis*, *Feltiella*, *Lestodiplosis*

4(1) On gymnosperms (excluding *Ephedra*) 5
On angiosperms or *Ephedra*. 26

5(4) On Taxodiaceae (bald cypress family). 6
On other conifers 7

6(5) In seeds. . . . . *Sequoiomyia*
In complex galls on needles (figs. 107, 108). . . . . *Taxodiomyia*

7(5) On Cupressaceae (juniper family). 8
On Pinaceae (pine family) 10

8(7) In cones . . . . *Janetiella*
In buds or tips of branchlets 9

9(8) Terminal segment with 1 pair of stubby, dorsally recurved papillae (as in fig. 69) . . . . *Contarinia*
Terminal segment with all papillae setiform, similar (fig. 47) . . . . *Oligotrophus*, *Walshomyia*

10(7) Spiracles of 8th abdominal segment greatly enlarged, projecting beyond terminal segment (figs. 86, 88); in droplet or mass of resin in cones, twigs, or needle bases. . . . . *Cecidomyia*
Spiracles of 8th abdominal segment not appreciably larger than other spiracles, not projecting over terminal segment; not immersed in resin 11

11(10) In cones 12
In needles or shoots 18

12(11) Terminal segment with lobes (figs. 41, 84); free between cone scales 13
Terminal segment without lobes other than enlarged papillar bases (figs. 68, 115); free between cone scales or in plant tissue 14

13(12) Integument rough; spatula with 2 anterior teeth; 8th abdominal segment with 2 setose dorsal papillae (fig. 84) . . . . *Resseliella*
Integument smooth; spatula with 1 anterior tooth (fig. 40); 8th abdominal segment with 4 asetose dorsal papillae (fig. 41) . . . . *Camptomyia*

14(12) Terminal segment with middle pairs of papillae stubby, outer pairs with setae of unequal length (figs. 115, 120). . . . . . . *Clinodiplosis*, *Hyperdiplosis*
Terminal segment with all papillae similar or with only 1 pair stubby (figs. 19, 72) 15

15(14) Terminal segment with 1 pair of stubby papillae (fig. 72) 16

Terminal segment with papillae uniform (as in fig. 19 or 41) 17

16(15) In seeds of spruce .... *Plemeliella*

In cone scales of Douglas-fir (figs. 72, 73) .... *Contarinia*

17(15) Eighth abdominal segment with 4 asetose papillae (as in fig. 41); free between cone scales .... *Asynapta*

Eighth abdominal segment with 2 setose papillae (as in fig. 19); in cone tissue .... *Kaltenbachiola*, "*Mayetiola*"

18(11) On fir or spruce 19

On pine 22

19(18) In needles of fir 20

In shoots of spruce 21

20(19) Integument mostly smooth; terminal segment with 1 pair of stubby papillae (fig. 83) .... *Paradiplosis*

Integument rough; terminal segment with all papillae setiform (as in figs. 19, 34) .... *Dasineura*

21(19) In enlarged apical buds .... *Dasineura*

In shoots at base of needles .... "*Mayetiola*"

22(18) Without spatula; on *Pinus edulis* and *P. monophylla* 23

With spatula; on other pines 24

23(22) Gregarious; in spindle-shaped needle gall (fig. 352) .... *Pinyonia*

Solitary; in cells at base of shortened needles (figs. 353–355) .... *Janetiella*

24(22) Integument rough; terminal segment with caudal lobes bearing a stubby papilla at apex and another papilla at base (fig. 84) .... *Resseliella*

Integument mostly smooth; terminal segment without lobes or the lobe a simple extension of a papilla (figs. 68, 69, 80) 25

25(24) Terminal segment with pair of stubby papillae closely adjacent, connected by sclerotized area (fig. 80) .... *Thecodiplosis*

Terminal segment with pair of stubby papillae separated, not situated on common sclerotized area (figs. 68, 69) .... *Contarinia*

26(4) On Poaceae (grass family) 27

Not on grasses 37

27(26) Last instar a nonfeeding stage, encased in brown, seedlike, second-instar integument (fig. 7) .... *Mayetiola*

Last instar a feeding stage, not encased in second-instar integument 28

28(27) Two groups of 3 lateral papillae present on each side of spatula (figs. 19, 36); on or in inflorescence 29

Four scattered lateral papillae on each side of spatula (fig. 67); usually in stems, one genus in seeds 32

29(28) Terminal segment with the 2 middle pairs of papillae stubby, outer pairs with greatly unequal setae (figs. 115, 119) ...... *Clinodiplosis*, *Sitodiplosis*

Terminal segment with all papillae with setae of approximately same length (figs. 19, 66, 70) 30

30(29) Integument smooth; papillar setae inconspicuous (figs. 21, 70) ....
.... *Contarinia sorghicola* species group

Integument rough; papillae conspicuously setose (figs. 19, 89, 98) 31

31(30) Terminal segment lobed (figs. 89, 98); in seeds .... undescribed genus of Cecidomyiidi

Terminal segment unlobed (as in fig. 19, 34); in florets......... *Dasineura*

32(28) Terminal segment lobed (figs. 59, 62) 33

Terminal segment unlobed, evenly rounded (fig. 66).................. 34

33(32) Terminal segment lobes smooth (fig. 62)................... *Edestochilus*

Terminal segment lobes rough (fig. 59) .... *Chilophaga*

34(32) In seeds; spatula with 2 anterior teeth.................... *Edestosperma*

In stems; spatula with 3 anterior teeth (fig. 67) or absent.............. 35

35(34) In leafy, broom gall at base of stem; spatula absent.......... *Astictoneura*

In apparently unmodified stems; spatula present or absent 36

36(35) Spatula absent; in *Elymus*............................ *Hybolasioptera*

Spatula present (fig. 67); on other grasses.................. *Calamomyia*

37(26) Thoracic segments with 6 lateral papillae in 2 groups of 3 (occasionally 1 papilla of 1 of the 2 groups missing) (figs. 19, 34)................... 38

Thoracic segments with 4 or fewer lateral papillae on each side of midline (figs. 60, 64, 129) 58

38(37) Eighth abdominal segment with setiform ventral papillae (as in fig. 19) 39

Eighth abdominal segment with asetose ventral papillae (as in fig. 24) 43

39(38) Terminal papillae stubby, borne on 2 lobes (fig. 39); on Juncaceae ....*Procystiphora*

Terminal papillae not stubby; if setae short, terminal segment evenly rounded (fig. 34); on other plants 40

40(39) Prothorax with ventral papillae setiform; inquilines in galls made by *Asphondylia* spp.. *Trotteria*

Prothorax with ventral papillae asetose (fig. 19); mostly gall makers.... 51

41(40) In stem or bud galls of willow (but not among outer leaves of conelike or cabbagelike galls); spatula short (fig. 42) or long (fig. 43) .... *Rabdophaga*

Not in stem or bud galls of willow (except occasionally among outer leaves forming conelike or cabbagelike galls); spatula most often clove-shaped (fig. 36) 42

42(41) Inquiline in various cecidomyiid galls .... *Camptoneuromyia*

Usually primary plant-feeders, a few species inquilinous (figs. 34–36) .... *Dasineura*

43(38) Terminal segment with setae of 1 pair of outer papillae more than twice as long as setae of remaining papillae (fig. 115)...................... 44

Terminal segment with setae of all papillae approximately equal (figs. 78, 84) 45

44(43) Terminal segment with 3 pairs of stubby papillae .... *Ametrodiplosis*

Terminal segment with 2 pairs of stubby papillae (figs. 115, 118) .... *Clinodiplosis*, *Parallelodiplosis*

45(43) Terminal segment with 1 pair of stubby, dorsally recurved papillae (figs. 72, 78, 83)... 46

Terminal segment with or without stubby papillae; if with, stubby papillae not dorsally recurved (figs. 82, 109) 51

46(45) Terminal segment with 2 or 3 pairs of stubby or near-stubby papillae (figs. 78, 82, 111) 47

Terminal segment with 1 pair of stubby papillae (figs. 74, 85) 49

47(46) On leaf galls of oak (figs. 111, 112)...................... *Macrodiplosis*

On leaf galls of poplar or plum 48

48(47) In leaf galls of poplar (fig. 85)............................ *Harmandia*

In marginal leaf roll of plum (figs. 78, 79).................. *Putoniella*

49(46) Eighth abdominal segment with dorsocaudal extension bearing both dorsal papillae (figs. 130–133) .... *Schizomyia*

Eighth abdominal segment without dorsocaudal extension 50

50(49) Terminal segment with 2 lobes, each bearing a papilla at end and one near base (fig. 84) .... *Resseliella*

Terminal segment without lobes (as in fig. 72)..... *Contarinia, Prodiplosis*

51(45) Terminal segment strongly cleft, pleural and dorsal abdominal papillae on lobes (fig. 82); in bud galls of St. John's-wort (*Hypericum*) .... *Zeuxidiplosis*

Terminal segment not deeply cleft 52

52(51) Terminal segment divided into 2 lateral lobes, 4 papillae situated at the end of each (fig. 109) 53

Caudal segment usually convex; if segment lobed, only 4 papillae present (fig. 99) 55

53(52) Integument smooth (figs. 109, 110); in leaf galls of *Amelanchier* and *Crataegus*... *Blaesodiplosis*

Integument rough; on other plants.. 54

54(53) Inquiline in galls of *Schizomyia* and other gall midge genera; teeth of spatula pointed (fig. 122)................................. *Youngomyia*

Fungus-feeder; teeth of spatula rounded (fig. 124) .... *Karshomyia*

55(52) Integument mostly smooth 56

Integument rough 57

56(55) In buds of birch.. *Apagodiplosis*

In stem galls of Cyperaceae (sedge family) (figs. 99, 100) .... *Planetella*

57(55) In curled leaf margins of *Robinia* (black locust) (fig. 95) .... *Obolodiplosis*

In leaf blister galls of *Coccoloba* (figs. 96, 97) .... *Ctenodactylomyia*

58(37) Terminal segment diminutive, less than half as wide as preceding segment (figs. 28, 128); spatula usually robust, with 4 anterior teeth, the surrounding area usually pigmented (fig. 129) .... *Asphondylia*

Terminal segment tapering gradually from preceding segment, more than half as wide (fig. 19); spatula, if present, with 1 or 2 teeth, the surrounding area not pigmented.. 59

59(58) On Asteraceae.. 60

On other plants.. 67

60(59) In seeds of *Achillea* (yarrow) .... *Ozirhincus*

Not on *Achillea*.. 61

61(60) In blisters on leaves and green stems on Lactuceae........... *Cystiphora*

Not in blister galls on Lactuceae 62
62(61) In tube-shaped bud galls or spherical leaf galls on *Helianthus* 63
Not on *Helianthus* 64
63(62) In tube-shaped bud galls on *Helianthus* (fig. 113) .... *Olpodiplosis*
In spherical leaf galls on *Helianthus* (figs. 101, 102) .... *Pilodiplosis*
64(62) In blister galls on leaves and green stems of Astereae (figs. 60, 61) .... *Asteromyia*
Not in blister galls on leaves and green stems of Astereae 65
65(64) In seeds of various Asteraceae .... *Neolasioptera*
Not in seeds of Asteraceae 66
66(65) Spatula narrow, only slightly widened anteriorly (fig. 64) .... *Neolasioptera*
Spatula greatly widened anteriorly (fig. 65) .... *Lasioptera, Neolasioptera*
67(58) In stem swellings or birch catkins 68
In or on leaves (simple spots, convex swellings, or complex galls) 71
68(67) In birch catkins (figs. 57, 58) .... *Semudobia*
In stem swellings 69
69(68) On willow (fig. 44) .... *Lygocecis*
On other plants 70
70(69) Spatula narrow, only slightly widened anteriorly (fig. 64) .... *Neolasioptera*
Spatula greatly widened anteriorly (fig. 65) .... *Lasioptera, Neolasioptera*
71(67) On oak, hickory, or willow 72
On other plants 75
72(71) In oak leaf galls (figs. 135, 136) .... *Polystepha*
On hickory or willow 73
73(72) On willow .... *Iteomyia*
In hickory leaf galls 74
74(73) On sticky spot (figs. 105, 106) .... *"Cecidomyia" glutinosa*
In complex galls .... *Caryomyia*
75(71) On maple or hackberry 76
On other plants 77
76(75) On maple; on spot on leaf surface (figs. 22, 32) .... *Acericecis*
On hackberry; in complex gall (figs. 37, 38) .... *Celticecis*
77(75) On *Viburnum* or grape 78
On other plants 79
78(77) On *Viburnum* (figs. 45, 46) .... *Sackenomyia*
On grape .... *Janetiella*
79(77) On boxwood or Chenopodiaceae 80
On other plants 81
80(79) On boxwood (figs. 103, 104) .... *Monarthropalpus*
On Chenopodiaceae (*Sarcobatus*) .... *Protaplonyx*
81(79) Terminal segment elongate-attenuate (fig. 134) .... *Bruggmannia*
Terminal segment short, rounded 82
82(81) Terminal segment with 8 long-setose papillae (as in fig. 63) .... *Meunieriella*
Terminal segment with 2 short-setose papillae .... *Stephomyia*

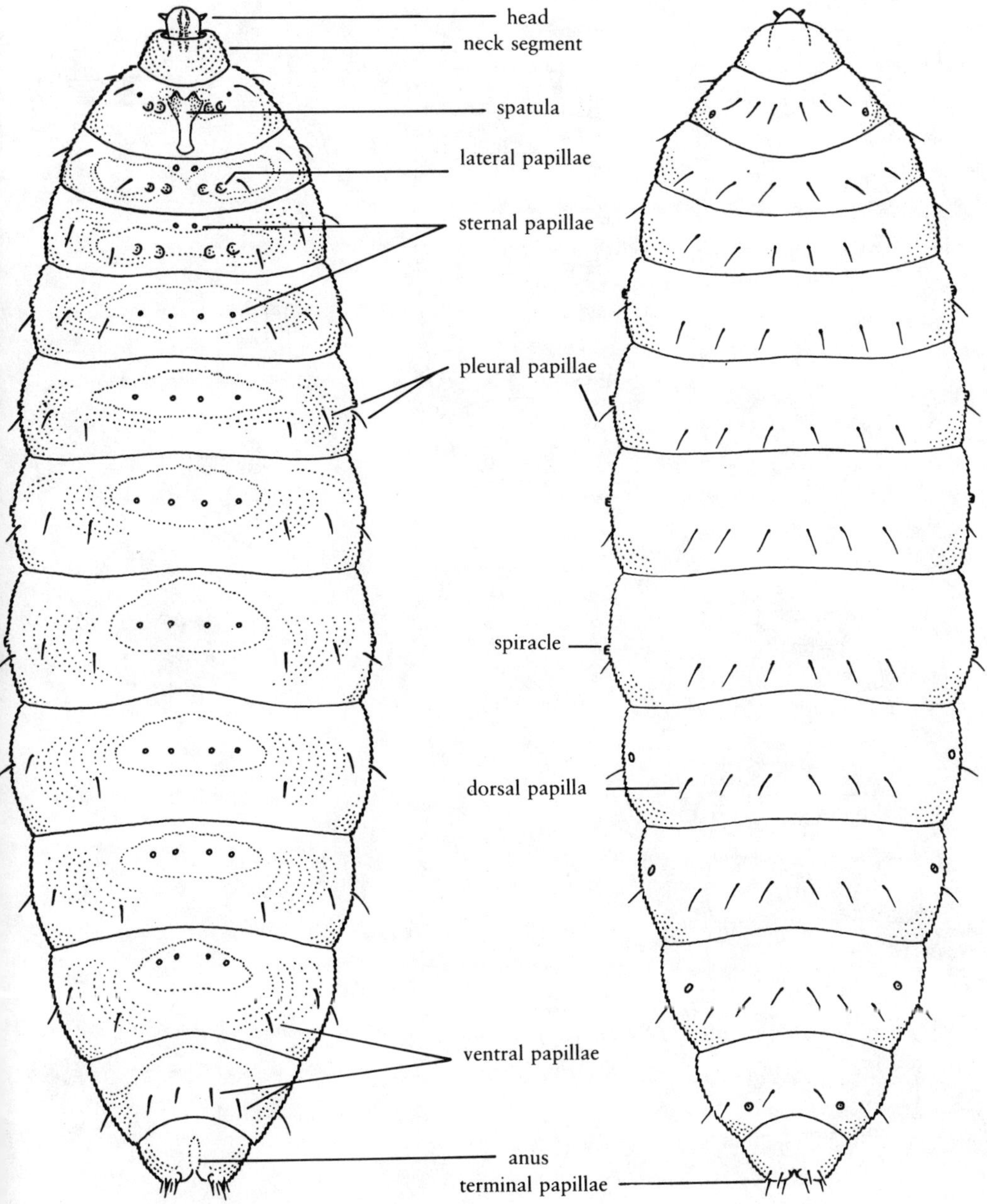

Figure 19. *Dasineura gleditchiae* larva, 50×. Ventral view on left, dorsal on right.

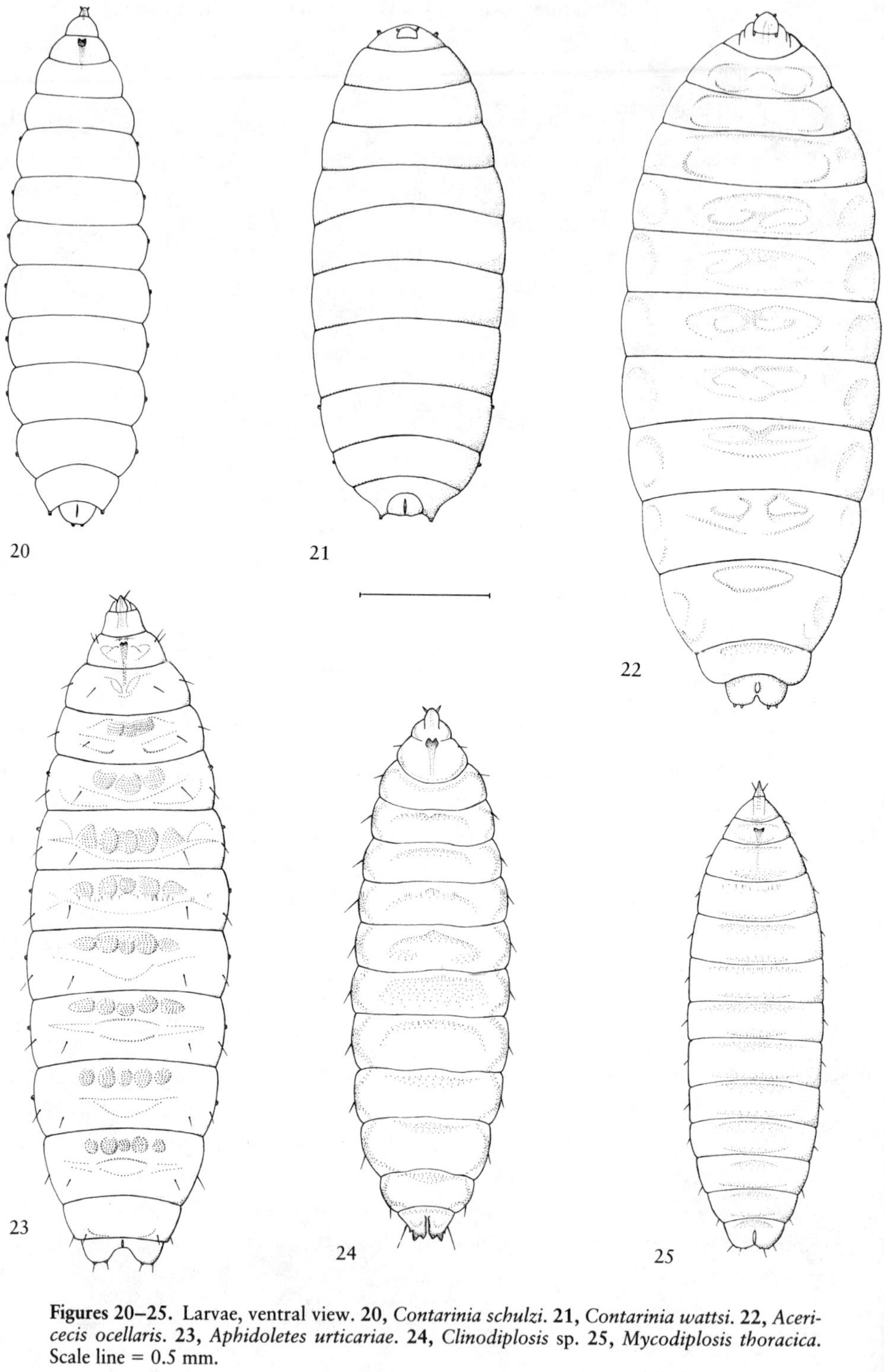

Figures 20–25. Larvae, ventral view. **20,** *Contarinia schulzi.* **21,** *Contarinia wattsi.* **22,** *Acericecis ocellaris.* **23,** *Aphidoletes urticariae.* **24,** *Clinodiplosis* sp. **25,** *Mycodiplosis thoracica.* Scale line = 0.5 mm.

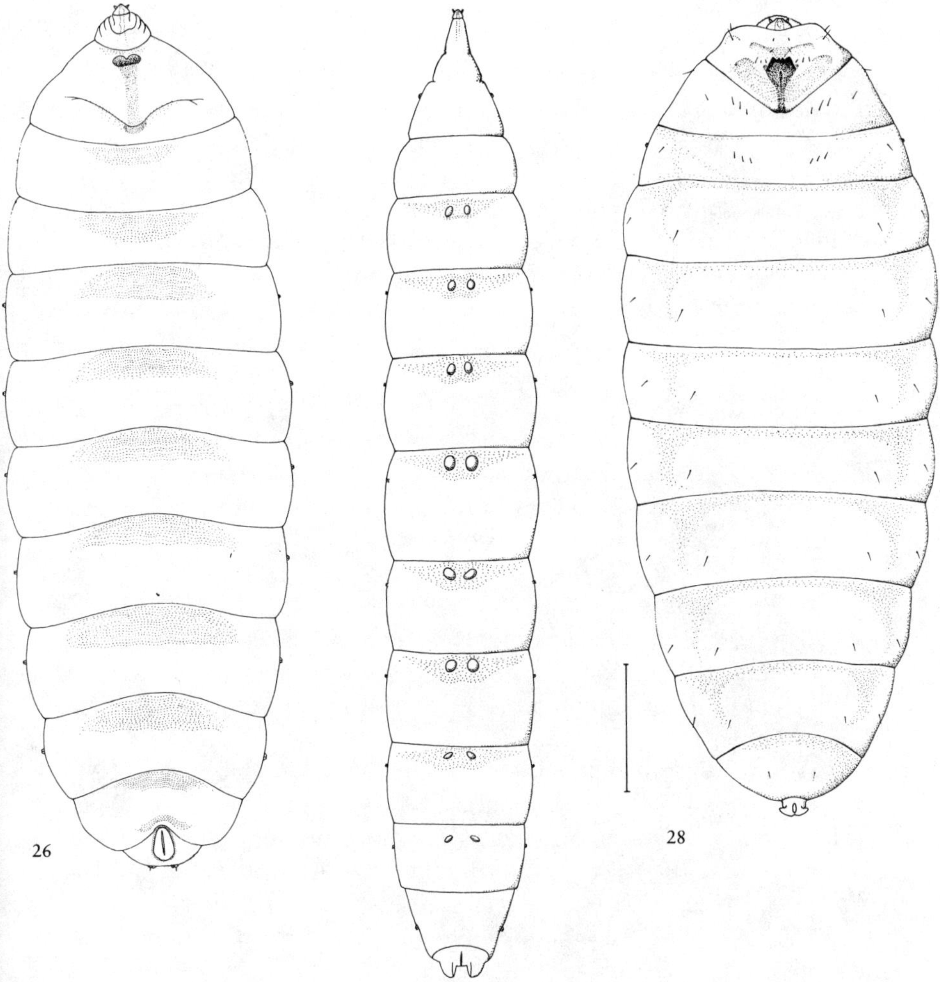

Figures 26–28. Larvae, ventral view. **26,** *Contarinia pyrivora.* **27,** *Asynapta hopkinsi.* **28,** *Asphondylia atriplicis.* Scale line = 0.5 mm.

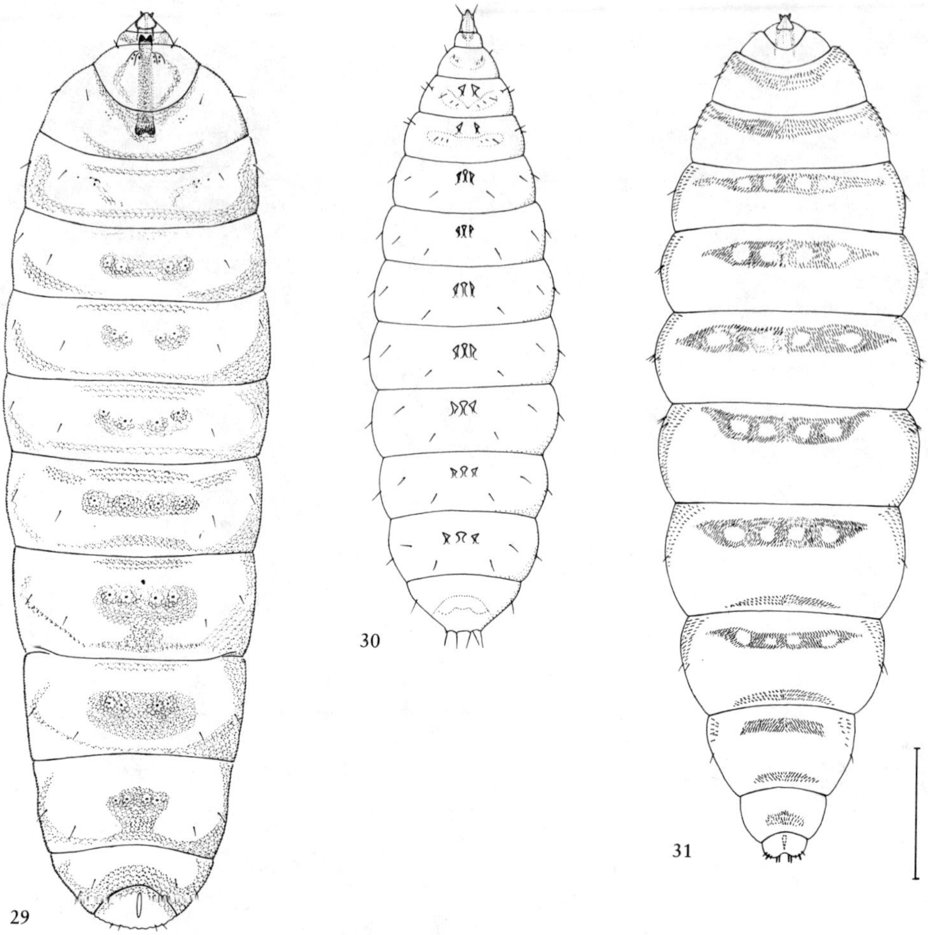

Figures 29–31. Larvae, ventral view. **29,** *Neolasioptera fontagrensis.* **30,** *Lestodiplosis* sp. **31,** *Cecidomyia resinicola.* Scale line = 0.5 mm.

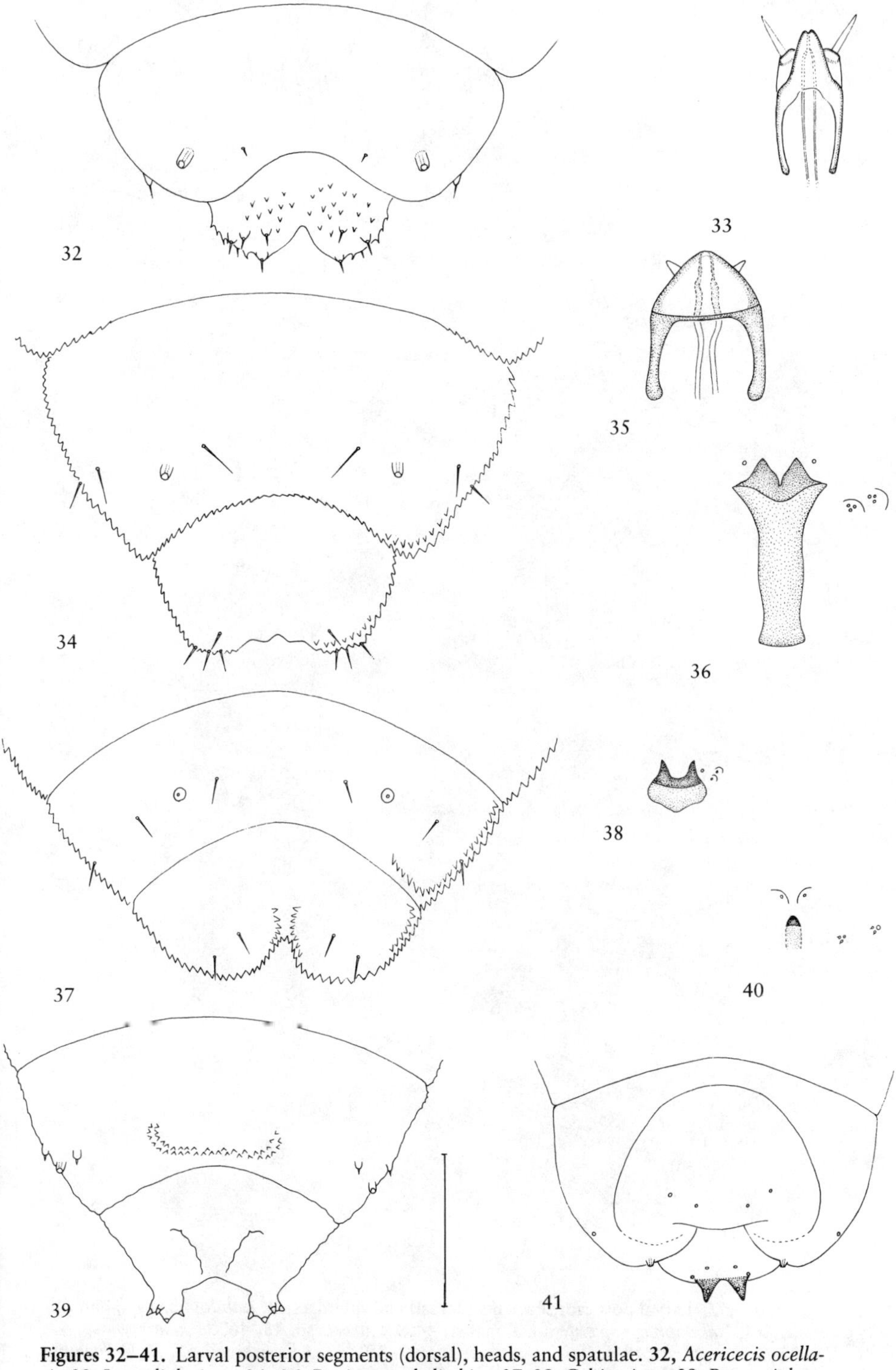

Figures 32–41. Larval posterior segments (dorsal), heads, and spatulae. **32,** *Acericecis ocellaris*. **33,** *Lestodiplosis* sp. **34–36,** *Dasineura gleditchiae*. **37, 38,** *Celticecis* sp. **39,** *Procystiphora gerardii*. **40, 41,** *Camptomyia pseudotsugae*. Scale line: 0.2 mm.

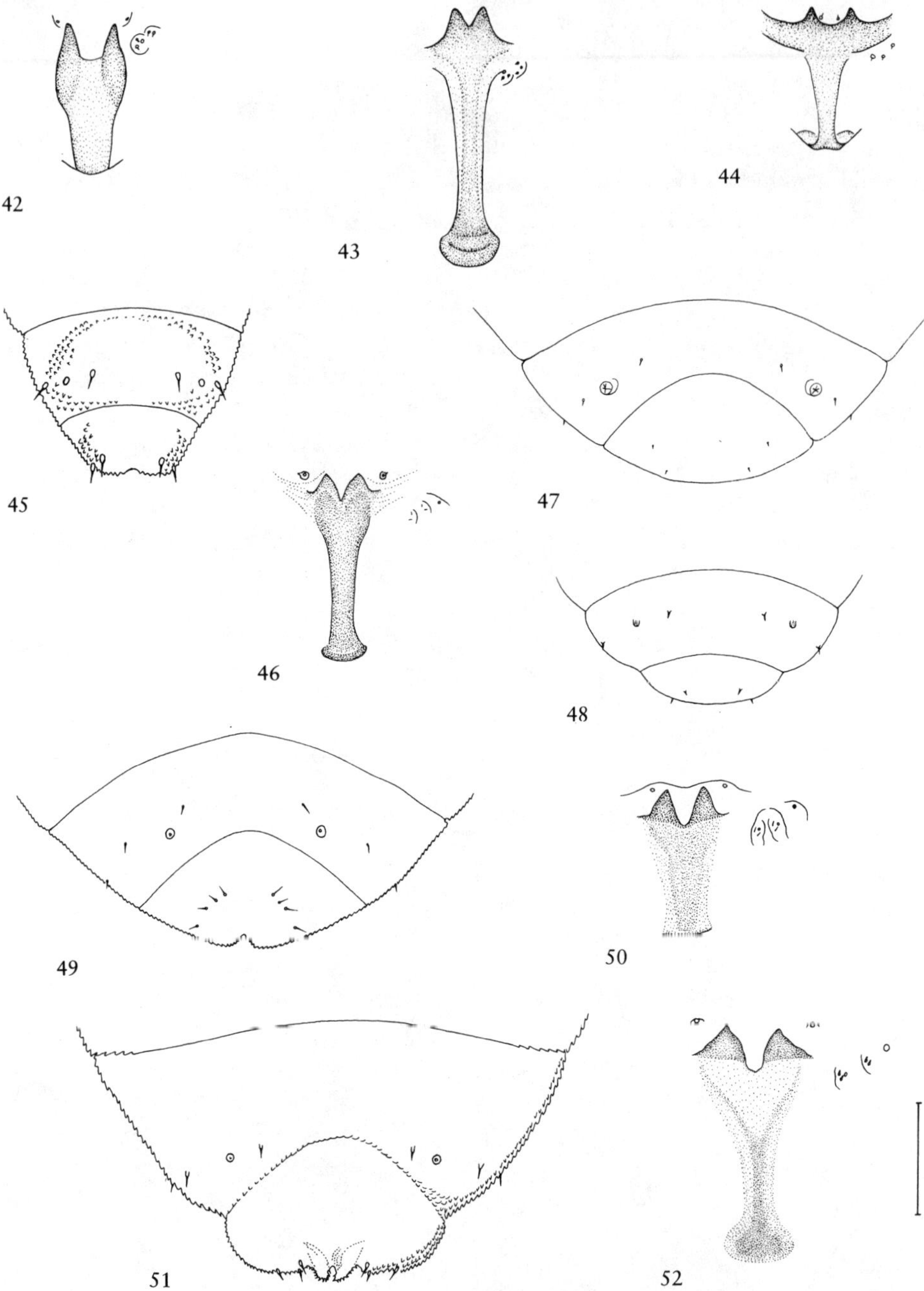

Figures 42–52. Larval posterior segments (dorsal) and spatulae. **42,** *Rabdophaga* sp., bud gall maker. **43,** *Rabdophaga* sp., stem gall maker. **44,** *Lygocecis* sp. **45, 46,** *Sackenomyia* sp. **47,** *Walshomyia cupressi*. **48,** *Janetiella coloradensis*. **49–50,** *Janetiella siskiyou*. **51, 52,** *Craneiobia tuba*. Scale line: 0.2 mm.

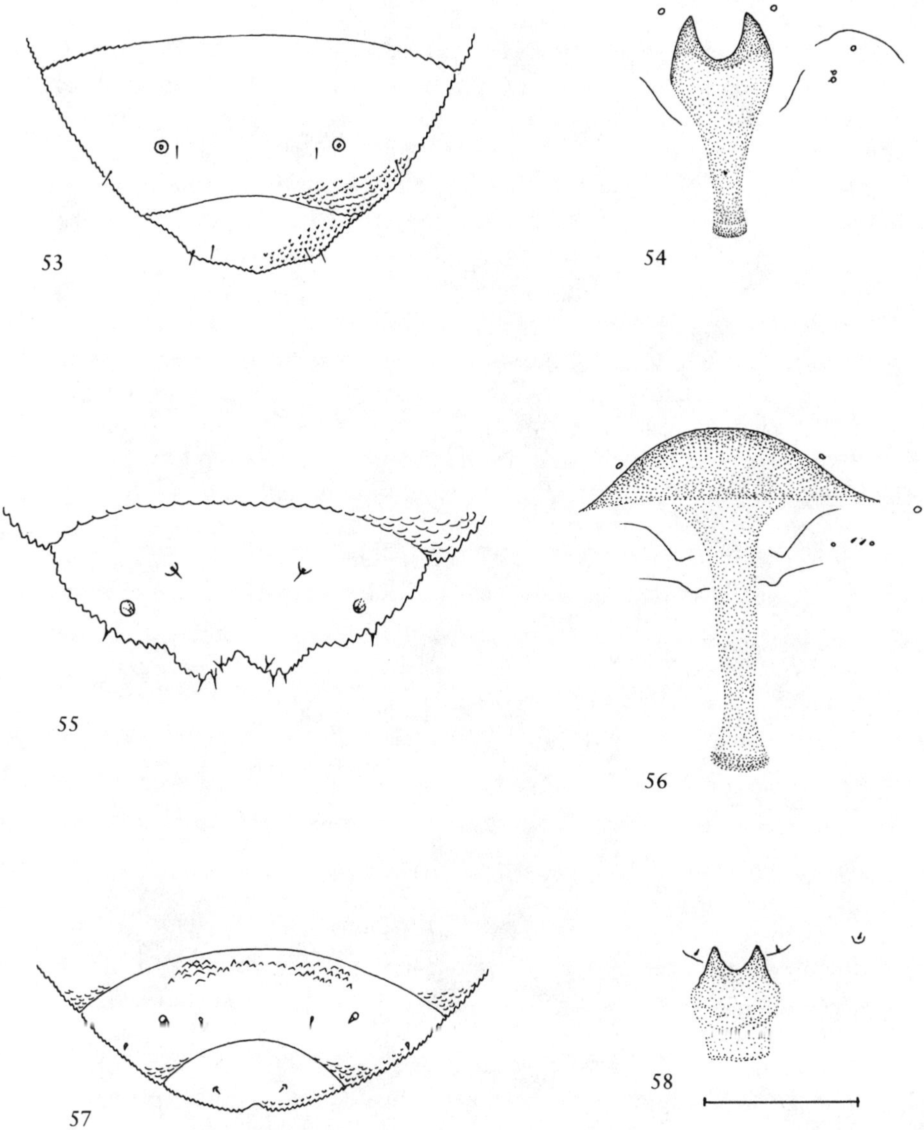

Figures 53–58. Larval posterior segments (dorsal) and spatulae. **53, 54,** *Janetiella* sp. from grape. **55, 56,** *Dasineura porrecta.* **57, 58,** *Semudobia tarda.* Scale line: larval posterior segments, 0.2 mm; spatulae, 0.1 mm.

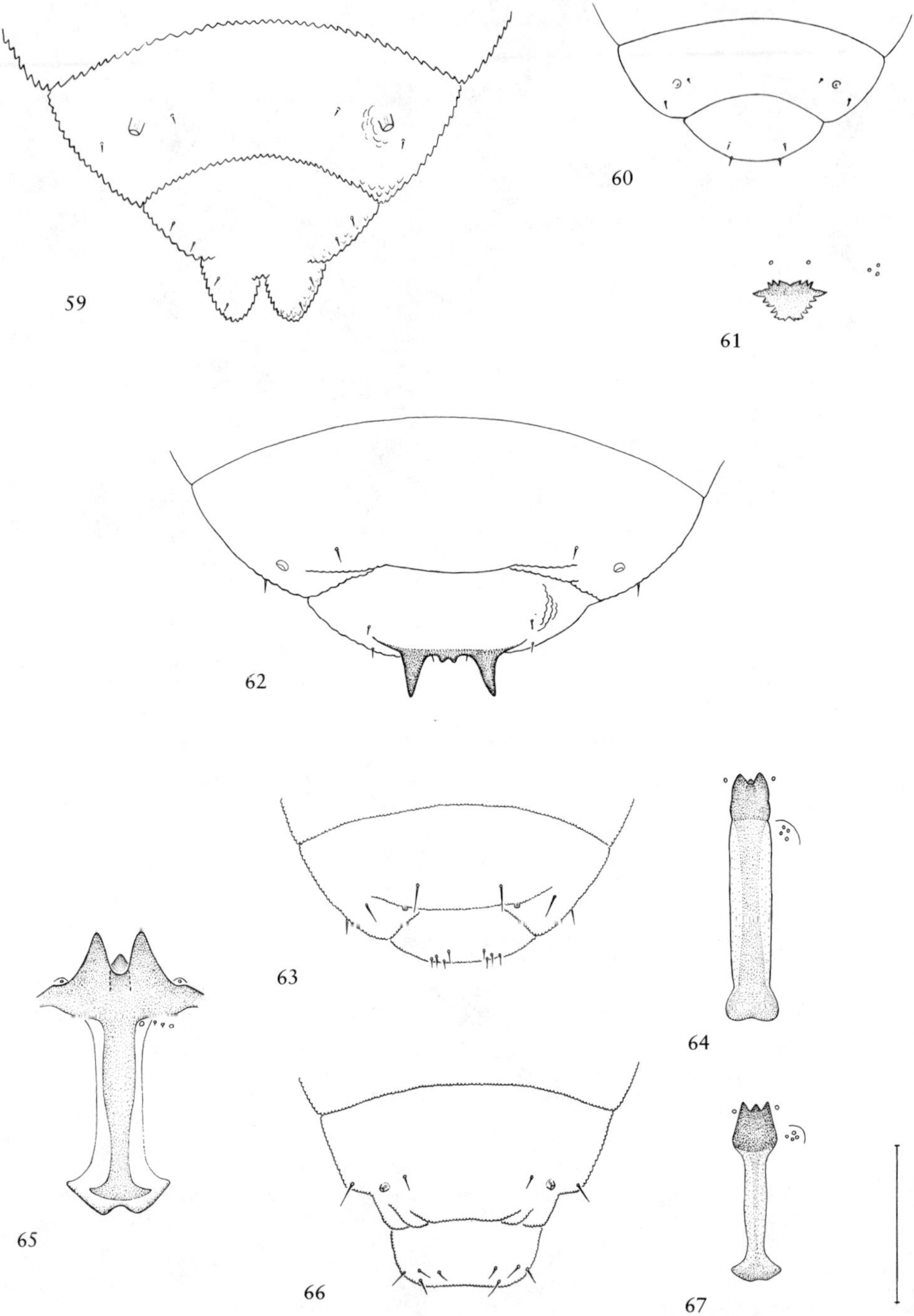

Figures 59–67. Larval posterior segments (dorsal) and spatulae. **59,** *Chilophaga gyrantis*. **60, 61,** *Asteromyia carbonifera*. **62,** *Edestochilus allioides*. **63, 64,** *Neolasioptera fontagrensis*. **65,** *Lasioptera solidaginis*. **66, 67,** *Calamomyia alterniflorae*. Scale line: 0.2 mm.

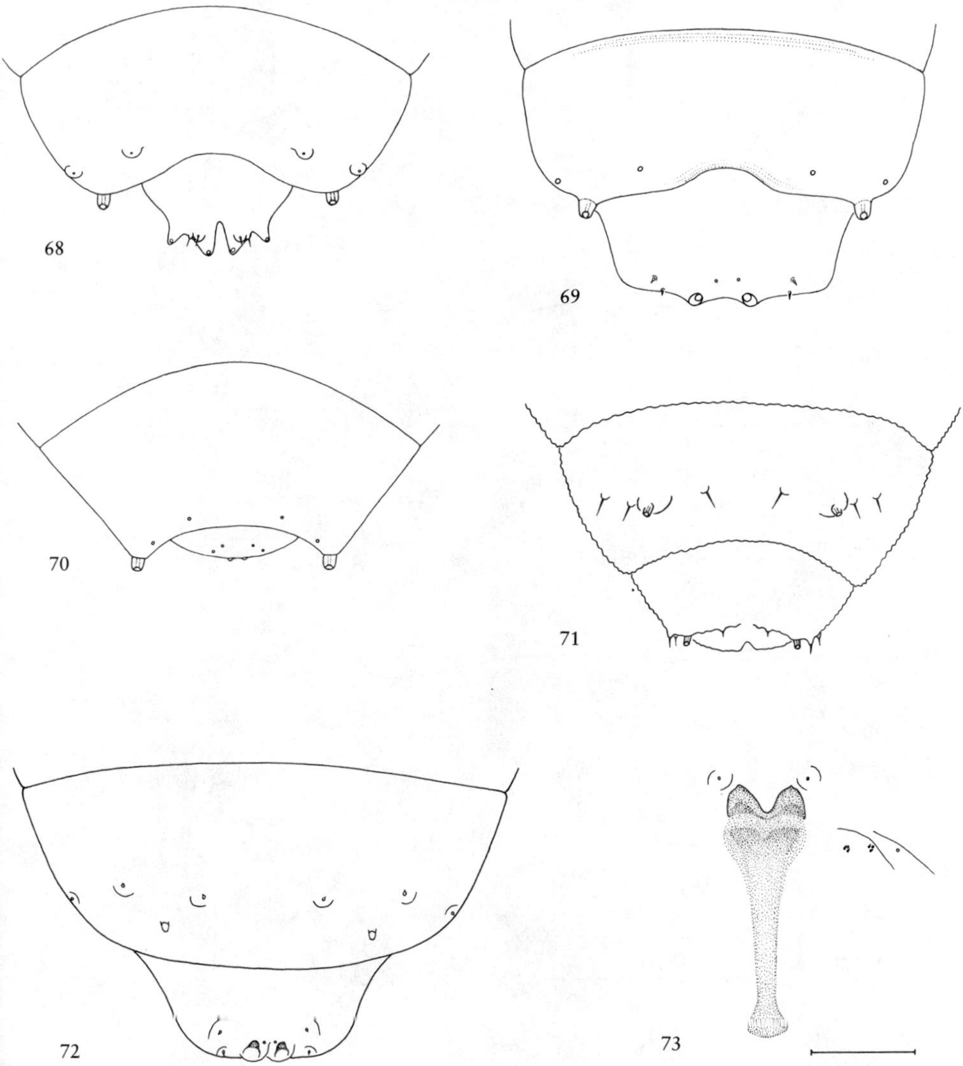

Figures 68–73. Larval posterior segments (dorsal) and spatula. **68,** *Contarinia baeri.* **69,** *Contarinia acuta.* **70,** *Contarinia wattsi.* **71,** *Contarinia clarkei.* **72, 73,** *Contarinia oregonensis.* Scale line: 0.2 mm.

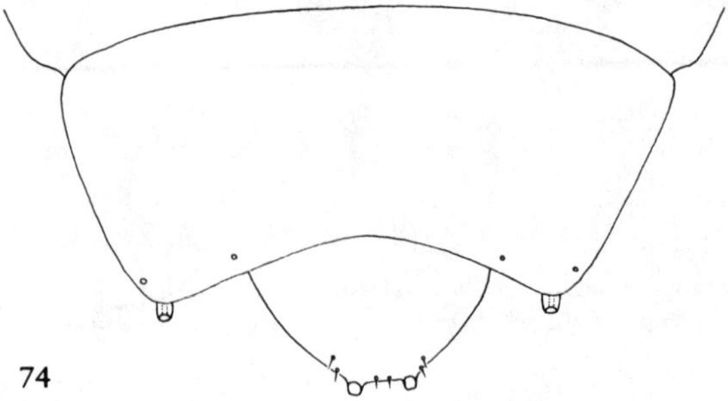

74

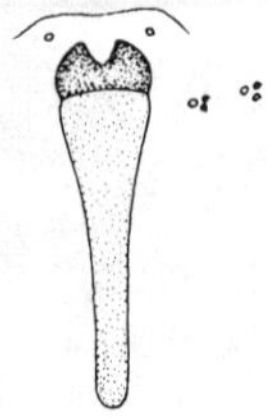

75

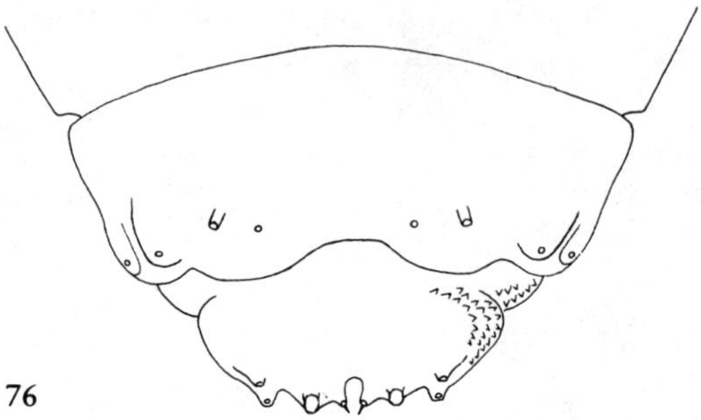

76

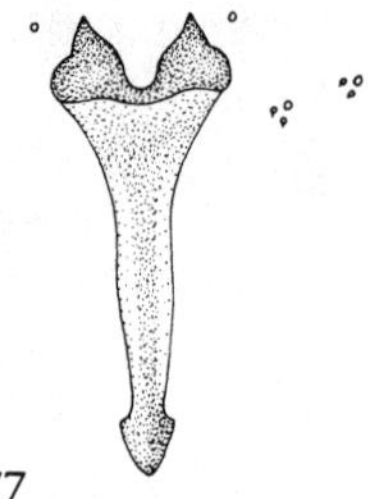

77

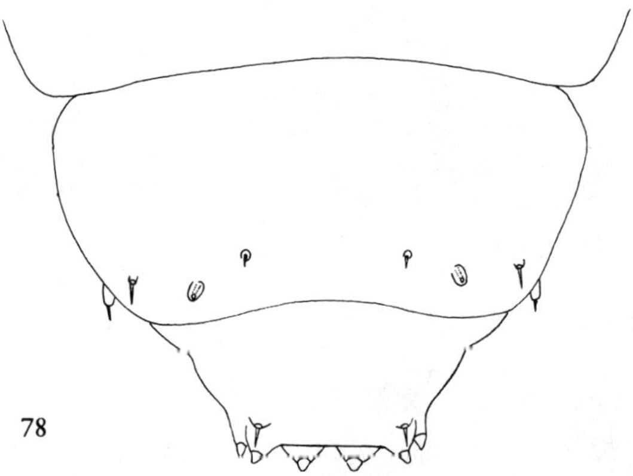

78

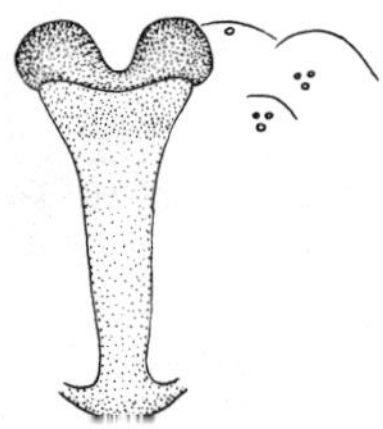

79

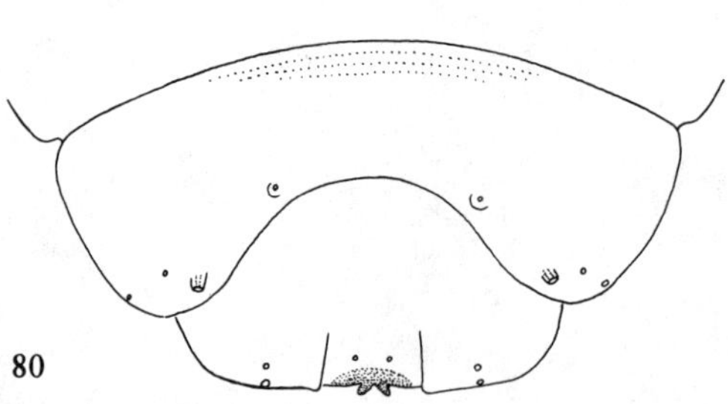

80

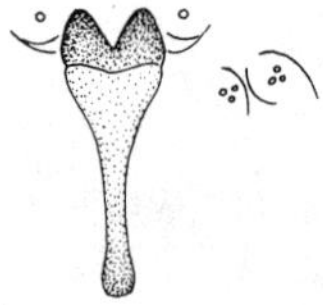

81

Figures 74–81. Larval posterior segments (dorsal) and spatulae. **74, 75,** *Contarinia* sp. near *schulzi*, from *Baccharis*. **76, 77,** *Contarinia nucicola*. **78, 79,** *Putoniella* sp. **80, 81,** *Thecodiplosis brachynteroides*. Scale line: 0.2 mm.

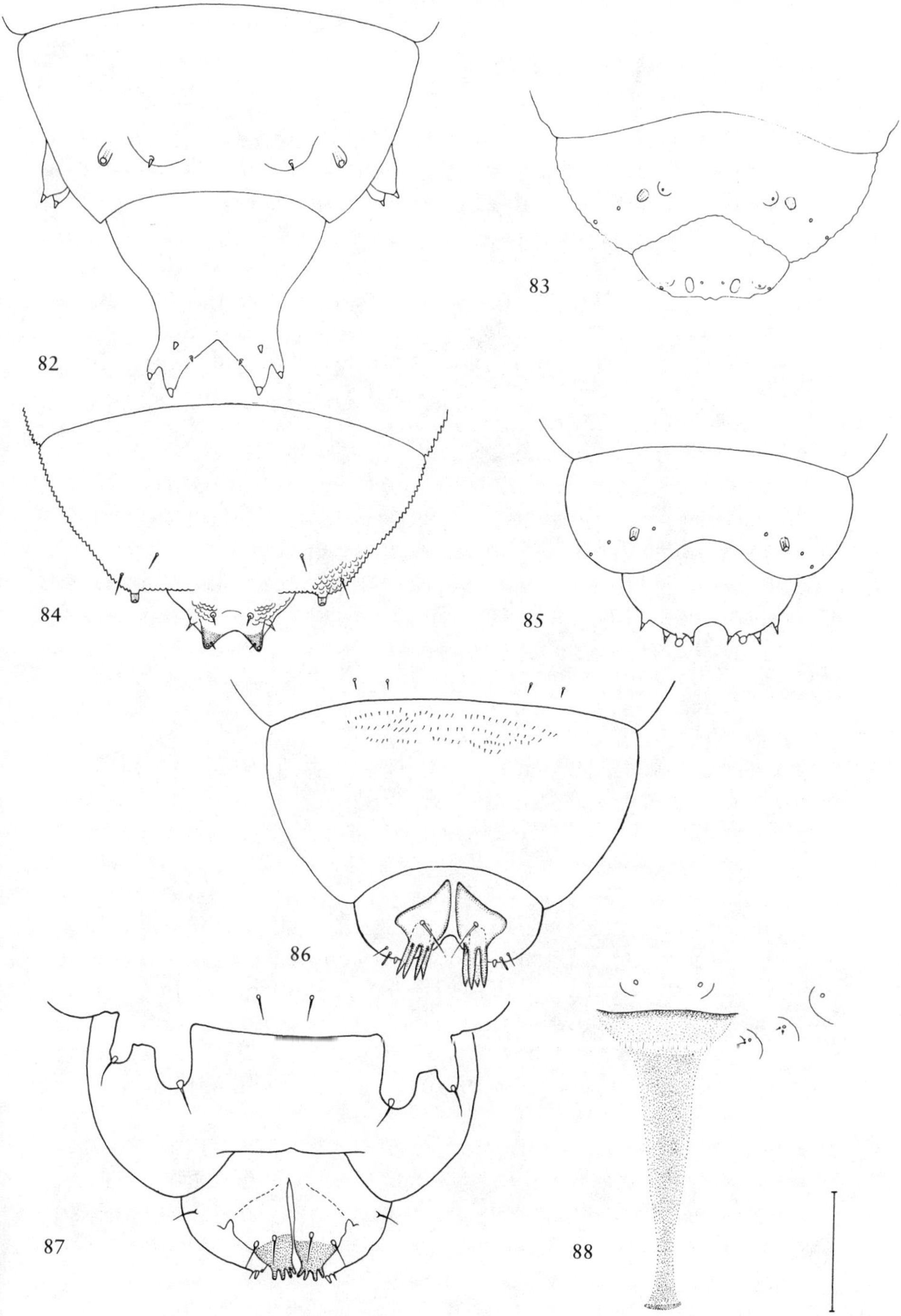

Figures 82–88. Larval posterior segments (dorsal) and spatula. **82,** *Zeuxidiplosis giardi.* **83,** *Paradiplosis tumifex.* **84,** *Resseliella conicola.* **85,** *Harmandia* sp. **86,** *Cecidomyia resinicola.* **87, 88,** *Cecidomyia candidipes.* Scale line: 0.2 mm.

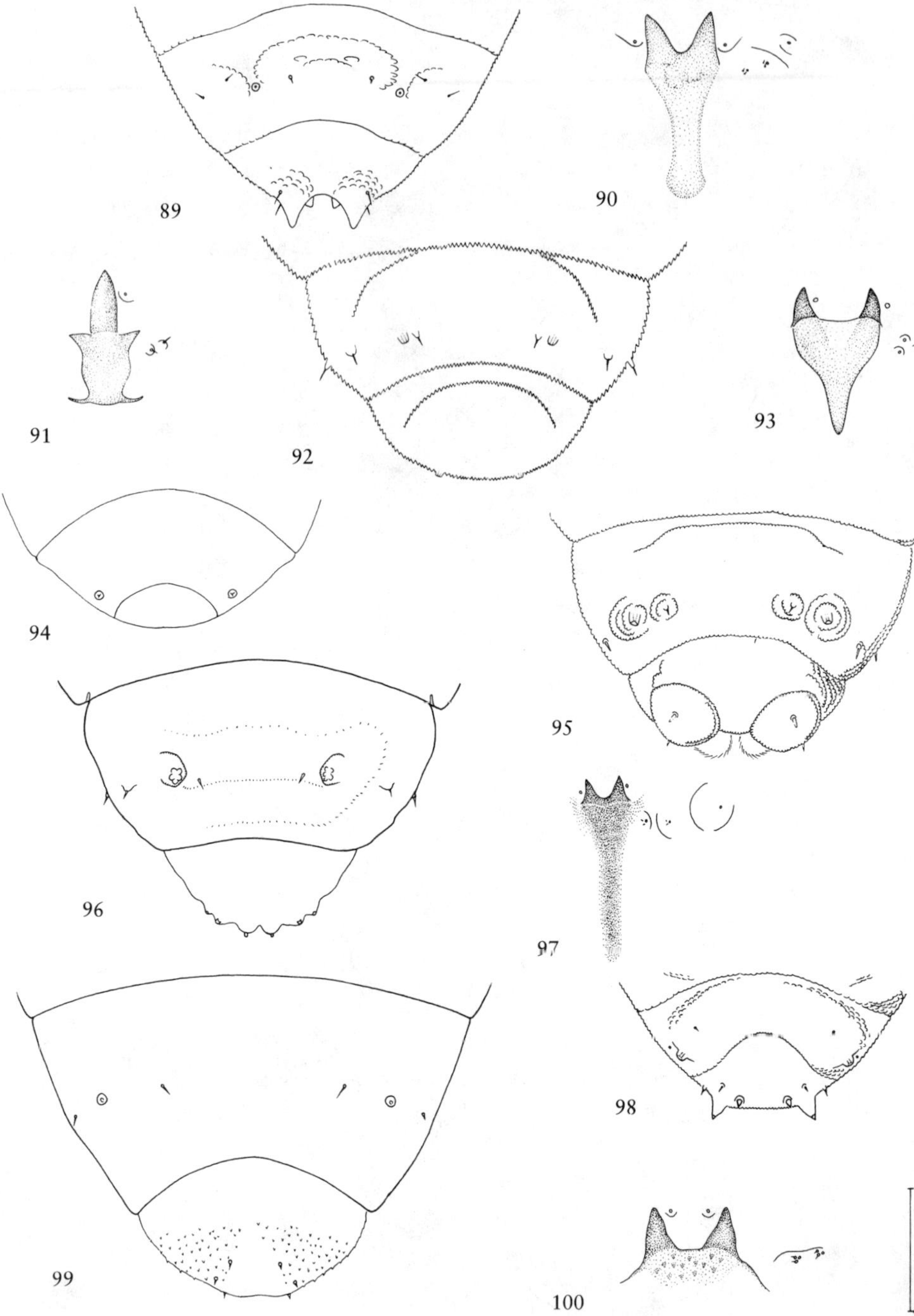

Figures 89–100. Larval posterior segments (dorsal) and spatulae. **89, 90,** Cecidomyiid on *Andropogon gerardii.* **91, 92,** *Caryomyia* sp. **93,** *Caryomyia* sp. **94,** *Pinyonia edulicola.* **95,** *Obolodiplosis robiniae.* **96, 97,** *Ctenodactylomyia watsoni.* **98,** Cecidomyiid on *Bouteloua curtipendula.* **99, 100,** *Planetella* sp. Scale line: 0.2 mm.

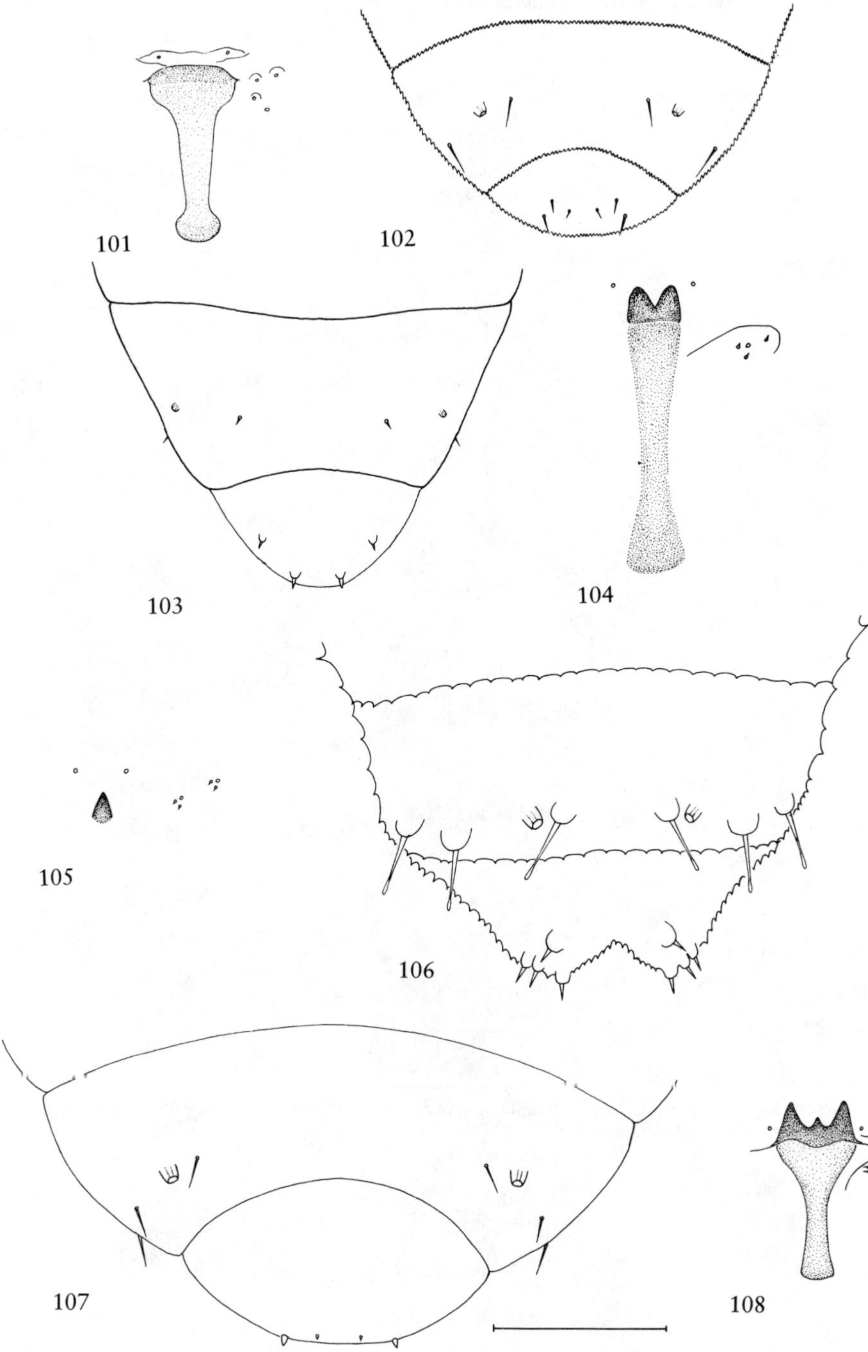

Figures 101–108. Larval posterior segments (dorsal) and spatulae. **101, 102,** *Pilodiplosis helianthibulla.* **103, 104,** *Monarthropalpus flavus.* **105, 106,** "*Cecidomyia*" *glutinosa.* **107, 108,** *Taxodiomyia* sp. Scale line: 0.2 mm.

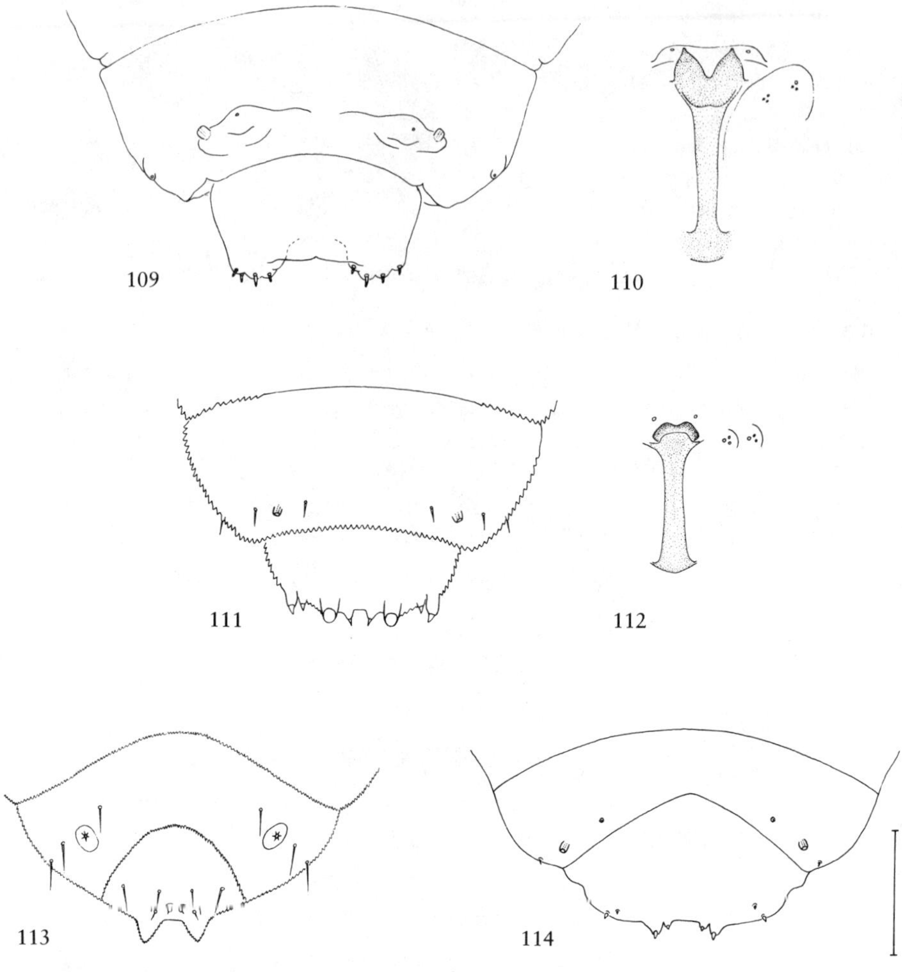

Figures 109–114. Larval posterior segments (dorsal) and spatulae. 109, 110, *Blaesodiplosis crataegifolia*. 111, 112, *Macrodiplosis* sp. 113, *Olpodiplosis helianthi*. 114, *Apagodiplosis papyriferae*. Scale line: 0.2 mm.

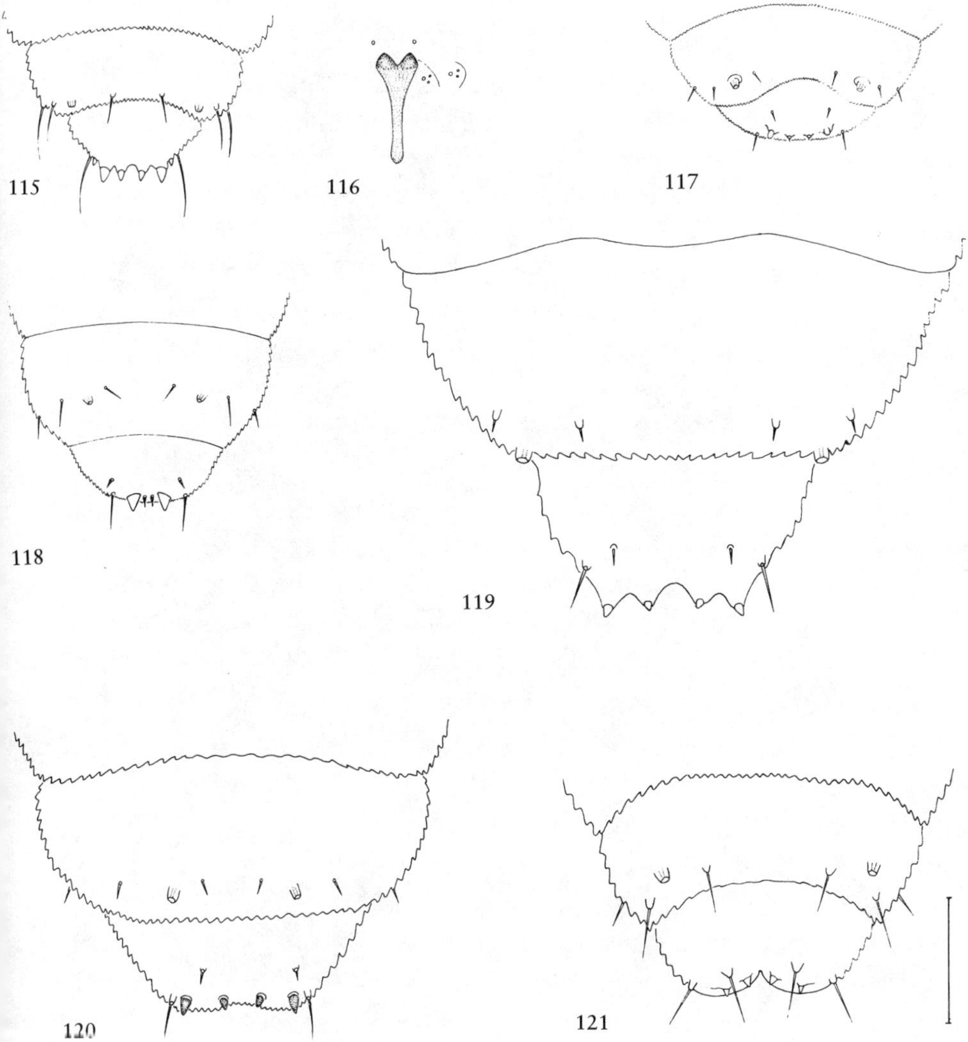

Figures 115–121. Larval posterior segments (dorsal) and spatula. **115, 116,** *Clinodiplosis* sp. **117,** *Clinodiplosis cattleyae.* **118,** *Parallelodiplosis subtruncata.* **119,** *Sitodiplosis mosellana.* **120,** *Clinodiplosis verbenae.* **121,** *Mycodiplosis thoracica.* Scale line: 0.2 mm.

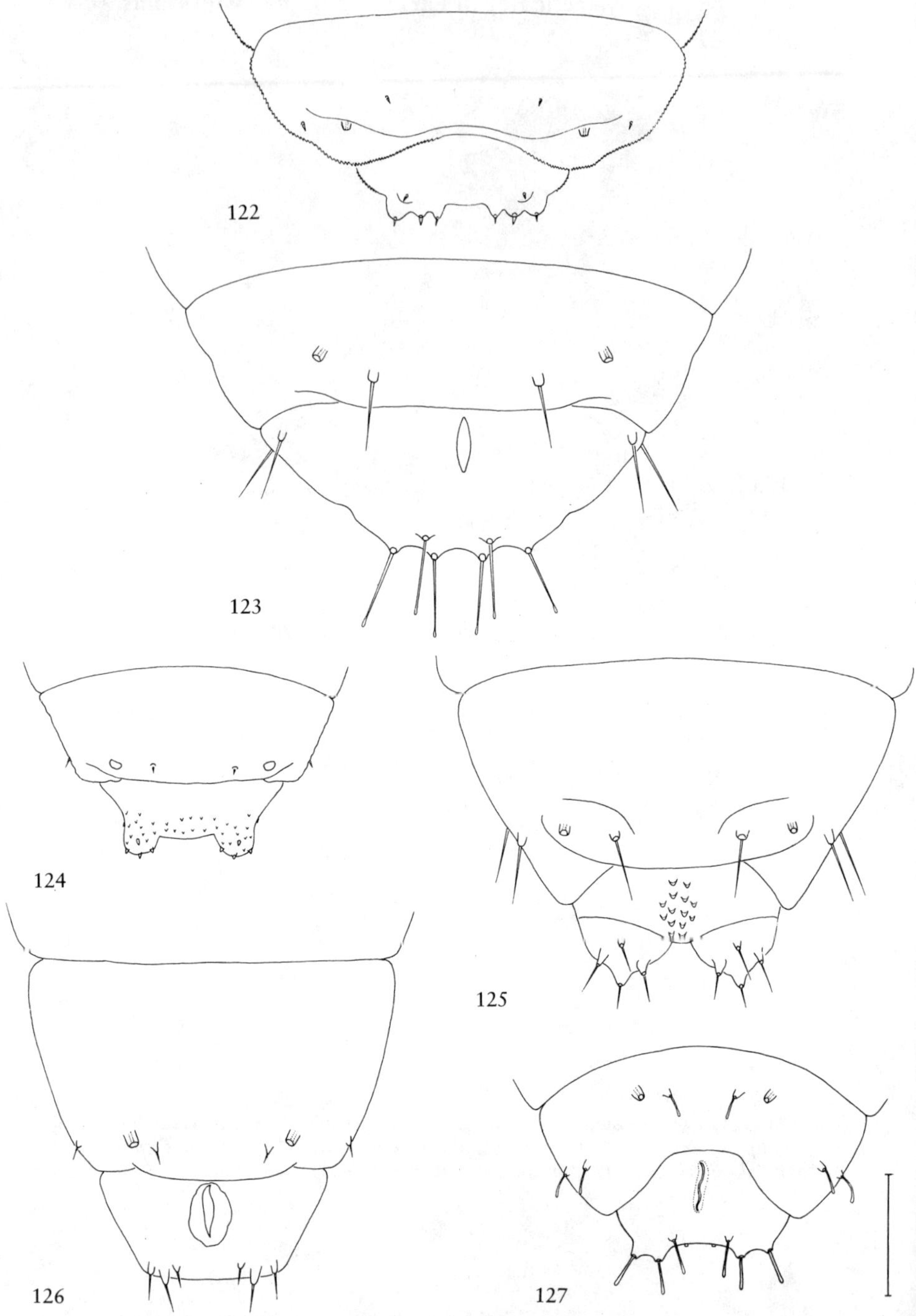

Figures 122–127. Larval posterior segments (dorsal). **122,** *Youngomyia podophyllae*. **123,** *Lestodiplosis* sp. **124,** *Karshomyia caulicola*. **125,** *Aphidoletes urticariae*. **126,** *Feltiella macgregori*. **127,** *Cartodiplosis nyssaecola*. Scale line: 122, 124, 125 = 0.2 mm; 123, 126, 127 = 0.1 mm.

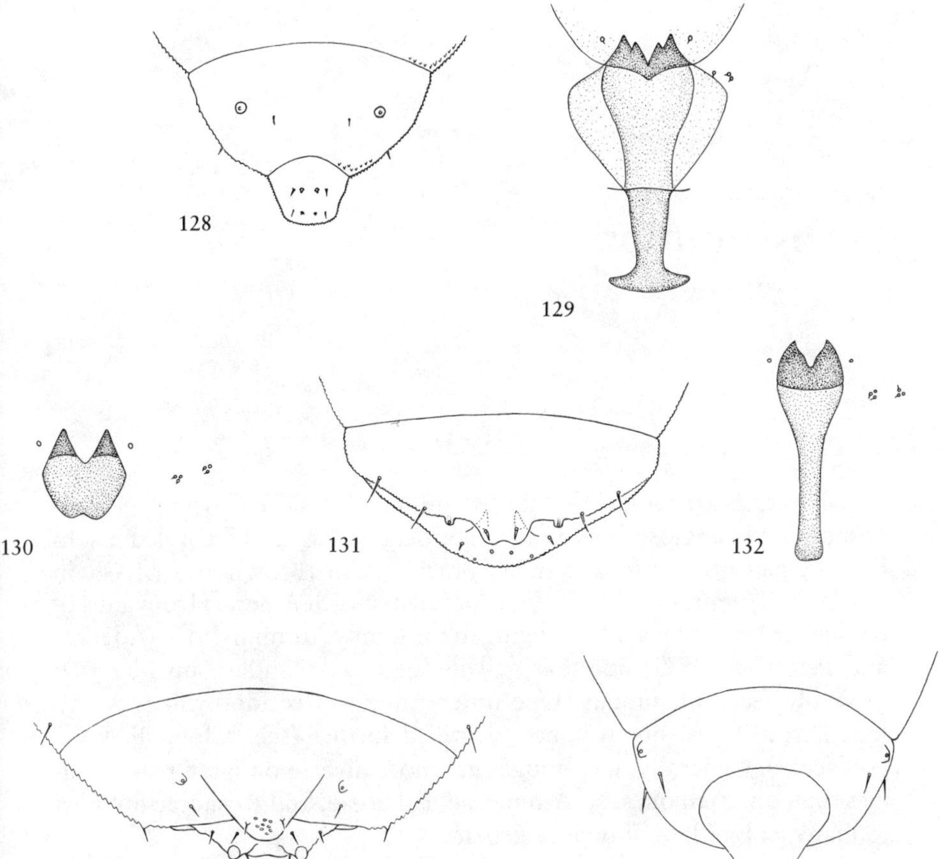

Figures 128–136. Larval posterior segments (dorsal) and spatulae. **128, 129,** *Asphondylia websteri.* **130, 133,** *Schizomyia vitispomum.* **131, 132,** *Schizomyia impatientis.* **134,** *Bruggmannia* sp. **135, 136,** *Polystepha pilulae.* Scale line: 0.2 mm.

5 Distribution

Hosts

Gall midges are widely distributed among plants. They occur on gymnosperms and angiosperms and on monocotyledons and dicotyledons. In Europe, gall midges feed even on bracken fern (Brown and McGavin 1982). A *Clinodiplosis* has been recorded in North America from galls of ostrich fern made by a fly belonging to the family Anthomyiidae (Aderkas and Peterson 1987), but that gall midge is a secondary invader that probably feeds on fungus. One other American cecidomyiid was collected from ferns, but was not identified further (see chapter 8, Polypodiaceae). Generally, gall midges are more diverse on large plant families than on small ones, so Asteraceae, Fabaceae, and Rosaceae are host to many gall midges of diverse genera.

No evidence exists for broad coevolution of plants and gall midges (Möhn 1961, Roskam 1985). The two species of *Dasineura* that roll the leaves of bracken in Europe are not especially primitive cecidomyiids, and there is no evidence that gall midges on conifers are more primitive than those on angiosperms.

It is obvious, however, that some groups of gall midges evolved on and are restricted to particular plant genera and tribes (Möhn 1961, Roskam 1985). Many groups of gall midges are restricted to particular hosts, e.g., *Caryomyia* on hickory, *Polystepha* on oaks, and *Mayetiola* on Festucoideae (Poaceae). The best-worked example of such an occurrence is of the genus *Semudobia* and its birch hosts (Roskam 1985, Roskam and Van Uffelen 1981).

Adaptation of gall midges to plants appears to have followed an ecological path rather than a phylogenetic one. A recent analysis by Roskam (1985) shows that gall midges are more numerous in plant families with woody representatives than in those that are herbaceous.

Woody plants generally live longer, are more diverse structurally, and are a more reliable resource than short-lived, herbaceous hosts. Möhn (1961) and Roskam (1985) show that there is some specialization among gall midges for certain plant organs. As examples, *Asphondylia* spp. are mainly on generative structures, flowers and buds, while *Lasioptera* spp. are mainly on vegetative structures, stems and leaves. There was possibly some preference for particular tissue in the original radiation of the groups (Roskam 1985).

Geography

Cecidomyiidae show several different patterns of geographic distribution. Many genera are shared between North America and Europe, as are their host plant genera. Species, except for recent immigrants, are rarely distributed across two continents, however. Other genera and an entire tribe are limited to the Americas regardless of the distribution of their hosts, and some genera and species groups are cosmopolitan.

Distribution patterns of gall midges can be explained by immigration and the history of continental plates. Immigration is attested to by the numerous established exotic weeds and pest insects brought by humans to North America in the past 500 years. Immigration is an ongoing, continuous process limited only by natural barriers. Eighteen thousand years ago, most of Canada and the northern tier of the United States were overlaid by glaciers. That those areas are now covered with vegetation is testament to the constant push made by living organisms against receding frontiers.

But older, more fundamental similarities between continents, particularly visible in the plants, animals, and geological formations shared between North America and Eurasia, are presumably due to past, very different, configurations of the earth's surface than exist today. The most likely reason North America and Eurasia share so many genera of gall midges is that these continents were still joined together about 50 million years ago rather than that each one of the similar groups of plants and animals migrated across straits or narrow land bridges (Gagné 1984, Noonan 1986). Alternatively, the reason that a given genus of gall midges is either Nearctic or Palearctic while its host plant is Holarctic might be either that the genus became extinct over part of its range or that it arose only after the barrier between the continents developed. The distribution of a genus or other group allows us to hypothesize about the age or origin of that group. Listed below are examples of several patterns of distribution as they relate to gall midges.

Recent Immigration through Commerce

It is important to know which species are recent immigrants spread through commerce so that they do not cloud the natural patterns. Generally, immigrants are restricted to alien hosts, although at least two, *Contarinia baeri* (Prell) and *Zeuxidiplosis giardi*, have been recorded from native American hosts. At least 30 plant-feeding cecidomyiids have been accidentally or purposely introduced into North America. One of them probably came from Japan, one is from Africa, and the remainder are from Europe. The most important of these include the Hessian fly, brought to America in wheat straw from Europe during the Revolutionary War (Fitch 1846), and the sorghum midge, which arrived in the nineteenth century with seed of Johnson grass, a close relative of sorghum, that was purposely introduced into the southern United States as a forage grass. Most of these immigrants were introduced with their hosts and remain restricted to them: some are innocuous or beneficial, if they feed on weeds, as do the two species of *Contarinia* that feed in the fruit of dock. Four exotic plant-feeding gall midges have been purposely introduced into North America for the control of various rangeland pests.

Many more exotic gall midges may occur here on weeds but have gone unnoticed. Only four years ago a characteristic leaf gall formed by *Dasineura glechomae* (Kieffer) on ground-ivy, *Glecoma hederacea*, was first noticed in Maryland by E. Sylvén, a Swedish scientist well acquainted with the gall in Europe.

Holarctic Distribution

It is almost a rule that a monophyletic group of gall midges restricted to a Holarctic host plant genus will be found in both North America and Eurasia. The synopsis of genera gives many examples, including *Rabdophaga* on willows, *Harmandia* on poplars, and *Kaltenbachiola* on spruce. *Celticecis* was until recently known only from North American hackberries, although Moser (1954) had predicted it would occur in the Orient. A species of *Celticecis* recently described from Japan (Yukawa and Tsuda 1987) and specimens in the U.S. National Museum of Natural History (USNM) of an undescribed species from Turkey sent to me by Moser, show that the genus is widespread in the Holarctic Region.

The genus *Semudobia* on birch provides an excellent illustration of Holarctic distribution because its interspecific and host relationships have been worked out so well (Roskam 1977, 1979). The genus consists of five species. *Semudobia skuhravae* Roskam, the only one native to both North America and Eurasia, has the most generalized habits and can live in either seeds or bracts. The remaining four species live only in seeds. Two of these four are restricted to North America. The remaining two were restricted to Eurasia until recently, when they were introduced into North America on ornamental birches. The two native American species are discontinuous, one in the West, the other in the East. Judging from an analysis of the external anatomy, Roskam (1985) hypothesized that the most generalized species is the sister group to the remaining four and has not changed because of its generalized host and biological needs. The two native North American species could have arisen from a common ancestor that became separated from its European counterpart when the two continents drifted apart. Subsequent cooling of North America and consequent isolation on eastern and western portions of the continent could have resulted in the division of the American species as well as the host birches. If evolution of the group did occur in this way, *S. skuhravae* is very old and has remained unchanged for at least 50 million years even though isolated on separate continents.

A more recent example of dispersal between North America and Europe prior to historic times is in *Mayetiola*. That genus, in the strict sense, contains 27 European species and one native American species, *Mayetiola ammophilae*. Another species in North America, the Hessian fly, was accidentally introduced here during historic times. *Mayetiola ammophilae* lives on American beachgrass, *Ammophila breviligulata*, a plant limited to the eastern North American coastline and Great Lakes. This grass is closely related to European beachgrass, *Ammophila arenaria* (L.) Link, which occurs along the northern European coastline. In Pleistocene times or earlier, the two species of grasses may have been one continuous species between Europe and North America, and the fly could have expanded its range from Europe. No *Mayetiola* species has yet been reported from European beachgrass, but one probably exists.

Relict Distribution

Some genera were at one time apparently much more widespread than they are now. *Sequoiomyia*, which contains one living species, on bald cypress in North America, was originally erected for a fossil species from a Miocene *Sequoia* found in Germany. The genus may occur on other extant Taxodiaceae also, but we know that it once ranged across the Holarctic. It presumably became extinct in Europe with the demise of its hosts there.

The Nearctic *Caryomyia*, a large, monophyletic group of species that form complex galls on hickory leaves (see fig. 310), may also have been more widespread at one time. It is presently known only from southeastern North America in a range congruent with hickories. Hickories now occur only in southeastern North America (about 11 species) and China (one species). If *Caryomyia* is ever found in China, we might suppose that at one time this genus of gall midges occurred in Europe, because hickories occur there as fossils.

A different situation occurs with gall midges of beech. Several complex leaf gall makers are known on this tree in Europe (Barnes 1951), and 22 such galls have recently been found in Japan (Tsuda 1982), yet to date none has been reported from American beech. Possibly some gall midges did occur in North America but became extinct during the glacial periods, when the range of beech in North America was severely restricted.

New World Distribution

The tribe Alycaulini, a large, diverse group of 188 described species in 17 genera, is restricted to North and South America. Its species occur on a wide array of plant genera. Many of these plants also occur in Eurasia but are free there of these particular gall midges. Since none of this large tribe occurs in the Old World, one can only suppose that the group appeared in North America after the mid-Eocene Euramerican separation. Possibly it arose in South America and migrated to North America after those continents were connected during the Cenozoic Era.

Another tribe, the Asphondyliini, is cosmopolitan, but shows its more primitive character states in genera restricted to tropical America, including Florida (Möhn 1961).

Worldwide Distribution

Some genera and species groups occur throughout the world. The *Contarinia sorghicola* species group occurs naturally on grasses on all continents, including Australia (Harris 1979). Australia has been isolated from the rest of the world for so long that it seems unlikely the immediate common ancestor of all the *C. sorghicola* group was present before the breakup of that continent and the rest of the world.

Asphondylia is a very large, cosmopolitan genus. Anatomical evidence indicates that American *Asphondylia* that feed on Fabaceae apparently come from a different ancestor within the genus than do the Eurasian species that feed on that family (Gagné and Wuensche 1986). Further, plant hosts of the genus are somewhat different in North America and Europe. Palearctic *Asphondylia* are found chiefly on Fabaceae, while in the New World, they are mainly on Asteraceae (Möhn 1961).

The Asphondyliini are richer in species in the New World than in the Old, but the reverse is true of the Lasiopterini. The latter tribe is depauperate in this hemisphere, where it is known from only 39 species on hosts scattered among many plant families and extends southward only to El Salvador. In the Eastern Hemisphere the tribe is known from 220 species, associated with very diverse plants. In fact in the Old World they also live in stems of grasses, whereas in North America, with the exception of *Hybolasioptera elymi* Gagné, a lasiopterine, that niche is taken over by nonlasiopterine gall midges. A plausible explanation for the distribution of the Lasiopterini is that they were widespread on Laurasia, the large northern continent of the Cretaceous. After the fragmentation of Laurasia, the species in North America evidently diversified less than did those in Eurasia. Alternatively, the tribe might have entered North America after the breakup of Laurasia, possibly across the Bering Land Bridge. *Hybolasioptera* is the only monophyletic genus of that tribe that is shared between America and Eurasia.

6 Galls

Galls as defined here include any predictable and consistent plant deformation that occurs in response to feeding or other stimulus by foreign organisms. This definition encompasses almost all damage done to plants by gall midges. It includes even leaf spots formed by *Acericecis ocellaris* on maple. Some gall midges that in aggregate form stem swellings may not show any enlargement when only one larva is in the stem. For convenience, that unenlarged stem is still considered a gall here.

Terminology

Comprehensive schemes for naming and classifying galls have been devised (Felt 1940, Küster 1911, Mani 1964, Wells 1921a). Available nomenclature for galls may be helpful in detailed studies of gall morphology but is unnecessary for purposes of this book, in which a simplified terminology is used. Here, galls are considered either *simple* or *complex*, terms roughly equivalent, respectively, to the terms *kataplasma* and *prosoplasma* used in other works. A simple gall, or kataplasma, is one that does not differ from normal plant tissue except that it is swollen or distorted. Simple galls include spots, swellings, folds, and invaginations: all the galls in figure 145. A complex gall, or prosoplasma, is the result of tissue that is completely reorganized into a fundamentally different shape from any structure normally found on a plant. Examples of complex galls are the leaf galls in figure 310f, i, l, and k.

In this book, a few phrases will often be used to characterize some of the simple galls. A *leaf spot* or *leaf blister* is an ovoid to circular area, with one or more larvae in a concavity on the underside of the leaf or inside the leaf tissue. If one or both sides is convex, the gall is called a blister instead of a spot. If the whole leaf is rolled to form a tube, the gall is a *leaf roll*; if only the margin of the leaf is rolled, it is a *marginal leaf roll*. A *vein fold* or *leaf fold* is an inpocketing of the leaf blade, along a vein or between veins, respectively, that does not affect the whole leaf.

Two terms occasionally used here for complex galls are *monothalamous* and *polythalamous*. The first refers to galls with a single larval cell that contains one or more larvae, the second, to galls with several discrete cells, each with a single larva.

Organisms That Form Galls

With a little practice, one can easily differentiate galls made by cecidomyiids from those made by other organisms. Most galls are made by gall midges, but an unknown, closed gall must usually be cut open to ascertain what kind of organism causes it. It is important to open more than one gall when trying to determine the gall maker; some may appear superficially to be made by a parasitoid hymenopteran that has completely destroyed the gall maker.

Cecidomyiidae

Simple cecidomyiid galls can be pulled apart to reveal one to several, legless, minute-headed larvae (fig. 137). The feeding area is smooth and clean; larvae produce no noticeable waste products. Complex cecidomyiid galls contain one or more ovoid or cylindrical cells, each usually inhabited by one larva. Full-grown larvae will often have a characteristic spatula (fig. 19), although that is lost in some genera. The inside surfaces of galls caused by *Asphondylia* and by most *Lasioptera* and *Neolasioptera* are lined with a white fungal mycelium.

Hymenoptera

The galls most easily mistaken for cecidomyiid galls are formed by legless larvae of Hymenoptera. These gall-forming hymenopterans belong mainly to the Cynipidae, but also to such families as Pteromalidae and Eulophidae. Like cecidomyiid larvae, the hymenopterous larvae typically produce no solid waste. The latter differ most from gall midges in having a conspicuous pair of opposable mandibles (fig. 140). Cynipids are responsible for most hymenopterous galls and occur mainly on oak,

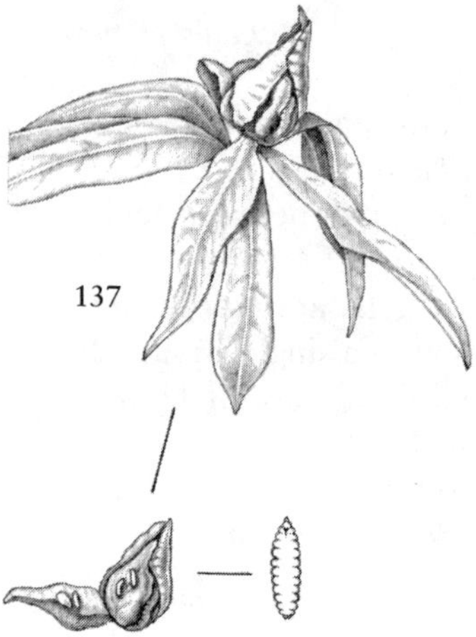
137

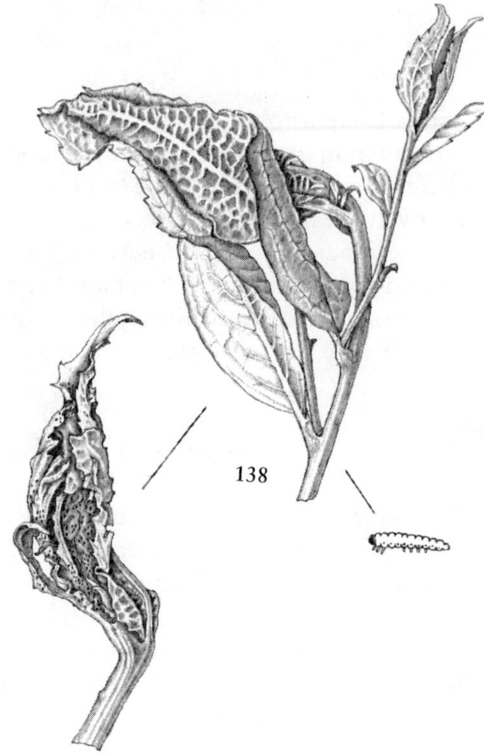
138

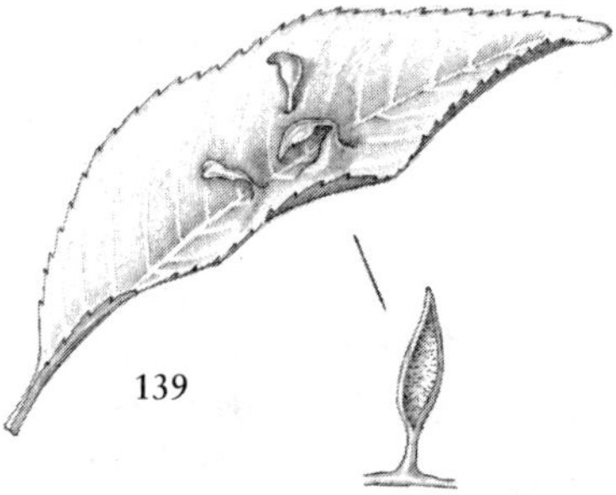
139

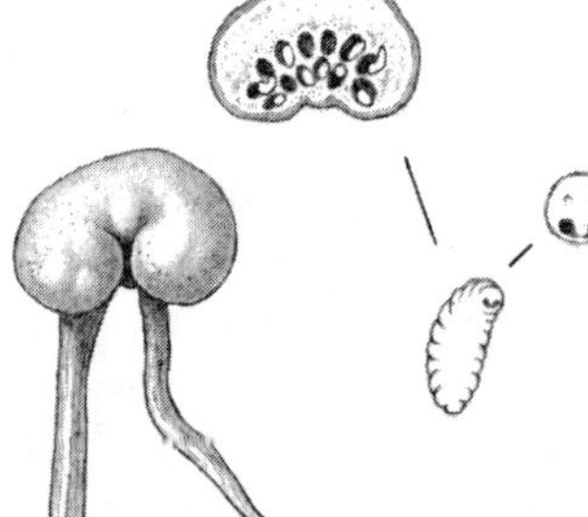
140

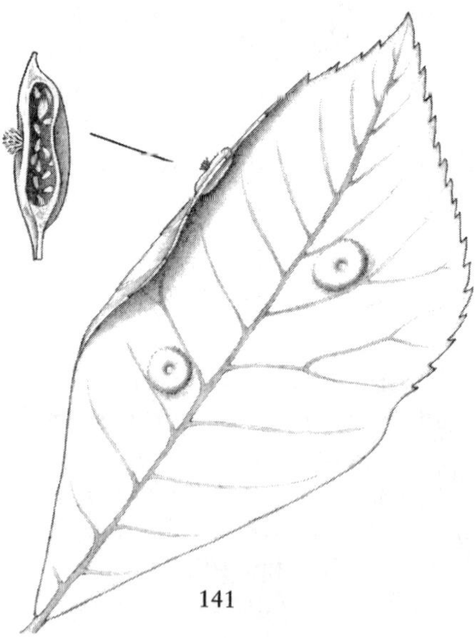
141

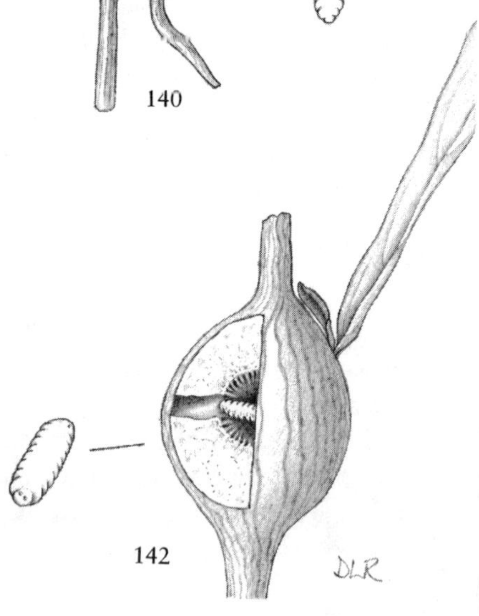

142

rose, and blackberry; on oaks they form complex galls very unlike the simple leaf blisters, folds, and rolls made by most gall midges on those trees. Figure 140 shows a many-celled gall formed by a pteromalid, *Hemadas nubilipennis* (Ashmead). Felt's (1940) *Plant Galls and Gall Makers*, although outdated, is still the most comprehensive beginner's guide for identification of these wasp galls.

Eriophyidae (Acarina)

Eriophyid mites form diverse galls on a great variety of plants. Their galls tend to be made up of greatly convoluted tissue or simple, globular, sometimes pointed outpocketings of leaves (fig. 139). The inner surface of a gall is highly convoluted and is often covered with fine hairs. Mites are gregarious, seldom as long as 1 mm, and have eight legs. The best reference to eriophyid mites is *An Illustrated Guide to Plant Abnormalities Caused by Eriophyid Mites in North America* by Keifer et al. (1982).

Tephritidae (Diptera)

Tephritid flies make galls chiefly on western North American asteraceous plants. A common eastern American gall on goldenrod is made by *Eurosta solidaginis* (fig. 142). Tephritid larvae are legless, like gall midge larvae, but are much larger and more robust than those of gall midges and have only two pairs of spiracles, an anterior pair and a large, conspicuous hind pair (fig. 142).

Agromyzidae (Diptera)

This large group of flies is chiefly responsible for flat, circular leaf-mines, although several form twig swellings or bud galls (fig. 218). The

Figures 137–142. Galls made by various groups of gall makers. All are drawn to actual size unless otherwise noted. **137,** Enlarged buds, *Euphorbia esula*, one with outer leaves removed to show larvae and cocoons; larva in 5×: *Dasineura* sp. near *capitigena* (Diptera: Cecidomyiidae). **138,** Composite leaf rolls, *Solidago rugosa*, one opened to show chewed leaves and frass; caterpillar in 2×: Lepidoptera: Olethreutidae. **139,** Leaf pouch galls, *Prunus serotina*, one in section, 2×: *Phytoptus cerasicrumena* Walsh (Acari: Eriophyidae). **140,** Kidney-shaped shoot galls, *Vaccinium angustifolium*, one in section; larva in 5×, head capsule in 10×: *Hemadas nubilipennis* (Ashmead) (Hymenoptera: Pteromalidae). **141,** Leaf blister galls, *Carya ovalis*, one gall in cross section, 3×; female and nymphs: *Phylloxera* sp. (Homoptera: Phylloxeridae). **142,** Globular stem swelling, partly sectioned, *Solidago canadensis*; larva in 2×: *Eurosta solidaginis* (Fitch) (Diptera: Tephritidae).

maggots always leave behind a trail of frass, distinguishing them from gall midges.

Other Gall Makers

Galls not caused by Diptera or legless Hymenoptera contain frass or waxy exudate in them. Most of the organisms that make these galls have legs; if not, as in some weevils (Coleoptera), they have prominent head capsules and mandibles. Several kinds of Homoptera make galls: Adelgidae form conelike galls on twigs of spruce; Pemphigidae, leaf galls on poplar; and Phylloxeridae, leaf galls on hickory (fig. 141), grape, and sour gum (fig. 256a). Other North American gall-making insects include Psyllidae (Homoptera) in galls on hackberry (fig. 430j); Tenthredinidae (Hymenoptera), or sawflies, in galls on willow and poplar; and a few Lepidoptera and Coleoptera in scattered stem and leaf galls (fig. 138) or mines. Felt's *Plant Galls and Gall Makers* (1940) is helpful for identifying most of these galls. *Insects that Feed on Trees and Shrubs* by Johnson and Lyon (1989) provides more recent references but does not have complete coverage of galls. Two articles by E. L. Smith (1968, 1970) give more information about Tenthredinidae.

Diversity

Cecidomyiids are responsible for a great variety of galls. Particular kinds of galls have evolved separately many times on unrelated plant groups. For example, spine-shaped galls that occur on grape, hackberry, and hickory are each made by a gall midge from a different tribe. Simple leaf spots or blister galls found on leaves of tulip poplar, magnolia, hazel, grape, and many others are made by an array of gall midge genera. Conversely, a group of closely related gall midges may be responsible for a wide variety of galls: all the galls in figure 157 are made by species belonging to a subgenus of *Rhopalomyia*.

Each cecidomyiid gall maker is responsible for a particular kind of gall, unlike some cynipid or phylloxerid species that may cause two very different types of deformation. But some cases are known of slight variations in cecidomyiid galls made by opposite sexes or distant populations of a gall midge. Galls made by females of the European *Mikiola fagi* (Hartig) are larger, more ovoid, and paler than those made by males (Coutin and Riom 1967); those of the Japanese *Masakimyia pustulae* are thick- or thin-walled in different parts of Japan, a difference that appears to be dependent on the genetic makeup of the gall midge (Sunose 1985).

Development

Many fine studies exist on the subject of cellular development of particular cecidomyiid galls (Jauffret 1970, 1973, Rohfritsch 1974, 1980, 1987a, 1987b, Rohfritsch and Shorthouse 1982, Rohfritsch and Skuhravý 1970, Solinas and Bucci 1982, West and Shorthouse 1982). Rapidly growing meristematic tissue of host plants reacts to larval salivary secretions. The larva of the Hessian fly, having migrated to a leaf node, settles down and begins to feed, its mandibles penetrating through several layers of cells. Cell growth ceases where the larval venter is in contact with the plant but continues around its periphery, resulting in a depression in which the larva develops (Painter 1951). The same process occurs in species that form complex galls that completely enclose the larva. If a larva is removed from the gall at any time during growth, the gall will stop growing. The rate of cell division around the affected area is greatly increased. Most galls are initiated on the external surface of the plant, and the tissue grows around the larva. Plant cells affected by salivary secretions change into cytoplasmically dense cells, the "nutritive tissue." The cuticle of these cells is thinner, and their contents are modified. The piercing mandibles of the larva break the cell walls, and the cell contents are sucked up by the larva (Solinas and Bucci 1982). Between these cells and normal plant tissue, vascular tissue is formed that intimately connects gall and plant organ (Rohfritsch and Shorthouse 1982).

Galls are in effect a new organ controlled by the gall maker, and cell proliferation is never disorganized. Rohfritsch and Shorthouse (1982) show how gall shape is in large part caused by the feeding pattern of the gall midge, whether the larva feeds at one spot or moves around the inside of the gall. In complex galls from which full-grown larvae exit, the eventual exit tunnel remains closed while the gall is growing, but spreads open near the time full-grown larvae are ready to leave (Rohfritsch and Shorthouse 1982).

Some cecidomyiids that lay eggs directly into plant tissue, notably asphondylias, may deposit a substance during oviposition that initiates gall formation. The indirect evidence for this is that larvae of many species of *Asphondylia* remain as quiescent first instars for many months, yet the seeds or other affected tissue do not grow normally.

Significance

Galls provide food and shelter and seem to protect gall makers from predators, but all three effects may be fortuitous. The ability of gall midges to invade plants has apparently developed many times and may be due to preadaptations that include extraintestinal digestion and salivary secretions with the ability to affect plant growth. Although many gall midges effectively obtain food from plants without becoming wrapped in plant tissue, it seems likely that there is some adaptive significance to the gall makers in persisting with this habit. Price et al. (1987) reviewed the various hypotheses that have been put forward to explain the significance of galls. They find some justification that galls lead to enhanced nutrition, create a more favorable microenvironment, and also possibly lend some protection from parasites.

Plants derive no known benefit from cecidomyiid galls, and most galls appear to be detrimental (West and Shorthouse 1982). Evidence shows that galls serve to trap plant energy and even manipulate the plant (Sunose and Yukawa 1979, Weis and Kapelinski 1984, Yukawa and Tsuda 1986). Effects of *Rhopalomyia solidaginis* (Loew) on goldenrod include increased stem production and decreased rhizome production and seed development (Hartnett and Abrahamson 1979). The apical bud gall of *Rabdophaga strobiloides* on willow asserts a dominance over lateral buds (Weis 1984). Galled twigs grow to greater girth than ungalled twigs, even when leaves are stripped from the galled twigs (Weis and Kapelinski 1984). When the European *Phragmites australis* grows in dry, mixed-plant situations, the internodes infested with the gall midge *Giraudiella inclusa* (Frauenfeld) become abnormally long (Tscharnke 1987). The effect is that infested plants became taller than surrounding plants and receive more sunlight.

Gall makers have their share of parasitoids, but there is evidence that galls afford some protection. For example, the hard excrescence of the fungal symbiont in a blister leaf gall on goldenrod inhibits attack by a parasitoid because the wasp's ovipositor cannot penetrate the mature gall (Weis 1982b).

7 Collecting, Rearing, and Preparing Gall Midges for Study

Gall midges can be caught in great numbers in flight and are sometimes the most numerous group of insects in passive flight traps. But plant-feeding cecidomyiids caught this way are of little value, unless one is looking for something particular. Sexes of flight-caught species are difficult to associate with one another. Even when they can be associated, the flies lack meaning outside the context of their hosts, biology, and immature stages. The work involved in preparing flight-caught specimens for study is often not worth the effort. The best study material is that found associated with plants, in galls or other damage.

Collecting Galls

When collecting to identify larvae or rear adults, the galls should be picked when larvae are fully grown. Most galls are best gathered in clear plastic bags, but paper bags are suitable when condensation on the sides of the containers is a problem, which is especially true when the galls cannot be processed until the next day. Mature larvae will usually have a spatula, and many kinds will leave the galls when disturbed or after their galls have been placed in a plastic bag for 10–30 minutes.

Larvae that normally leave the plant tissue to pupate are mainly those from leaf fold galls or that are free-living in buds or flowers. If larvae inside the galls are barely visible and have no spatula, they are probably in an early instar and will not reach maturity if the galls are picked then. Some species remain in the early larval stages for much of the summer, but most gall midge larvae live only a short time, two weeks or so, in the galls.

Samples of galls should be preserved. Hard galls on stems, most bud galls, and the majority of complex galls on leaves need no special treatment and can be kept dry. Any larvae inside the gall tissue will desiccate in place and can be left there until needed for future study. Galls on herbarium specimens more than 100 years old are still good sources of specimens. Flat galls such as leaf folds or blisters on leaves that might curl or shrivel can be dried between pages of books or in a plant press. The more succulent, large galls can be kept in alcohol or can be freeze dried to retain their shape. Since plant specimens in alcohol lose their color and freeze dryers are not generally available, an alternative is to photograph galls for the record. Another is to use silica gel to dry the tissue gradually.

Rearing Gall Makers

A few full-grown larvae from a collection should be preserved in alcohol or on slide mounts for future reference and identification. The remainder, either free or still within the galls, can be kept for rearing atop a moist mixture of one part clean sand and one part peat moss. The sand keeps the peat moss from binding, and the acidic peat moss retards fungal growth. The pot should be at least 10 cm in diameter but short enough to stand in a cardboard shoebox with 2–3 cm clearance between the top of the pot and the cover of the box. Plastic pots used by plant growers to package seedlings are suitable. Pack the soil more tightly at the very bottom of the pot to prevent larvae from descending all the way through the soil and out the drainage holes. At the end of the shoebox make a hole into which a lipped, 8–10 dram vial will fit snugly. Label both the pot and the shoebox as to contents. Keep the shoebox in a sheltered, cool place, and check the vial daily for emergence of gall midges. To avoid contamination, do not reuse boxes.

Open the box about once a week to check soil moisture and add water, if necessary. Remove any spiders or psocids that have taken up residence. When adult cecidomyiids emerge, they will usually fly into the vial. The vial with midges in it can be removed and quickly inverted into a killing jar, or the flies can be picked out with an alcohol-moistened artist's brush.

If adults do not emerge by autumn of the year in which larvae are collected, the larvae probably require diapause. When cold weather arrives, remove the pots from the shoeboxes and place the pots in a cold frame outdoors. In early spring place the pots into shoeboxes as described above.

Fall is a good time to collect stem galls. Bundle them together, label them, and put them in a cold frame, preferably in a container because much of the stem tissue may rot away during the winter.

A collection of one kind of gall or flower head will sometimes contain more than one species of cecidomyiid. This can usually be determined by examining the larvae closely, but inquilines sometimes occur in small numbers and can be missed. Even when larvae leave a host readily, it is wise to retain some of the plant tissue in case another species is present. For example, *Contarinia* and *Dasineura* larvae in flower heads of *Baccharis* will readily leave the plant when full-grown, but *Neolasioptera* larvae in the same heads will remain embedded in the plant tissue.

Mounting Specimens

A dissecting microscope should allow one to distinguish among some genera of cecidomyiids. For detailed study of larvae or adults, however, the specimens will need to be mounted on microscope slides for study under a compound microscope, preferably one with phase contrast. If a slide has to be made, it is best to make a good-quality, permanent one. This means that all the muscles should be dissolved.

To dissolve the muscles of a larva, first prick it once or twice with a fine insect pin. Then place the specimen in a small crucible with a solution of sodium hydroxide. It can be heated quickly on a hot plate or left in a solution for a longer period of time at room temperature. If a hot plate is used, allow the solution to boil gently for a minute or so. When the specimen is cleared, remove it with an eye dropper and place it in a porcelain spot plate. With fine forceps transfer the larva to a drop of acetic acid for a minute to neutralize the sodium hydroxide, and then transfer it successively to 70% alcohol, 95% alcohol, and creosote or clove oil, leaving it in each liquid for about five minutes. Specimens can be kept in creosote or clove oil overnight if necessary. Then transfer the larva to a drop of Canada balsam on a slide and lay a cover slip on it. Attach a label to each end of the slide: one label to record the host, other collection data, and the mounting medium, and the other to record the identification.

Adults are more difficult to prepare. Taking a specimen from alcohol, first remove the wings and place them directly into the creosote. Puncture the abdomen with a fine insect pin and place the body in the sodium hydroxide solution. When the specimen is cleared, follow the same process as for a larva. The wings can be mounted under one cover slip, the rest of the body under the other. Once on the cover slip in the drop of balsam, the head can be removed from the body and placed so that it can be seen in frontal view. If the specimen is a male, the terminalia should also be disconnected from the body and placed dorsoventrally. The remainder of the body is mounted laterally. When several specimens are available, it is a good idea to mount some with the head and terminalia still attached. Pupal skins can be mounted without clearing in the same way as wings, but they should be stored in alcohol long enough so that no air remains in the leg or antennal sheaths. A little practice is required to get the Canada balsam just thick enough so that the specimen is not distorted and yet not so thick that one cannot use the high power microscope objectives. An easier, alternate mounting medium is Euparal, which is miscible with 70% alcohol. With Euparal, two steps can be skipped, the 95% alcohol and the creosote. The optical properties of Euparal are adequate for most work.

Specimens stored in alcohol deteriorate with time, so mount an adequate sample on slides as soon as possible after collecting or rearing. Because many species are strikingly marked with colored scales that are lost when immersed in liquid, it is best to pin a series on double mounts or on card points. Most adults shrivel slightly, but the scale or color pattern will preserve well. Larvae and pupae can also be mounted on card points for long-term storage. These should be kept the same way and with the same care as for any pinned specimens. More general guidelines for preparing and maintaining a collection are given in Steyskal et al. (1987).

Plate 1. A-C, *Dasineura porrecta* on elm flowers. **A,** Eggs at base of flower bud. **B,** First larval instar crawling on developing samara. **C,** Third larval instar in pocket of malformed samara. **D-F,** *Dasineura tumidosae* on ash leaves. **D, E,** Galls. **F,** Gall opened to show gregarious larvae.

Plate 2. The tumor stem gall midge, *Asphondylia atriplicis,* on saltbush, *Atriplex canescens* (photographed by Bradford A. Hawkins, University of California, Riverside). **A,** Sectioned gall with third larval instar in cell. Note spatula (*arrow*). **B,** Sectioned gall with pupa. Note fungal lining in cell. **C,** Pupa cutting its way out. **D,** Pupa fully exserted. **E,** Adult emerging. **F,** Adult hardening before flight.

Plate 3. A, B, Needle galls on Virginia pine made by *Thecodiplosis brachynteroides*: gall in **B** broken longitudinally to show larvae in individual cells. **C,** Swollen blueberry infested with *Dasineura* sp. (photographed by Jerry A. Payne, Byron, Ga.). **D,** Vein fold gall on chinquapin made by *Macrodiplosis* sp. (photographed by Jerry A. Payne, Byron, Ga.). **E, F,** The holly berry gall midge, *Asphondylia ilicicola*; berry in longitudinal section to show aborted carpels, first larval instar, and fungus beginning to spread through the chamber.

Plate 4. A, B, Complex galls on inflorescence of *Laportea* made by *Dasineura* sp. **B,** Gall in section to show larval cell. **C, D,** Complex hickory leaf galls made by *Caryomyia* spp. **E,** Simple leaf spot galls on red maple made by *Acericecis ocellaris*. **F,** Complex needle galls on bald cypress made by *Taxodiomyia cupressi* (photographed by James E. Appleby, Champaign, Ill.).

8 Keys to Galls and Plant Damage

How to Use the Keys

This chapter is arranged by host family and genus, in alphabetical order. Plant names are indexed at the back of the book. Find the host genus in the plant index and refer to the page given in boldface. Before using a key, read any general notes under the plant family or generic name for additional general information. Keys are based primarily on the damage itself, but occasionally on the site of infestation or characters of the gall midge larva. The paragraph following the gall midge name summarizes what is known about the damage and the gall midge. The host(s) and distribution are given for each cecidomyiid, followed by any references for further reading. The original citation for a gall midge can be located in Gagné (in prep.). General information about any Nearctic genus can be found in the synopsis in chapter 3. To be more certain of an identification, compare larvae from the damage in question with a known example of the genus illustrated in chapter 4.

Records of Damage

The records of damage listed are thought to be complete and are based on reports in the literature and on collections in the U.S. National Museum of Natural History (USNM) and elsewhere. Felt (1940) listed most records of galls known through 1939, but some of his records were of incorrectly associated gall midges and galls (adults reared from galls were sometimes not the gall maker); many of these erroneous records are corrected here. Felt's book gave only a short description of the damage with no attribution to source, but his earlier *Key to American Insect Galls*

(Felt 1918a) gave a citation to a journal or a code number for many of these records. The code numbers referred to notes in Felt's unpublished notebooks and allowed me to associate particular damage with previously undetermined larvae that had been labeled only with a number.

Records published after 1939 are either cited in the text or traceable by looking up the species name in Gagné (in prep.). Records based on specimens in the USNM are given without attribution. Particular records can be consulted in that collection for additional information.

Only the currently valid name for a species is given in the keys. Synonyms are in the index and are cross-referenced to the currently valid name and the page on which it appears.

Plant Names

Family names all end with "aceae." Names formerly used that were not based on a genus, e.g., Compositae, are indexed and appear in parentheses after the correct family name in the headings. Host names accord as far as possible with the National List of Scientific Plant Names (Anonymous 1982). Plant names in quotation marks are as originally given. Common names of plant genera are listed in parentheses in the headings and in the index. Author names of plants are not given.

Past host records are not always correct, and doubtful ones are given in quotations. One should be aware that a collector could have misidentified a plant genus or that a generic concept has changed since the time of collection. An old record of "*Aster* sp." could now fit in one of many genera.

Distribution

The range of a species is drawn with a broad brush and is described from north to south and east to west. Only states or provinces are given, and Washington, D.C., is considered here part of Maryland and is identified as Maryland in the keys. Not all states or provinces are listed if a gall midge is widespread; the distribution of a species known from New York, Maryland, North Carolina, and Florida is given as New York to Florida. The species will very likely be found in states in between those where it has already been collected. Some species are likely to occur wherever their hosts occur, but this is not always true (Ehler 1982).

References

References are given when they contain more detail than is presented here. Only the more informative or more recent references are given. They will usually key a user into any additional literature.

Illustrations

Illustrations are presented for most kinds of damage. The figure number, in parentheses, immediately follows the description of the damage. Some galls, e.g., fusiform stem swellings, are not always illustrated even if examples were available because they have a similar appearance on many hosts. Plant parts are mostly shown at actual size. Some damage or details were drawn at 2×, as is noted in the figure legend. Most illustrations were made from fresh or preserved examples of damage. In some cases, when fresh or dried galls were not available, previous illustrations were redrawn to approximately actual size, and attribution for the original drawing is given in the figure legend.

Aceraceae

Acer (maple, box-elder)

Gall midges of box-elder, the compound-leaved maple, are distinctly different from those of the simple-leaved maples.

1a Gall on simple-leaved maples 2
1b Gall on compound-leaved maple, box-elder 6
2a Circular, usually red, leaf spot, each with one exposed larva on underside of leaf (145b, color plate 4E) . . . . *Acericecis ocellaris* (Osten Sacken)

Spots are about 5 mm in diameter, usually red, but sometimes green or yellow, and evident in spring as soon as leaves are fully expanded. Some new spots may develop in small numbers throughout the summer. Larvae, one under each spot, are stationary in a sticky substance on the underside of the leaf and feed only at the slightly concave central point to which their mandibles are attached. When fully grown, still in spring, the larvae drop to the soil, where they spin cocoons and overwinter as larvae. The leaf spot tissue turns brown after the larvae have dropped to the soil. Damage to the leaves appears to be slight, but occasional large infestations can cause leaves to curl. Hosts: Chiefly *A. rubrum*; also *A.*

pensylvanicum, *A. saccharinum*, *A. saccharum*, *A. spicatum*; typical spots occur on *A. circinatum* from Oregon and Washington, but larvae of the gall midge have not been reported. Distr.: eastern (and possibly throughout) North America. Ref.: Gagné (1983b).

2b Pouch, fold, or other distortion of leaf; the larva covered by leaf tissue 3

3a Yellow and red marginal leaf fold (145c) .... *Contarinia* sp. and *Dasineura* sp.

One or two larvae of either genus, but usually of *Contarinia*, are found in each of these folds, visible in early spring after the leaves are fully expanded. Larvae develop quickly and drop to the soil. Adults have not been reared. Host: *A. rubrum*. Distr.: Connecticut and New York to Maryland.

3b Leaf pouch or fold along veins, or greatly distorted leaf 4

4a Veins at junctures on underside of leaf expanded and compressed against the leaf surface (145d).. *Dasineura* sp.

Galls are visible only from the underside of the leaf. They appear in spring, and each contains one larva. Larvae drop to the ground when full-grown. Adults have not been reared. Host: *A. rubrum*. Distr.: Maryland.

4b Crumpled leaf or longitudinal vein pouch 5

5a Crumpled leaf; on silver maple, *Acer saccharinum* (144) .... *Dasineura aceris* (Shimer)

Damage usually occurs on expanding leaves, which grow distorted and crumpled. Young leaves are often killed. When the gregarious larvae are full-grown, they form a webbing over the folds, where they pupate. Several overlapping generations may occur over the summer. Host: *A. saccharinum*. Distr.: Ontario to Wisconsin, south to Maryland and Kansas. Ref.: Watson (1973).

5b Crumpled leaf or vein pouch; on other maples, chiefly *Acer rubrum* (144, 145a) .... *Dasineura communis* Felt

More than one distinct type of damage has been attributed to *D. communis*, but the various kinds may be caused by more than one species of gall midge. Expanding leaves may be crumpled as in figure 144, or full-grown leaves may have a longitudinal pouch gall (145a), usually along the principal veins. The pouches, usually situated next to a major vein, open to the upper leaf surface and may be long and narrow to short and pea-shaped. One or more larvae may occur in a pouch. Full-grown larvae escape through the opening on the top surface of the leaf and drop to the soil. Young leaves are often killed, but older leaves with pouch galls do not appear severely affected. Hosts: *A. pensylvanicum*, *A. ru-*

brum, *A. saccharum*. Distr.: Maine to Wisconsin south to Maryland and Missouri.

6a Rough, irregular twig swelling .... cecidomyiid

The swelling is reportedly 2–4 cm long (Felt 1940). Host: *A. negundo*. Distr.: unspecified.

6b Bud or leaf galls 7

7a Globular pouch gall along veins (143c) .... *Contarinia* sp.

These reddish, inconspicuous galls appear in June on the upper side of fully expanded leaflets and result from an inpocketing of the leaflet. Each gall contains a single larva. Adults have not been reared. Host: *A. negundo*. Distr.: Maryland.

7b Folded or swollen leaflet or swollen leaf bud (143a, b) .... *Contarinia negundinis* (Gillette)

Damage occurs in early spring when leaves are expanding. The feeding site of the larvae, in the bud at the base of developing petioles or, more commonly, the leaflets, evidently depends on the tissue available when eggs are laid. The white larvae are gregarious and can be very numerous within a swelling. Occasionally, more thickly swollen galls have separate pockets. Larvae develop quickly and drop to the ground, where they spin a cocoon. Unlike most gall midges that overwinter in the ground, *C. negundinis* pupates in the fall instead of the following spring. Host: *A. negundo*. Distr.: Massachusetts to Saskatchewan, south to Virginia and Illinois. Ref.: Wilson (1966).

Amaranthaceae

Amaranthus (amaranth)

1a Enlarged, aborted fruit .... *Asphondylia amaranthi* Felt

Infested fruit are sometimes only slightly swollen. Pupation occurs inside the fruit. The gall midge is probably multivoltine. Hosts: *Amaranthus blitoides*, *A.* "*spinosus*". Distr.: Florida and Texas; Tamaulipas, Mexico. Ref.: Painter (1935).

1b Swollen stem .... *Neolasioptera palmeri* (Felt)

The original report of this damage (Cockerell 1902) indicated that the swelling was inconspicuous. Only the larva is known, and it is tentatively placed under *Neolasioptera*. Host: *A. palmeri*. Distr.: New Mexico. Ref.: Cockerell (1902).

143

144

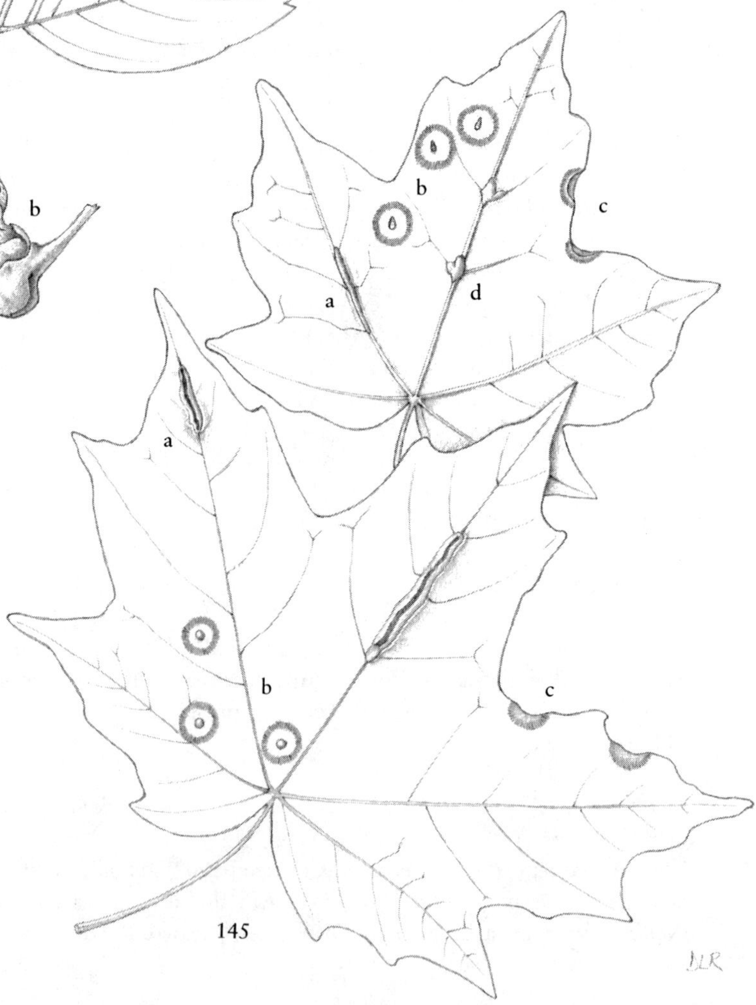

145

Amaryllidaceae

Agave (agave)

Discolored area on leaf .... *Neolasioptera martelli* Nijveldt

Larvae live in tunnels under the leaf surface. Pupation occurs in the plant. Affected tissue dies after adults emerge. Host: *Agave* sp. Distr.: Chapingo, Mexico. Ref.: Nijveldt (1967).

Anacardiaceae

Rhus (poison ivy, poison oak, sumac)

1a Swollen, aborted flower buds (146) .... *Asphondylia integrifoliae* Felt

Infested flowers are about twice the diameter of normal flowers. One larva occurs per flower and pupates there. Adults emerge from the current year's flowers, so there may be another generation on a different part of the plant. Hosts: *R. integrifolia*, *R. trilobata*. Distr.: Utah and California. Ref.: Felt (1916).

1b Gall on other plant parts 2
2a Leaf gall 3
2b Stem, aerial rootlet, or subterranean bud gall.......................... 4
3a Curled leaf margin .... *Contarinia* sp., *Dasineura* sp.

Margins of newly expanded leaves are tightly curled from beneath. *Contarinia* larvae were taken in New York in June, *Dasineura* larvae in Maryland in early May. Larvae of both species are white. At least the *Dasineura* sp. is gregarious. Host: *R. toxicodendron*. Distr.: New York and Maryland.

3b Leaf blister .... cecidomyiid

Blisters were reported as variable in size and attributed to a cecidomyiid (Maze 1949). Host: *R. toxicodendron*. Distr.: Pennsylvania.

4a Tuberlike swelling on aerial rootlet (147)........ *Dasineura rhois* (Coquillett)

Figures 143–145. Galls of Aceraceae. **143,** *Acer negundo*: **a,** swollen, folded leaflets and **b,** swollen bud: *Contarinia negundinis*; **c,** globular pouch gall: *Contarinia* sp. **144,** Distorted leaf, *Acer saccharinum*: *Dasineura aceris*. **145,** Lower leaf surface (at top) and upper surface (at bottom), *Acer rubrum*: **a,** pouch galls: *Dasineura communis*; **b,** leaf spots: *Acericecis ocellaris*; **c,** rolled leaf margin: *Contarinia* sp. and *Dasineura* sp.; **d,** swollen vein gall: *Dasineura* sp.

Each swelling contains one larva. Larvae overwinter in the galls, and adults emerge in spring. Host: *R. toxicodendron*. Distr.: New York to Virginia.

4b Stem swelling or subterranean bud gall. 5

5a Stem swelling . . . . cecidomyiid

Reports of this damage (Felt 1911b, 1940) possibly referred to the rootlet gall of *Dasineura rhois* in 4a. Host: *R. toxicodendron*. Distr.: unspecified.

5b Mossy, subterranean bud gall. . . . . cecidomyiid

This gall was further described only as 1/2 inch in diameter (Felt 1940). Host: *R. toxicodendron*. Distr.: unspecified.

Apiaceae (Umbelliferae)

Angelica

Tapered stem swelling . . . . *Neolasioptera* sp.

Swellings are 4–5 cm long. Larvae from the galls are tentatively identified as *Neolasioptera*. Host: *A. venenosa*. Distr.: New York.

Sanicula (black snakeroot)

Spherical stem and leaf enlargement . . . . cecidomyiid

Galls are small, reportedly about 2 mm in diameter. Felt (1940) identified the gall maker as "?*Asteromyia*." His specimens are evidently lost, but it is doubtful they belonged to *Asteromyia*, a genus restricted to a tribe of Asteraceae. Host: *Sanicula* sp. Distr.: unspecified.

Zizia (golden Alexander)

Tapered stem swelling . . . . *Lasioptera ziziae* Felt

Swellings are about 4 cm long. Larvae overwinter in the stems, and adults emerge in spring. Host: *Z. aurea*. Distr.: Illinois.

Apocynaceae

Apocynum (dogbane)

1a Oval, discolored stem depression .... *Neolasioptera apocyni* (Felt)

Larvae live in individual cells just below the stem surface. An adult was reared from stems in August, so there may be more than one generation of this species each year. Host: *A. androsaemifolium*. Distr.: New Jersey. Ref.: Weiss and West (1921).

1b Enlarged flower bud or deformed seed pod 2
2a Enlarged flower bud.................................... *Contarinia* sp.

Felt (1918a, 1921) attributed the damage to *Clinodiplosis apocyni* (Felt), but a contarinia collected simultaneously from the same damage was probably responsible. The contarinia was reared in August from larvae taken the same year, so this species probably has more than one generation per year. Host: *A. androsaemifolium*. Distr.: New York.

2b "Pocket gall" on seed pods.................................. cecidomyiid

The damage, attributed to a cecidomyiid by Maze (1949), was not described further. Host: *A. androsaemifolium*. Distr.: Pennsylvania.

Aquifoliaceae

Ilex (holly)

1a Aborted berry (149, color plate 3E, F)........... *Asphondylia ilicicola* Foote

Ovules of infested berries do not develop. Berries grow to almost normal size but remain green through winter, making the trees less desirable for ornamental purposes. The usually single larva (but occasionally two or three per berry) overwinters in the first instar. In spring, larvae grow quickly, pupate in the berries, and emerge when the new flowers open. Females lay eggs in the young fruit when the flower petals are still present. Hosts: *I. glabra*, *I. opaca*, *I. vomitoria*. Distr.: Connecticut and Pennsylvania, south to North Carolina. Ref.: Highland (1964).

1b Swollen, folded leaves (148)................................ cecidomyiid

Host: *I. vomitoria*. Distr.: North Carolina. Ref.: Wells (1921b).

Nemopanthus (mountain holly)

Enlarged bud .... *Asphondylia ilicoides* Felt

Infested buds are ovoid, green, 5 mm long, 3 mm in diameter, and are reportedly inhabited by several larvae, presumably in individual cells. Galls were collected in mid-June, and adults emerged shortly after. Host: *N. mucronatus*. Distr.: New York.

Aristolochiaceae

Aristolochia (Dutchman's pipe)

Leaf blister .... "*Cecidomyia*" *hageni* Aldrich

This species is known only from a description of its damage and from the larvae being brick red and agile. The circular blisters are about 4 mm in diameter on leaf blades, but are larger on petioles. Full-grown larvae escape through a hole in the upper surface of the blister. Host: *A. macrophylla*. Distr.: Massachusetts. Ref.: Hagen (1881).

Asclepiadaceae

Asclepias (milkweed)

1a Rolled leaf or swollen midrib fold .... *Contarinia* sp.

The damage was described as a swollen midrib fold in Felt (1940) but as a rolled leaf in his unpublished notebook. Larvae were taken in July and adults reared in August, so this species probably has more than one generation per year. A lestodiplosis was also reared from the damaged leaves. Host: *A. incarnata*. Distr.: New York.

1b Tapered node swelling .... *Neolasioptera asclepiae* Felt

Swellings usually contain two larvae, one on each side of the septum of the joint. Larvae overwinter in the gall; adults emerge in spring. Host: *A. incarnata*. Distr.: New York. Ref.: Felt (1918a).

Figures 146–156. Galls of Anacardiaceae, Aquifoliaceae, and Asteraceae. **146,** Bud galls, *Rhus integrifolia*: *Asphondylia integrifoliae*. **147,** Aerial rootlet swellings, *Rhus toxicodendron*: *Dasineura rhois*. **148,** Swollen, folded leaves, *Ilex vomitoria*: cecidomyiid (redrawn from Wells 1921b). **149,** Berries, *Ilex glabra*: **a,** normal berries (red), one in cross section to show normally developed seeds; **b,** berries (still green in spring) infested by *Asphondylia ilicicola*, one in cross section to show larval cell of *A. ilicicola* and aborted or poorly developed seeds. **150,** Stem swelling, *Ageratina altissima*: *Asphondylia eupatorii*. **151,** Flower galls, *Ageratina altissima*: *Schizomyia eupatoriflorae* (redrawn from Beutenmüller 1907). **152,** Bud gall, *Ambrosia artemisiifolia*: *Asphondylia ambrosiae*. **153,** Leafy bud galls, whole and in cross section, *Ambrosia psilostachya*: *Rhopalomyia ambrosiae*. **154,** Bud galls, *Ambrosia psilostachya*: *Contarinia partheniicola*. **155,** Swollen stem, *Ambrosia trifida*: *Neolasioptera ambrosiae*. **156,** Bud galls, *Antennaria plantaginifolia*: *Asphondylia antennariae*.

146

147

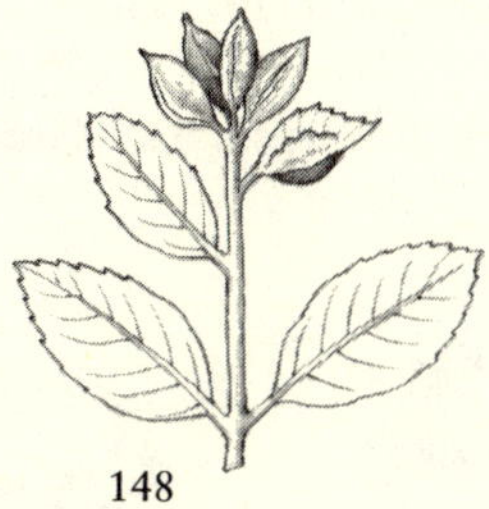

148

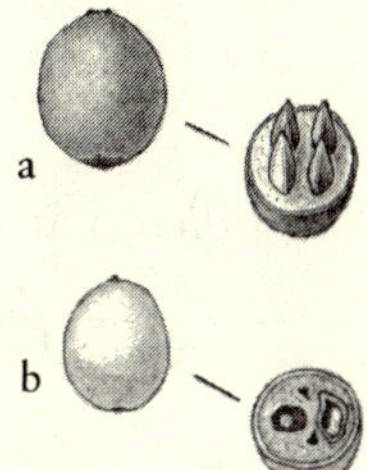

149

150

151

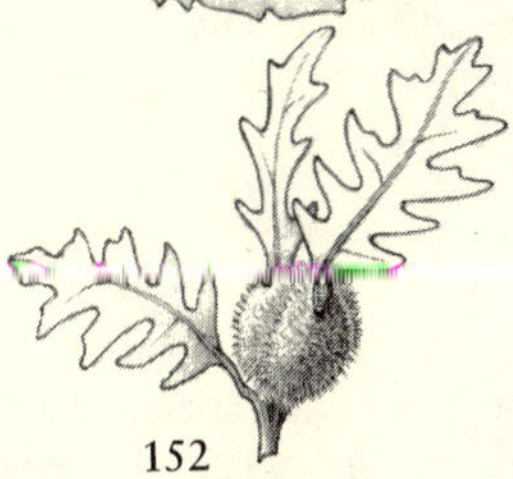

152

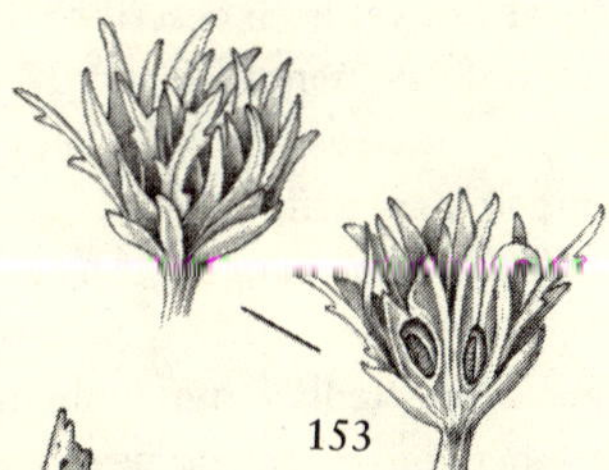

153

154

155

156

DLR

Asteraceae (Compositae)

Achillea (yarrow)

1a Damaged seed . . . . *Ozirhincus millefolii* (Wachtl)

This species was introduced into North America, probably during colonial times, in the seeds of its host. The gall midge is multivoltine and can be reared in great numbers from flower heads in summer. Adults are notable for their elongate mouthparts, flattened heads, and long necks, evidently modified for taking liquids from the narrow florets of the host. Host: *A. millefolium*. Distr.: coast to coast in North America, probably wherever the host occurs; Europe. Ref.: Rübsaamen and Hedicke (1926–1939).

1b Ovoid, white-downy, lateral bud gall. . . . . *Rhopalomyia* sp.

This 7-mm-long gall was questionably referred to *Rhopalomyia millefolii* Loew, a European species, by Felt (1918a, 1940), but the galls from North America or any specimens that might have been reared from them are evidently lost. Host: *A. millefolium*. Distr.: unspecified. Ref.: Rübsaamen and Hedicke (1926–1939).

Ageratina

1a Terminal, globular, succulent flower bud gall (151) . . . . *Schizomyia eupatoriflorae* (Beutenmüller)

Galls can be found in late summer. The red, active larvae overwinter in the soil. Adults have not yet been described. Host: *A. altissima*. Distr.: Ontario to North Carolina. Ref.: Beutenmüller (1907).

1b Leaf mine or stem gall 2
2a Oval, yellowish leaf mine . . . . cecidomyiid

The mine is reportedly 8 mm in diameter. The galls and specimens are evidently lost, and there have been no new reports since Felt (1918a). Host: *A. altissima*. Distr.: unspecified.

2b Stem gall 3
3a Gall succulent, just below terminal leaves (150). . . . *Asphondylia eupatorii* Felt

Galls appear in late summer and are 8–15 mm long and 5–10 mm in diameter, usually green, often elongate-tapered but sometimes globular, and polythalamous. Pupation occurs in the galls, and adults emerge in late summer. Eggs are probably laid on the following year's buds close to the ground. Host: *A. altissima*. Distr.: New York and Maryland.

3b Gall hard, woody, near upper part of flowering stems .... *Neolasioptera eupatorii* (Felt)

Galls are tapered, 15 mm long, and 10 mm in diameter. Larvae live in tunnels and overwinter as third instars in the galls. Adults emerge from the gall in spring. Host: *A. altissima*. Distr.: New York. Ref.: Felt (1918a).

Ageratum

Swollen leaf vein .... *Isolasioptera eupatoriensis* Möhn

Adults were reared in August. Host: *A. adenophorum*. Distr.: Morelos, Mexico; El Salvador. Ref.: Möhn (1964b).

Ambrosia (ragweed)

1a Tapered stem swelling (155).................. *Neolasioptera ambrosiae* Felt

This species is based on a New York series from giant ragweed, *A. trifida*. Specimens from Maryland on *A. artemisiifolia* differ somewhat, so more than one species may be involved. This species is probably univoltine in the Northeast. Larvae are orange, single or gregarious. Single larvae do not always cause a swelling of the stem. Hosts: *A. artemisiifolia*, *A. confertiflora*, *A. dumosa*, *A. trifida*. Distr.: New York to South Carolina, Nebraska to Texas, and California.

1b Leaf swelling or bud galls 2

2a Globular leaf swelling.................................... *Schizomyia* sp.

Swellings are thin-walled and monothalamous, slightly discolored, about 5 mm in diameter. When full-grown, the active, orange larvae break out of the gall and drop to the soil to pupate. Adults have not been reared. Host: *A. artemisiifolia*.

2b Bud galls 3

3a Green, leafy bud gall (153) .... *Rhopalomyia ambrosiae* Gagné

This species causes a foreshortening of the stem. The gall has green leaves growing out of its apex. Pupation occurs in the gall. Adults were reared in June. Specimens from Texas are somewhat different from the type series taken on *A.*

artemisiifolia in Florida and may be a different species. Hosts: *A. artemisiifolia*, *A. psilostachya*. Distr.: Florida and Texas. Ref.: Gagné (1975a).

3b White-fuzzy, ovoid, bud or flower galls. 4

4a Larval cells lined with a white fungus (152). . . . *Asphondylia ambrosiae* Gagné

Galls are whitish, hairier than the normal stem (especially in California specimens), and, because they are polythalamous, are generally larger than those of the following species. Pupation takes place inside the gall. In Florida, adults have emerged in September, January, and February. Hosts: *A. artemisiifolia*, *A. dumosa*, *A. monophylla*. Distr.: Florida, Texas, and California. Ref.: Gagné (1975a).

4b Larval cells green, not lined with a white fungus (154) . . . . *Contarinia partheniicola* (Cockerell)

Galls are 5–8 mm long, 4–5 mm wide and are found on axillary buds, along stems, on young leaves, on flowers, and along the rachis of the inflorescence. Adults have been reared in spring and fall. Hosts: *A. chamissonis*, *A. confertiflora*, *A. dumosa*, *A. eriocentra*, *A. psilostachya*, and *Parthenium* (q.v.). Distr.: Florida, and Texas to California.

Antennaria (everlasting, pussy-toes)

1a Ovoid, long-woolly bud gall. . . . . *Rhopalomyia pilosa* Felt

Galls are white and enclose several hard, thin-shelled cavities, each with one larva. The host was only tentatively identified as an *Antennaria*. *R. pilosa* is distinct from *R. antennariae*, a species definitely known from *Antennaria*. Host: ?*Antennaria*. Distr.: Northwest Territories.

1b Imbricated, short-woolly leaf bud gall. 2

2a Some leaves forming gall recurved at tips . . . . *Rhopalomyia antennariae* (Wheeler)

Galls are polythalamous and develop in May. Adults reportedly emerge from the galls early the next spring. Host: *A. plantaginifolia*. Distr.: Maine to New York, and Wisconsin.

2b Leaves forming gall not recurved at tips (156) . . . . *Asphondylia antennariae* (Wheeler)

Galls are reportedly more elongate than those of *R. antennariae*, and each contains one larva in a hard, central cell. The adult emerges in early spring, reportedly after hibernating in the gall (Wheeler 1889). Perhaps it hibernates as a young larva, and its growth is completed in spring, as occurs in many other *Asphondylia* spp. Host: *A. plantaginifolia*. Distr.: Maine and Wisconsin.

Anthemis

Larva in flower head .... *Clinodiplosis* sp.

A typical *Clinodiplosis* larva, evidently the basis for the record of "*Cecidomyia* sp." in Felt (1911b), is in the Felt collection. This species is probably not a primary plant-feeder and is not necessarily specific to this host. Host: *Anthemis* sp. Distr.: New York.

Arctium (burdock)

Larva in flower head .... cecidomyiid

This record is based on yellowish larvae (Felt 1911b:453) that are apparently no longer extant. No particular damage to the fruit was noticed. Host: *Arctium* sp. Distr.: unspecified.

Artemisia (sagebrush)

The 32 cecidomyiids keyed here all belong to the subgenus *Diarthronomyia* of *Rhopalomyia*. Their larvae overwinter in living plant tissue, but the instar in which they overwinter varies among species. Eggs are laid on the plant surface, and the hatching larvae crawl to an appropriate site, where they soon become embedded in the plant tissue. Pupation occurs within the larval chamber and lasts about two weeks. Pupae break through the gall surface to the exterior with the help of the antennal horns and frontal projections. All species except two have one generation per year. The remaining two are bivoltine. Some species are common, others, quite rare. Six other described species from the western United States, five from that same subgenus and one from *Asphondylia*, were reared from *Artemisia* but are not keyed here because their galls were not described in enough detail. These are *Rhopalomyia alticola* (Cockerell), *R. coloradella* (Cockerell), *R. erii* (Felt), *R. floccosa* (Felt), *R. tridentatae* Rübsaamen, and *Asphondylia artemisiae* Felt. In addition, an *Asphondylia* sp. from spongy bud galls was reported on *Artemisia tridentata* in Wyoming (Fronk et al. 1964), but the galls and specimens cannot be located. No asphondylias were found on *Artemisia* during a concerted three-year collecting effort in Idaho (Jones et al. 1983). All the species in the key below are treated in Jones et al. (1983), so that reference will not be repeated after each species.

1a Gall on leaves 2
1b Gall on other plant parts 15
2a Gall globular to lobed, soft, spongy 3
2b Gall cylindrical or capsular, hard 6
3a Gall surface with short trichomes or none, red to green 4
3b Gall surface with long trichomes, surface color not visible 5
4a Gall surface with short trichomes, red to green (157m, n) . . . . *Rhopalomyia pomum* Gagné

Galls are irregularly lobed and polythalamous to globular and monothalamous, and are 8–45 mm in diameter. This is probably the most common gall on big sagebrush. Galls occur on the persistent leaves. In Idaho they begin to grow in autumn, but greatest growth is in spring. Adults emerge in May to mid-July, depending on elevation. Host: *A. tridentata*. Distr.: Montana to Washington south to Nebraska and Utah.

4b Gall surface bare (157o) . . . . *Rhopalomyia calvipomum* Gagné

Galls are irregularly lobed and polythalamous to globular and monothalamous, smaller than those of *R. pomum*, with a diameter of 8–20 mm. As with *R. pomum*, galls appear on the persistent leaves. Growth begins slowly in autumn, but the galls are more readily seen and grow quickly in spring. Adults emerge in April and May in Idaho. Host: *A. tridentata*. Distr.: Idaho and Nevada.

5a Gall on *Artemisia tridentata* (157l) . . . . *Rhopalomyia hirtipomum* Gagné

Galls are ovoid and monothalamous, but several may grow close together on the same leaf. They are smaller than those of *R. pomum*, with a diameter of 8–15 mm. Galls begin growth in autumn but enlarge quickly to full size in spring. Adults emerge in April in Idaho. This gall is probably the one that Felt (1918a) ascribed to *Rhopalomyia navasi* Tavares, which makes a somewhat similar gall on terminal or lateral buds on an artemisia in Spain and North Africa. Host: *A. tridentata*. Distr.: Idaho and Washington.

5b Gall on *Artemisia* "*gnaphaloides*" . . . . *Rhopalomyia gnaphalodis* Felt

---

Figures 157–160. Galls of Asteraceae: *Artemisia tridentata*, all formed by *Rhopalomyia* spp. **157a,** Bud gall with cross section in 2×: *R. anthoides*; **b,** node gall: *R. nucula*; **c,** bud galls, *R. obovata*; **d,** bud gall: *R. lignitubus*; **e,** bud gall: *R. cramboides*; **f,** ephemeral leaf galls: *R. ampullaria*; **g,** ephemeral leaf gall: *R. culmata*; **h,** bud gall: *R. lignea*; **i,** leaf galls: *R. tubulus*; **j,** leaf galls, one in 2×: *R. hirtibulla*; **k,** bud gall, one in section: *R. conica*; **l,** leaf galls: *R. hirtipomum*; **m,** leaf gall: *R. pomum* (in section); **n,** leaf galls: *R. pomum*; **o,** leaf gall: *R. calvipomum*. **158,** Bud gall: *R. gossypina*. **159,** Node gall: *R. medusirrasa*. **160,** Node gall: *R. medusa*.

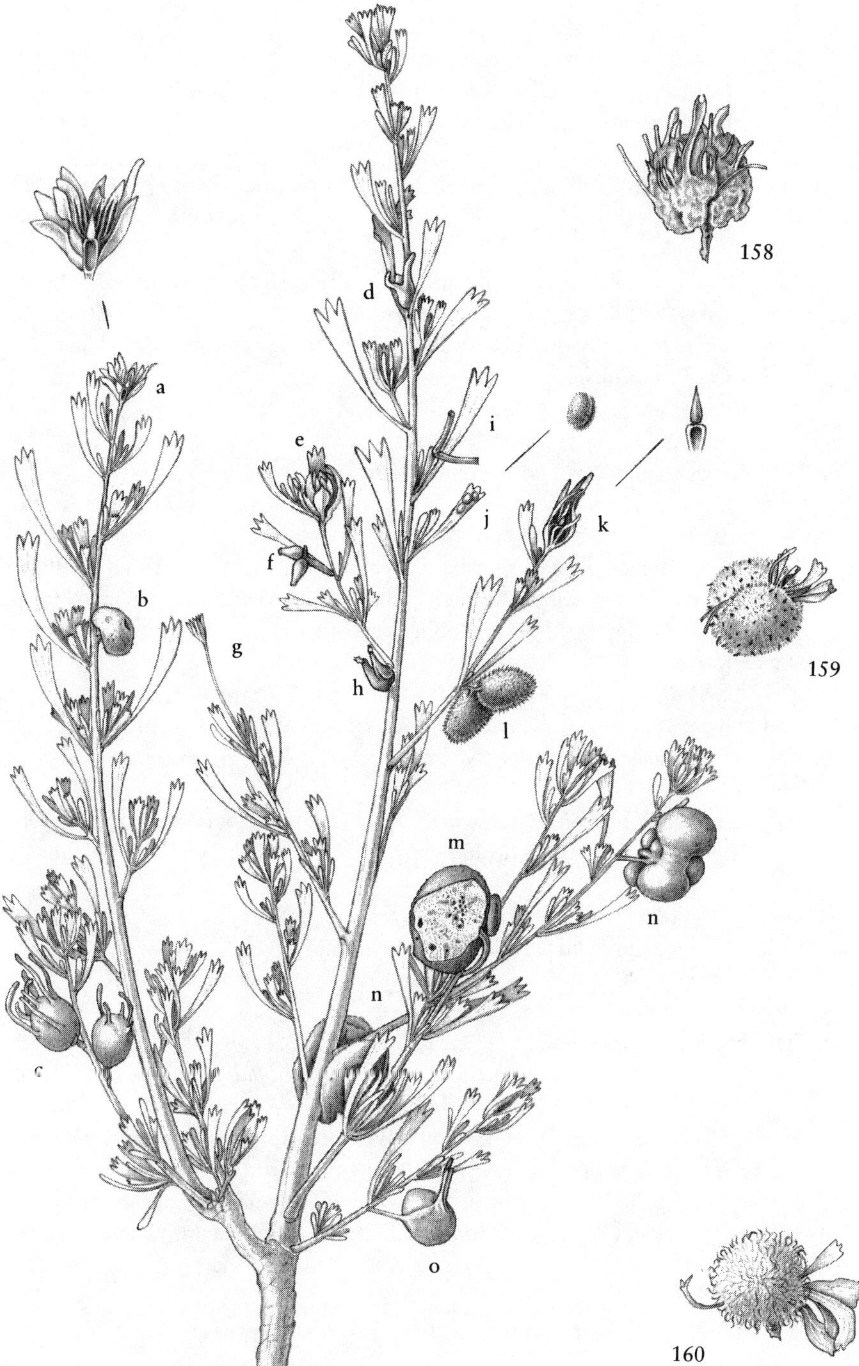

157

158

159

160

Galls are similar to those formed by *R. hirtipomum*, but further comparison of the two species is not possible because males of *R. gnaphalodis* are unknown. Host: *A. "gnaphaloides."* Distr.: Colorado.

6a Gall a flowerlike swelling at end of long, narrow, deformed ephemeral leaf (157g) . . . . *Rhopalomyia culmata* Gagné

The gall, about 5 mm long at the apex of a 10 mm long stalk, occurs only on ephemeral leaves that develop in spring. Adults emerge in spring. The site of the presumptive overwintering generation of this species is unknown. Host: *A. tridentata.* Distr.: Idaho.

6b Gall sessile 7

7a Gall elongate-cylindrical, not tapered at apex (157i) . . . . *Rhopalomyia tubulus* Gagné

The 3–6 mm long gall has a single larval chamber at its base and a thin septum near the apex through which the adult escapes. Adults emerge in spring in Idaho and presumably lay eggs on developing persistent leaves. Host: *A. tridentata.* Distr.: Idaho.

7b Gall convex at apex 8

8a Gall ovoid, widest at midlength, wrinkled . . . . *Rhopalomyia rugosa* Gagné

Galls are 4 mm long by 2–3 mm wide. These were found only once, in spring on persistent leaves. Adults emerge in spring. Host: *A. tridentata.* Distr.: Idaho.

8b Gall hemispherical or cone shaped 9

9a Gall conical or pyriform 10

9b Gall hemispherical 13

10a Gall 2–7 mm long 11

10b Gall less than 2 mm long 12

11a Gall conical, on *Artemisia tridentata* (157f) . . . . . *Rhopalomyia ampullaria* Felt

These galls occur mainly on ephemeral leaves, occasionally on stems and flower bracts. This gall midge has two generations per year, the overwintering one developing in galled flowers, the summer generation in the conical galls on stems and ephemeral leaves. Host: *A. tridentata.* Distr.: Montana and Washington to Utah.

11b Gall pyriform, on *Artemisia frigida* (162) . . *Rhopalomyia betheliana* Cockerell

Galls are 3–4 mm long, 2 mm in basal diameter. They occur also on flowers and possibly on other plant parts. Adults emerge in early spring. Host: *A. frigida.* Distr.: Colorado.

12a Gall on *A. "californica"* (161) .... *Rhopalomyia clinata* Gagné

Adults emerge in spring. Host: *A. "californica."* Distr.: California.

12b Gall on *Artemisia "heterophylla"* (163)...... *Rhopalomyia occidentalis* (Felt)

Adults emerge in spring. Host: *A. "heterophylla,"* possibly also *A. tridentata.* Distr.: California.

13a Gall protruding from both sides of leaf...... *Rhopalomyia tumidibulla* Gagné

Galls are about 3 mm in diameter and appear on persistent leaves. Adults emerge in spring. Host: *A. tridentata.* Distr.: Idaho.

13b Gall protruding from one leaf surface only.............................. 14
14a Gall covered with trichomes similar to those on leaf .... *Rhopalomyia brevibulla* Gagné

Galls are hemispherical, 1 mm wide and 1 mm in diameter. Adults emerge in March from galls on persistent leaves. Host: *A. tridentata.* Distr.: Montana to Utah.

14b Gall covered with trichomes longer than on leaf (157j) .... *Rhopalomyia hirtibulla* Gagné

Galls are 1 mm wide and 1 mm in diameter. Adults emerge in March from galls on persistent leaves. Host: *A. tridentata.* Distr.: Montana to Utah.

15a Gall on branch tips or buds 16
15b Gall at nodes, in stems, or on flowers 25
16a Gall woody, a swollen bud with no leaf vestiges 17
16b Gall made up of a rosette of leaves or with distinct leaf vestiges........... 18
17a Gall tubular, open and flared at apex (157d) .. *Rhopalomyia lignitubus* Gagné

Galls are monothalamous, 10–19 mm long, 4–6 mm in diameter at base, and develop in spring at the onset of plant growth. A single larval chamber lies at the base of the gall. Host: *A. tridentata.* Distr.: Idaho.

17b Gall an irregular aggregate of roughly cylindrical galls, closed and blunt at apex (157h) .... *Rhopalomyia lignea* Gagné

Galls are evidently made up of leaf rudiments, are 8 mm long, 4–6 mm at base across their widest diameter. Host: *A. tridentata.* Distr.: Idaho.

18a Gall spherical, soft to leathery, with short leaf vestiges growing from surface of gall (157c).................................. *Rhopalomyia obovata* Gagné

Galls are 5–10 mm diameter with one to several larval cells inside. Mature or overwintering galls are leathery and yellow-brown. The larva is full-grown in fall; the adult emerges in spring. Host: *A. tridentata*; similar galls occur on *A. arbuscula* and *A. tripartita*. Distr.: Idaho and Washington to Nevada.

18b Gall a rosette or globular and cottony.................................. 19
19a Gall globular with long, dense, cottony trichomes and many leaf vestiges... 20
19b Gall a rosette with long and short trichomes 21
20a Gall on *Artemisia tridentata* (158) .... *Rhopalomyia gossypina* Gagné

Gall 8–15 mm high, 8–16 mm diameter; one to several capsules at base on axis of malformed stems. Galls develop in late spring and when mature resemble balls of cotton. Adults emerge in late summer; this species probably overwinters as young larvae. Host: *A. tridentata.* Distr.: Idaho.

20b Gall on *Artemisia canadensis*.............. *Rhopalomyia alticola* (Cockerell)

This species is known only from pieces of a broken gall in the USNM. Host: *A. canadensis*. Distr.: Colorado.

21a Inner whorl of gall leaves distinctly shorter than outer.................. 22
21b Whorls of leaves of similar length but may differ in shape................ 23
22a Larval capsule smooth, tapered at its apex, prominent at center of rosette (157k) .... *Rhopalomyia conica* Gagné

Gall development begins in spring. Larvae are full-grown in fall and adults emerge in spring. Host: *A. tridentata*; similar galls occur also on *A. arbuscula*, *A. nova*, *A. tripartita*. Distr.: Idaho and Washington to Nevada.

22b Larval capsule rounded, concealed by leaves of rosette (157a) .... *Rhopalomyia anthoides* Gagné

Although this gall is similar to that of *R. florella*, flies of the two species are quite different. The gall is 4–15 mm in diameter and 5–17 mm long. One to 49 capsules can be found in a gall. Adults emerge in early spring and overwinter as larvae. Host: A. *tridentata*. Distr.: Idaho and Washington to Utah and Nevada.

23a Outer whorls of gall leaves as wide as long, inner whorl with more linear leaves (157e).................................. *Rhopalomyia cramboides* Gagné

The outermost leaves of the whorl are as wide as long, the inner leaves oblong. Adults emerge in spring. New galls develop on new stem growth. Host: *A. tridentata*. Distr.: Idaho.

23b All whorls of gall leaves similar in shape 24
24a Leaves of rosette widest at base, tapering to strong point .... *Rhopalomyia florella* Gagné

One generation occurs per year. Adults emerge in early spring. Galls are 12–15 mm diameter, 8–15 mm high. Host: *A. tridentata*. Distr.: Idaho.

24b Leaves of rosette linear, slightly constricted apically .... *Rhopalomyia mammilla* Gagné

Galls are polythalamous and about 6 mm diameter at base to 5 mm long. Adults emerge in September. This species evidently overwinters as a young larva. Host: *A. tridentata*. Distr.: Idaho.

25a Flower gall.. 26
25b Node or stem gall.. 27
26a Gall conical, inside unopened flower bud of normal size; on *Artemisia tridentata* ....*Rhopalomyia ampullaria* Felt

Flower parts inside involucral bracts are modified to form a hard capsule, densely covered with long, white trichomes. Galls are 3–4 mm long. See remarks under entry 11a of this key.

26b Gall pyriform, larger than normal bud size; on *Artemisia frigida* .... *Rhopalomyia betheliana* Cockerell

Adults emerge in early spring. Galls are 3–4 mm long, 2 mm in basal diameter, and occur also on leaves as in figure 162. Host: *A. frigida*. Distr.: Colorado.

27a Aerial or subterranean node gall 28
27b Stem swellings.. 31
28a Subterranean, ovoid node gall (164) *Rhopalomyia subhumilis* Gagné

Gall midges emerge in late June. One generation occurs per year. Host: *A. ludoviciana*. Distr.: Utah. Refs.: Gagné (1978a), Ranasinghe (1977).

28b Aerial node gall.. 29
29a Gall ovoid to pyriform, solid .... *Rhopalomyia nucula* Gagné

This species overwinters as full-grown larvae; adults appear in spring. The gall is covered with white or gray down, is 3–4 mm in diameter, 3–10 mm long, and has 1–11 larval chambers. Host: *A. tridentata*. The same kind of gall occurs on *A. nova*. Distr.: Montana and Washington to Nevada and Utah.

29b Gall a spherical mass of fleshy, filamentous, modified leaves.............. 30
30a Modified leaves with long trichomes (159) .. *Rhopalomyia medusirrasa* Gagné

Galls are white to gray due to the long trichomes covering the modified leaves. The globular mass is 6–16 mm in diameter. This species overwinters as first-instar larvae. These grow quickly in spring, and adults emerge in the same season. Host: *A. tridentata.* Distr.: Idaho and Nevada.

30b Modified leaves lacking trichomes (160) .... *Rhopalomyia medusa* Gagné

Galls are green to red, or mixed green and red, 8–30 mm in diameter, and begin development in fall but grow mainly in spring. The cecidomyiid overwinters as early-instar larvae, and adults appear in spring. One generation occurs per year. Hosts: *A. arbuscula*, *A. cana*, *A. nova*, *A. tridentata*, *A. tripartita.* Distr.: Montana to Nevada.

31a Conical gall on stem .... *Rhopalomyia ampullaria* Felt

These conical galls occur also on ephemeral leaves, stems, and flower bracts. See remarks under entry 11a of this key.

31b Stem swelling 32
32a Trichomes of swelling no longer than on normal stem ....
.... *Rhopalomyia tumidicaulis* Gagné

The swollen area, usually on a stunted stem with foreshortened internodes, is 5–23 mm long and contains one to several larval capsules. Adults emerge in late summer. The first sign of gall development is in spring, so the species probably overwinters as early-instar larvae. Host: *A. tridentata.* Distr.: Idaho.

32b Trichomes of swelling forming a thick, white, cottony mat ....
.... *Rhopalomyia hirticaulis* Gagné

The remarks under entry 32a apply here also. Host: *A. tridentata.* Distr.: Idaho.

Aster (aster)

In addition to the species keyed below, two others were reared from galls that were not described in adequate detail. They are *Rhopalomyia astericaulis* Felt, reared from an ovoid stem gall on *Aster novaeangliae* in Massachusetts, and *Rhopalomyia asteriflorae* Felt, reared from abnormally bunched or stunted heads of *Aster paniculatus* in New York.

1a Smooth or rough, hard stem swelling 2
1b Leaf or bud gall.. 3
2a Smooth, tapered swelling near apical branches (165)...... *Neolasioptera* spp.

Swellings are usually found on thin, more apical branches, are 10–25 mm long, and usually have a single larva in a tunnel along the plant's main axis. Larvae overwinter in the galls, and adults emerge in late spring. Three species have been described from such galls, although these have not yet been critically examined and may actually be one species. Such galls have been found from New England and Ontario to North Carolina and Missouri on many more asters than listed for the three species below.

N. albitarsis (Felt). Host: *A. cordifolius*. Distr.: New York.
N. punicei (Brodie). Host: *A. puniceus*. Distr.: Ontario.
N. ramuscula (Beutenmüller). Hosts: *A. dumosus*, *A. patens*. Distr.: North Carolina.

2b Rough, uneven swelling, usually near base of stem (as in 189) .... *Lasioptera asterspinosae* White

Galls may completely or only partially encircle the stem. They are brown, polythalamous, and spongy. Galls occur from ground level to 1 m above the ground. Larvae overwinter in the galls, and adults emerge in spring. Host: *A. spinosus*. Distr.: Texas. Ref.: White (1950).

3a Circular, variously colored leaf blister (168) or soft, black or purple swelling of twig of inflorescence (as in 193)......................... *Asteromyia* spp.

Asteromyia clarkeae (Felt) and *A. laeviana* (Felt) form the same kind of galls and have hosts in common. A characteristic fungus is associated with the galls but may be only a benign associate. The fungus eventually develops into a black cast surrounding the larvae. Both gall midge species have several generations per year beginning on the lower cauline leaves in spring and progressing upward through summer. The blisters are 4–10 mm in diameter, or in length if they occur on twigs, and the color varies with the host. As examples, those on *A. laevis* are yellow to light gray, those on *A. ericoides*, purple to black. Hosts: *Aster* spp. Distr.: throughout North America. Ref.: Gagné (1968a).

3b Bud gall 4
4a Gall composed of a rosette of leaves 5
4b Nonleafy bud gall or modified flower 6
5a Loose rosette of wide, short leaves (170, 171) .... *Asphondylia recondita* (Osten Sacken)

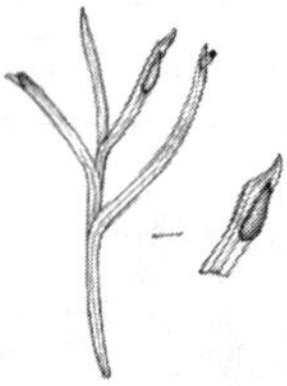
161

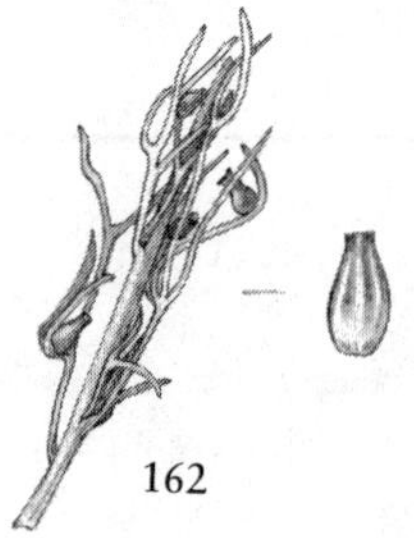
162

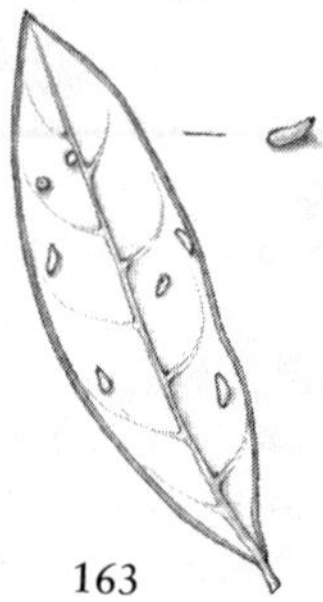
163

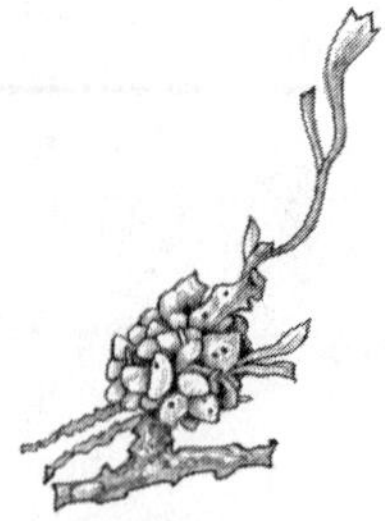
164

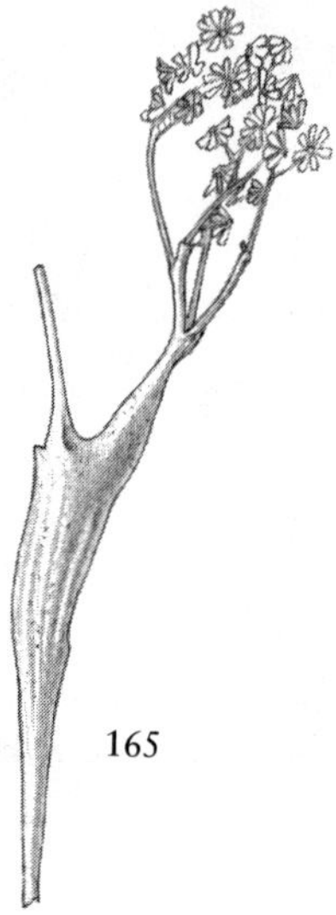
165

166

167

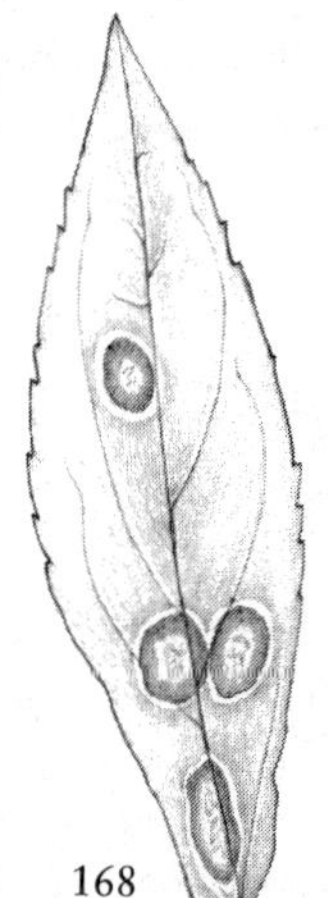
168

169

170

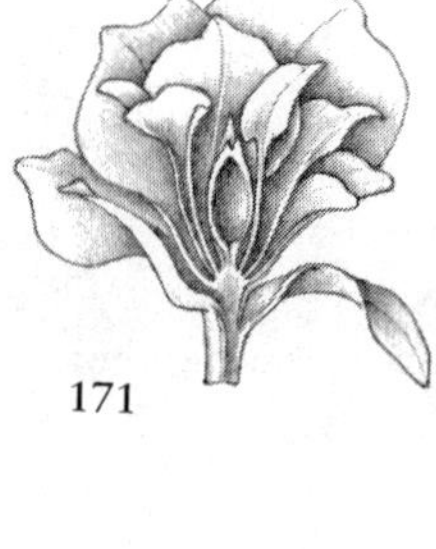
171

DLR

Galls are 10–20 mm long and occur at the tips of branches. The single larval cell is at the center of the gall. Adults emerge from the gall in summer. Hosts: *A. novaeangliae*, *A. patens*, *A. paternus*. Distr.: New York to North Carolina.

5b Catkinlike growth, a foreshortened stem, with narrow leaf tips along length of gall (167) .... *Rhopalomyia strobiligemma* (Stebbins)

Galls are polythalamous. Adults have emerged from the galls in September. Host: *A. paniculatus*. Distr.: Massachusetts and New York.

6a Gall ovoid-tapered, at base of stem or on apical bud 7
6b Galls among flowers or modified flowers 8
7a Gall at base of stem............................ *Rhopalomyia weldi* Felt

Galls were found in early spring and were described as soft, about 10 mm long and 5 mm wide, single or clustered, and purplish. Adult midges emerged the same spring. Host: *A. macrophyllus*. Distr.: Illinois.

7b Gall on apical bud (233) .... *Asphondylia* sp.

The brownish galls occur in midsummer at stem apices. Adults emerged in mid-August. It is possible that this species makes a gall on the flower head later in the season. Host: *A. subulatus*. Distr.: Maine.

8a Gall ovoid, white-haired, larger than florets (166)......... *Rhopalomyia* spp.

Galls are sessile and widest at base, 4–7 mm long by 3–5 mm in diameter, one-celled, ovate, pointed at apex, and green with white down. Adults have been reared from the galls in fall. The three species listed are too incompletely known to determine whether they are synonyms. In addition to the localities listed below, similar galls have been taken on an aster in Nebraska.

R. crassulina Cockerell. Host: *A. crassulus*. Distr.: Colorado.
R. gemmaria (Stebbins). Host: *A. ericoides*. Distr.: Massachusetts.
R. lateriflorae Felt. Host: *A. lateriflorus*. Distr.: Massachusetts and New York.

Figures 161–171. Galls of Asteraceae. **161,** Leaf galls, one in 2×, *Artemisia "californica"*: *Rhopalomyia clinata*. **162,** Leaf galls, one in 2×, *Artemisia frigida*: *Rhopalomyia betheliana*. **163,** Leaf galls, one in 2×, *Artemisia heterophylla*: *Rhopalomyia occidentalis*. **164,** Subterranean node galls, *Artemisia ludoviciana*: *Rhopalomyia subhumilis*. **165,** Stem swelling, *Aster patens*: *Neolasioptera ramuscula*. **166,** Bud galls, one in 2×, *Aster crassulus*: *Rhopalomyia crassulina*. **167,** Bud gall, *Aster ericoides*: *Rhopalomyia strobiligemma*. **168,** Leaf blisters, *Aster laevis*: *Asteromyia laeviana*. **169,** Floret gall, section through flower head, *Aster novaeangliae*: *Rhopalomyia* sp. **170, 171,** Flower head galls, one in cross section, *Aster patens*: *Asphondylia recondita* (redrawn from Beutenmüller 1907).

8b Gall no longer than normal floret. 9

9a Single, deformed, lateral floret. . . . . *Rhopalomyia* sp.

Galls are 2–4 mm long and 2 mm in diameter. Adults probably emerge from galls in autumn. Host: *A. undulatus.* Distr.: Massachusetts. Ref.: Felt (1940).

9b Ovoid gall among flowers (169) . . . . *Rhopalomyia* sp.

Galls were found among normal flowers in October. Adults were reared in October. Host: *A. novaeangliae.* Distr.: Maryland.

Baccharis

In addition to the species in the key below, two others are known from baccharis. A potatolike swelling, 1.2–3.5 mm long and about 1.2 mm wide has been attributed to an *Asphondylia* sp. and an unspecified stem gall to *Rhopalomyia baccharis* Felt. Both were from *Baccharis pilularis*, but the locality was not specified (Felt 1940). *Rhopalomyia baccharis* is distinct from *R. californica*, which causes the terminal, spongy galls keyed below.

1a Larva on or in flowers 2

1b Larva on other plant parts. 3

2a Larva free in flowers . . . . *Contarinia* sp. near *schulzi* Gagné and *Dasineura* sp.

Larvae of the contarinia are white, those of the dasineura reddish. Both live among developing florets and drop to the soil when full-grown. Specimens collected in October in Maryland emerged the following August at the time *Baccharis* was coming into bloom. Hosts: *B. glomeruliflora, B. halimifolia.* Distr.: Maryland, Florida, and Texas.

2b Larva in swollen flower receptacle . . . . *Neolasioptera* sp.

Larvae live in individual, spherical cells in the receptacles, where they pupate. Two generations occur per year. Overwintering larvae diapause until late the following summer when the hosts come into bloom. Adults have elongate mouthparts, apparently modified for feeding in flowers. Hosts: *B. glomeruliflora, B. halimifolia,* and *B. neglecta.* Distr.: Maryland, Florida, and Texas; Dominican Republic. Ref.: Gagné and Boldt (in press).

3a Bud gall, usually terminal, lobed and succulent (172) . . . . *Rhopalomyia californica* Felt

This species is multivoltine. Eggs are laid in terminal growth, and the hatching larvae crawl between bud scales where the buds begin to form. Host: *B. pilularis*;

similar galls occur on *B. glutinosa* in Oklahoma, but it is not known whether *R. californica* is responsible for that gall also. Distr.: California and Oregon; also Australia, where it has been successfully introduced to control *B. halimifolia*. Refs.: Ehler (1982, 1987), Tilden (1951).

3b Leaf or stem swelling 4
4a Black, circular leaf blister (174). . . . . *Asteromyia* sp. near *gutierreziae* Felt

Galls become hard with age. Only the larva of this species is known. It is probably multivoltine. Host: *B. angustifolia*. Distr.: Texas.

4b Stem swellings 5
5a Relatively soft, irregular stem swellings, often on upper stem (173) . . . . *Neolasioptera lathami* Gagné

This species may be multivoltine because the galls occur on lower and upper parts of the stem. Galls sometimes produce stunting in seedlings. Eggs are laid mostly in stem terminals, but some in leaf axils. Host: *B. halimifolia*, *B. neglecta*, *B. sarothroides*. Distr.: New York to Florida, and Texas and Arizona. Ref.: Diatloff and Palmer (1987).

5b Hard, woody, tapered, stem swellings at base of plant . . . . *Neolasioptera baccharicola* Gagné

This species is probably multivoltine. Adults were reared from galls in spring. Host: *B. halimifolia*. Distr.: Virginia. Ref.: Gagné (1971a).

Bebbia

Galls on flower heads . . . . *Asphondylia* sp.

The damage of this gall midge was not specified further. Adults were reared from November to January. Host: *B. juncea*. Distr.: California.

Bidens (beggar-ticks)

1a Modified, aborted floret (175) . . . . *Asphondylia bidens* Johannsen

Adult midges were reared from November to January. Host: *B. pilosa*. Distr.: Florida. Ref.: Needham (1946).

1b Stem gall 2
2a Slender, woody, tapered stem swelling . . . . *Neolasioptera* sp.

Swellings were reportedly about 13 mm long (Felt 1940). The gall maker is unknown and is only tentatively placed in *Neolasioptera.* Host: *Bidens* sp. Distr.: New York.

2b Soft, lateral, globular outgrowth of stem (176) .... *Asphondylia* sp.

Galls are monothalamous. This is a species distinct from *A. bidens.* Host: *Bidens* sp. Distr.: Pennsylvania and Illinois.

Borrichia

Basally joined, swollen pair of leaves at branch apex (178) .... *Asphondylia* sp.

Galls are green and succulent and contain one to several cells, each with a larva or pupa. Adults emerge from the swellings in November and December. Host: *B. frutescens.* Distr.: North Carolina, Florida, and Mexico.

Chondrilla (rush skeletonweed)

Blister on leaf or green stems (179) .... *Cystiphora schmidti* (Rübsaamen)

This gall midge was introduced into North America from Europe in 1975 to aid in controlling rush skeletonweed, an immigrant range pest. Leaf blisters are usually purplish. Leaves with many galls shrivel and die. Females insert eggs directly into leaf tissue by means of the piercing ovipositor. This species is multivoltine. Host: *C. juncea.* Distr.: Idaho, Oregon, and California; Europe and Middle East. Refs.: Harris (1975), Caresche and Wapshere (1975), and Sobhian and Andres (1978).

Chrysanthemum (chrysanthemum)

Cone shaped galls on leaves, stems, and buds (177) ....
.... *Rhopalomyia chrysanthemi* (Ahlberg)

Figures 172–179. Galls of Asteraceae. **172,** Succulent bud galls showing exit holes and pupal exuviae, *Baccharis pilularis*: *Rhopalomyia californica.* **173,** Stem gall showing exit holes, *Baccharis halimifolia*: *Neolasioptera lathami.* **174,** Leaf blisters, *Baccharis angustifolia*: *Asteromyia* sp. **175,** Normal floret (left) and infested floret (right), *Bidens pilosa*: *Asphondylia bidens.* **176,** Globular stem swellings, *Bidens* sp.: *Asphondylia* sp. **177,** Conical, recumbent leaf galls, one in 2×, *Chrysanthemum* sp.: *Rhopalomyia chrysanthemi.* **178,** Succulent leaf galls, *Borrichia frutescens*, one with pupa in exit hole, the other in section to show larval cells: *Asphondylia* sp. **179,** Leaf blisters, *Chondrilla juncea*: *Cystiphora schmidti.*

172

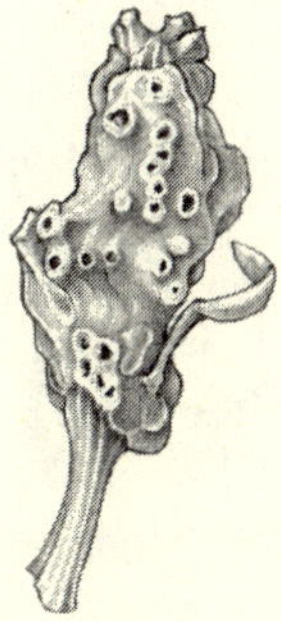

173

174

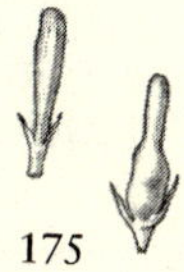

175

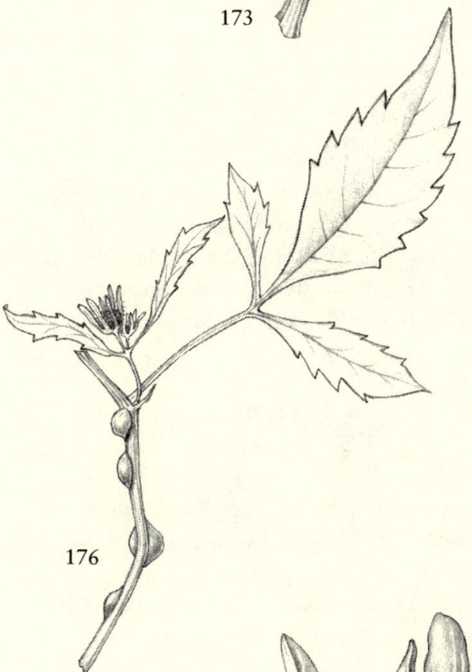

176

177

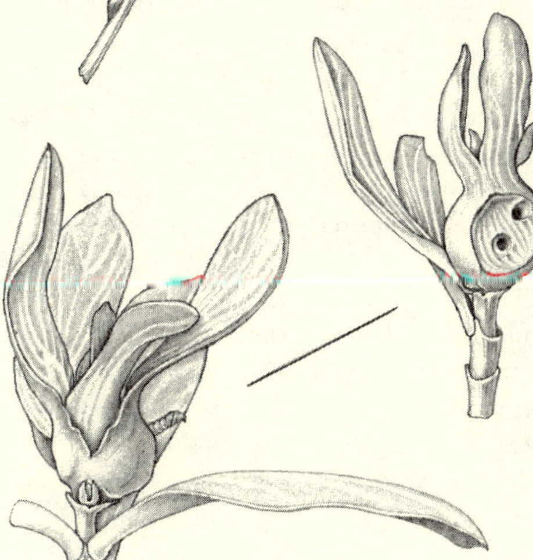

178

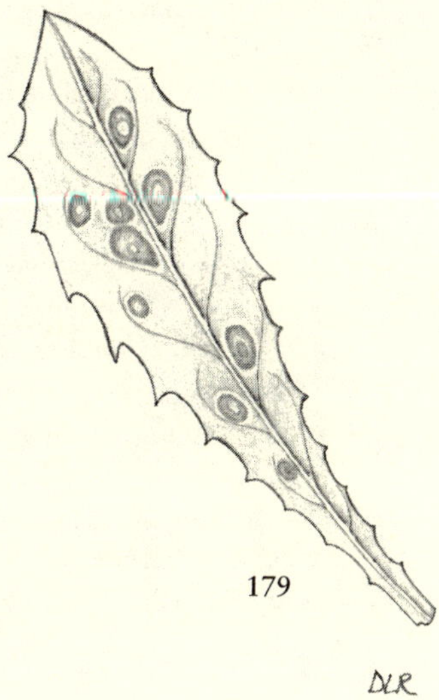

179

DLR

Severe infestations can cause the plants to become deformed and bear a reduced number of blooms. At one time this multivoltine species was a serious pest on chrysanthemums, but it is rarely heard of now, probably due to chemical control. This species was probably an immigrant from Asia and transported with cultivated chrysanthemums. It arrived in North America probably as early as 1900. Hosts: *Chrysanthemum* spp., including cultivars. Distr.: cosmopolitan, wherever chrysanthemums are grown, although no longer common in North America. Ref.: Barnes (1948b).

Chrysothamnus (rabbitbrush)

Many rabbitbrush galls are caused by gall midges; several kinds of galls represented in the USNM collection, e.g., figure 183, cannot be assigned to particular species and are not keyed here. Some described species of gall midges are only tentatively assigned to particular galls in this key. Some rhopalomyias on rabbitbrush have a close association with galls made by Tephritidae or form galls very similar to those of fruit flies.

1a Elongate-elliptical, discolored blister on leaf or stem (180) .... *Asteromyia chrysothamni* Felt

Galls are buff to blue-gray. Larvae are surrounded by a symbiotic fungus that becomes hard and black with age. Hosts: *C. albidus*, *C. nauseosus*, *C. greenei*, *C. parryi*, *C. "plattensis,"* *C. viscidiflorus*, and *Haplopappus* spp. (q.v.). Distr.: widespread in western North America: Saskatchewan to Texas and west. Ref.: Gagné (1968a).

1b Bud gall or lateral outgrowth of stem 2
2a Bud gall 3
2b Lateral, nondetachable outgrowth of stem 5
3a Gall covered exteriorly by overlapping leaves joined together at gall apex (182) .... *Asphondylia* spp.

A. bigeloviaebrassicoides (Townsend). This gall seems to fit the original description of this species, but is only tentatively placed under this name. The species was described only from galls that are now lost. Galls in the USNM fit the description and contain *Asphondylia* pupae in a central cell. Townsend's description pointed out that the larval cell was covered with white tomentum, probably a reference to the fungal associate typical of *Asphondylia* galls. Host: *Chrysothamnus* sp. Distr.: New Mexico.
A. chrysothamni Felt. This species also is only tentatively associated with the gall illustrated. Galls were evidently not kept. Host: *C. graveolens*. Distr.: Utah.

3b Gall cottony or smooth, leafless 4
4a Gall smooth, ovoid, without leaves incorporated in surface (181) . . . . *Rhopalomyia bigelovioides* Felt

Host: *Chrysothamnus* sp. Distr.: California.

4b Gall cottony, irregularly globular, with recurved leaf tips emerging from surface (183, 184). . . . . *Rhopalomyia* spp.

R. bigeloviae (Cockerell). This species reportedly came from a large, woolly gall up to 12 mm long typical of that made by an *Aciurina* (Tephritidae). Host: *Chrysothamnus* sp. Distr.: Colorado.
R. bigeloviaestrobiloides (Townsend). This is known only from the original description of the gall, which may be the same as that made by *R. utahensis*. Host: *Chrysothamnus* sp. Distr.: New Mexico.
R. utahensis Felt. Galls develop quickly in early spring with the onset of leaf growth. Adult midges emerge from the galls in April and May. One generation occurs per year. Eggs or first-instar larvae enter diapause until the following spring. Host: *C. nauseosus*. Distr.: Idaho and Utah. Ref.: Wangberg (1981).

5a Globular, resinous swellings (185) . . . . *Rhopalomyia glutinosa* Felt

Swellings are monothalamous. Host: *C. nauseosus*. Distr.: unspecified and Utah.

5b Conical swellings of same texture as stem and with cottony fibers filling exit (186) . . . . *Rhopalomyia chrysothamni* Felt

Host: *C. nauseosus*. Distr.: Utah.

Cirsium (thistle)

1a Larva in swollen stem . . . . *Lasioptera* sp.

Adults were reared from stems in August. Host: *C. proteanum*. Distr.: California.

1b Larva in flower head 2
2a Larva free in flower head . . . . *Dasineura gibsoni* Felt

This species lives among the florets and can severely check seed set. Adults were reared in June from overwintering larvae and in September from flower heads, so the species is at least bivoltine. It is possibly synonymous with *Dasineura com-*

positarum Kieffer or *Dasineura cirsii* (Rübsaamen), both of which are known from *Cirsium* flowers in Europe. Hosts: C. *arvense*, C. *vulgare*, both introduced species. Distr: Massachusetts to Ontario, Indiana, and British Columbia.

2b Larva in plant tissue .... *Neolasioptera* sp.

Adults were reared from flower heads in July. Host: C. *proteanum*. Distr.: California.

Conoclinium

Larva in flower head .... *Asphondylia* sp.

Adults were reared from unspecified damage in seed heads in April in Florida. This species could possibly be *Asphondylia corbulae* Möhn, known from El Salvador. Host: C. *coelestinum*. Distr.: Florida.

Conyza (horseweed)

1a Larva in flower head or bud gall 2
1b Larva in leaf or stem 3
2a Larva free in flower head .... *Contarinia* sp.

Larvae were found crawling out of flower heads in great numbers in October. Host: C. *canadensis*. Distr.: Maryland.

2b Larva in hairy bud gall .... *Rhopalomyia* sp.

The gall is apical, 6–12 mm in diameter. Brandhorst (1977), who found this gall, attributed it to *R. erigerontis* Felt, which is known from *Erigeron*, but specimens from the two plants have not been compared. Host: C. *canadensis*. Distr.: Nebraska.

Figures 180–195. Galls of Asteraceae. **180–186,** Galls of *Chrysothamnus* sp. **180,** Leaf gall: *Asteromyia chrysothamni*. **181,** Bud gall: *Rhopalomyia bigelovioides*. **182,** Bud gall: *Asphondylia bigeloviaebrassicoides*. **183,** Bud gall: *Rhopalomyia utahensis*. **184,** Node gall: *Rhopalomyia ?utahensis*. **185,** Resinous stem swellings: *Rhopalomyia glutinosa*. **186,** Stem swellings with hair in openings: *Rhopalomyia chrysothamni*. **187–189,** Galls of *Conyza canadensis*. **187,** Leaf blisters: *Asteromyia modesta* Felt. **188,** Petiole swellings: *Asteromyia modesta*. **189,** Stem swelling with emergence holes and pupal exuviae: *Neolasioptera erigerontis*. **190,** Bud gall, *Erigeron "fragilis"*: *Rhopalomyia erigerontis*. **191,** Enlarged flower bud (at lowest fork), *Eupatoriadelphus purpureus*: *Dasineura purpurea*. **192,** Stem galls, one in section, *Eupatorium hyssopifolium*: *Neolasioptera perfoliata*. **193,** Stem blisters, *Gutierrezia sarothrae*: *Asteromyia gutierreziae*. **194,** Rosette bud gall, *Ericameria palmeri*: *Rhopalomyia ericameriae*. **195,** Fuzzy stem swelling, *Gnaphalium* sp.: *Asphondylia* sp.

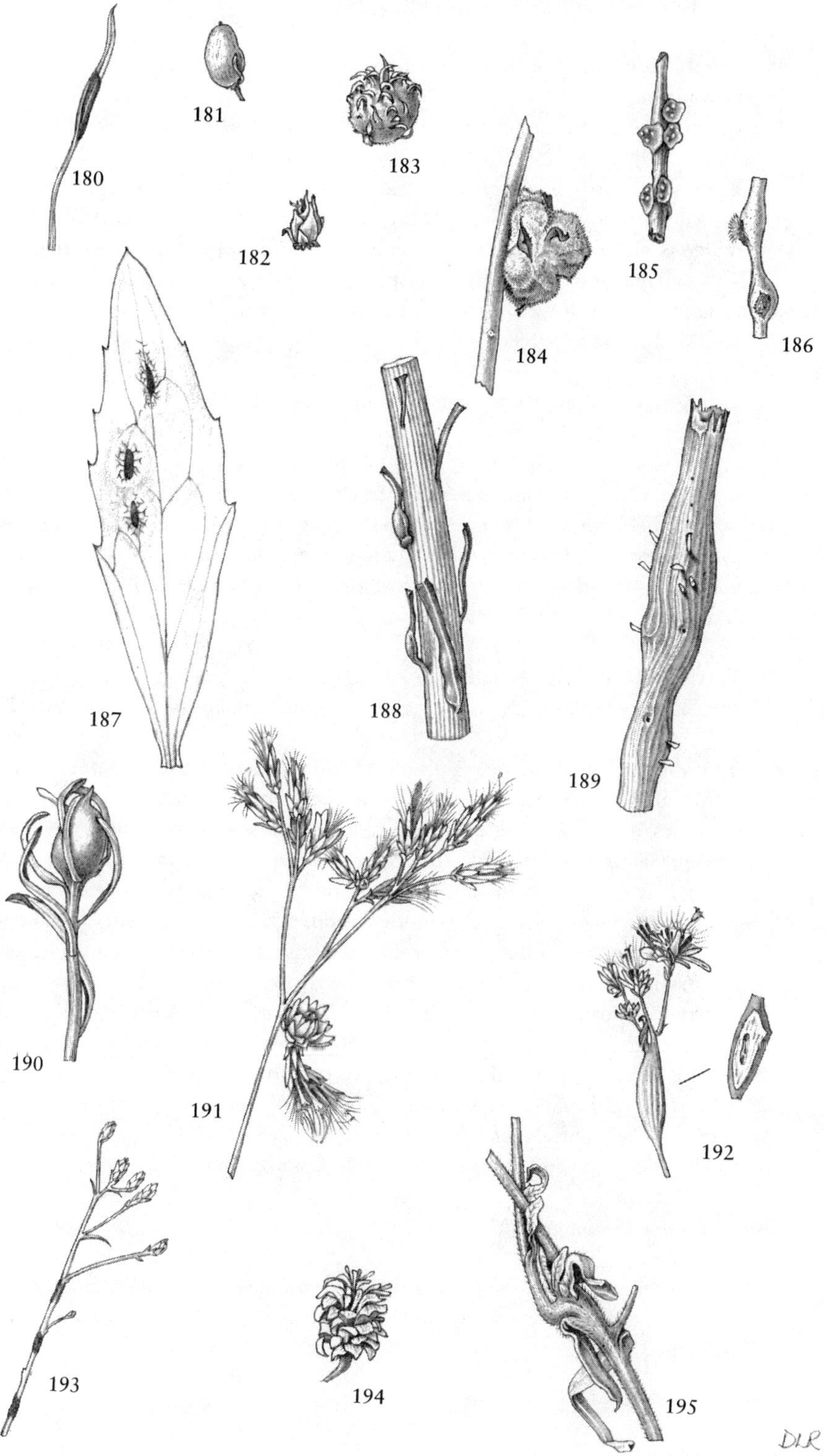
180
181
183
182
184
185
186
187
188
189
190
191
192
193
194
195
DLR

3a Larva in leaf blister or petiole swelling 4
3b Larva in stem.. 5
4a Green to purplish, simple leaf blister (187) .... *Asteromyia modesta* (Felt)

Blisters are often inconspicuous unless discolored. No fungus appears to be associated with these galls, unlike galls of most other asteromyias. This gall midge is multivoltine, each succeeding generation appearing on newer leaves. This gall maker may be a different species from the *A. modesta* in 4b. Hosts: *C. canadensis* as well as *Erigeron* and *Grindelia*. Distr.: Massachusetts to Maryland. Ref: Gagné (1968a).

4b Enlarged, calloused, winter-persistent base of petiole (188) .... *Asteromyia modesta* (Felt)

Galls are readily seen along most of the stem length in autumn after leaves have fallen. This species is either the overwintering form of *Asteromyia modesta* or a different species. Full-grown larvae are orange and overwinter as larvae in the galls. Adults emerge in spring. Host: *C. canadensis*. Distr.: New York and Maryland. Ref.: Gagné (1968a).

5a Larva in stem swellings, either rough and usually to one side of stem (189) or smooth, tapered ..... *Neolasioptera eregeroni* (Brodie) and *N. erigerontis* Felt

Two different kinds of stem galls occur on this host: a large, more or less soft, pithy gall near the base of or midway up the stem and a smaller, harder gall on branches of the inflorescence. The first kind appears in summer and is variable in size but usually many centimeters long and twice the normal diameter of the stem; it is polythalamous and can contain as many as 50 larvae. The other appears much later, in fall, is monothalamous and usually about 10 mm long by 5 mm in diameter. Adults from at least the large galls emerge from the hosts in both summer and spring. Larvae from the two kinds of gall show a few minor differences, but adults have not been reared from the inflorescence gall. If the two galls are shown to be made by two separate species, Brodie's name, *N. eregeroni*, would be available for the inflorescence gall. Brodie wrote that some larvae escaped from the galls, which is not known in *Neolasioptera*, but the adults he reared do belong to *Neolasioptera*. Host: *C. canadensis*. Distr.: Massachusetts to Ontario, south to Maryland, and Nebraska and California.

5b Larva in pith in apparently normal stems .... *Resseliella* sp.

Some infested stems showed black discoloration externally. One to 20 bright red, gregarious larvae lived in tunnels in the pith. Host: *C. canadensis*. Distr.: Maryland.

Coreopsis (tickseed)

"Succulent gall" .... *Asphondylia* sp.

This species is known only from adults in the USNM. The gall was not described further. Host: *C. palmata*. Distr.: Probably midwestern United States.

Encelia

1a Leafy bud gall .... *Asphondylia enceliae* Felt

The gall is monothalamous, 1 cm long, and 4 mm in diameter. Adults were reared in February. Host: *E. californica*. Distr.: California.

1b Conical, thick-walled, lateral bud gall .... *Rhopalomyia enceliae* Felt

The gall is 6 mm long, 3 mm in diameter, gray, and its apex is rounded or flared. Galls are monothalamous but occasionally confluent and may deform the stem. Host: *E. californica*. Distr.: California.

Erechtites (fireweed)

1a Irregular stem enlargement .... *Neolasioptera* sp.

Swellings are 2–5 cm long and 1 cm in diameter. Since adults were reared in March and August, this species has at least two generations per year. Host: *E. hieraciifolia*. Distr.: Florida.

1b Flower head gall .... *Asphondylia* sp.

The gall was not described further. Host: *Erechtites* sp. Distr.: unspecified.

Ericameria

1a Leafy rosette bud gall (194).................. *Rhopalomyia ericameriae* Felt

Adults were reared in April. Host: *E. palmeri*. Distr.: California.

1b Swollen bud................................ *Prodiplosis falcata* Gagné

Adults were reared in May. This species is evidently multivoltine. Hosts: *E. ericoides*; also *Eriogonum* sp. (Polygonaceae). Distr.: California. Ref.: Gagné (1986b).

Erigeron (see also *Conyza*) (fleabane)

1a Oval leaf blister (as in 187) .... *Asteromyia modesta* (Felt)

Blisters are usually green and inconspicuous but occasionally show some purple. Galls of this species do not have the associated fungus of other *Asteromyia* spp. Hosts: *E. strigosus, E.* sp. Known also from *Conyza*, *Grindelia*, and *Solidago* (q.v.). Distr.: Maryland and Minnesota.

1b Bud or flower gall.. 2
2a Apical bud gall (190) .... *Rhopalomyia erigerontis* Felt

Galls are whitish-downy. Host: *E. "fragilis."* Distr.: California.

2b Swollen, deformed calyx............................... *Rhopalomyia* sp.

Galls are polythalamous and can severely distort the flower heads. Adults emerged in early August. Host: E. *salsuginosus*. Distr.: Washington.

Eupatoriadelphus (Joe-Pye weed)

Enlarged, aborted flower (191) .... *Dasineura purpurea* Felt

Larvae live among slightly swollen, imbricated leaves of aborted flower buds. Adults were reared in September, so the species is at least bivoltine. Host: *E. purpureus*. Distr.: Massachusetts to Virginia.

Eupatorium (see also *Ageratum, Conoclinium,* and *Eupatoriadelphus* for some species formerly in *Eupatorium*)

1a Larva in flower ....
.... *Contarinia perfoliata* Felt and *Neolasioptera eupatoriflorae* (Felt)

C. perfoliata. Larvae crawl about among the developing flowers and may prevent seed set. Host: *E. perfoliatum*. Distr.: New York.

N. eupatoriflorae. The exact site of the larvae of this species is unknown, but adults generally resemble those neolasiopteras that live in achenes of Asteraceae.

Host: *E. hyssopifolium*, *E. perfoliatum*. Distr.: New York and Maryland. Ref.: Felt (1918a).

1b Larva in stem or leaf galls 2
2a Woody, tapered stem swelling (192) .... *Neolasioptera perfoliata* (Felt)

Swellings appear near the upper part of the flowering stems. Larvae overwinter in the stems. Hosts: *E. hyssopifolium*, *E. perfoliatum*. Distr.: Massachusetts to Virginia.

2b Hemispherical leaf swelling .... cecidomyiid

Swellings are thin-walled, about 5 mm in diameter, and show equally on both leaf surfaces. Larvae exit on underside of leaf through a hole lined with hairs. This gall is only tentatively identified as one made by a cecidomyiid; larvae had apparently only recently dropped to the soil when the galls were found in late September. Host: *E. perfoliatum*. Distr.: Maryland.

Euthamia (grass-leaved goldenrod)

This genus is host to several species of *Asphondylia*, among other gall midges. Adults of the asphondylias have black, velvety scales on the abdomen and banded white and black scales on the legs. These and the species on *Solidago* and *Aster* evidently form a natural group.

1a Elongate, ribbed, tapered, sometimes pediceled, complex gall on leaf, stem, or flower (196c, f) 2
1b Gall generally leafy or a simple deformation 3
2a Gall pediceled (196c) .... *Rhopalomyia pedicellata* Felt

Galls are monothalamous and green to purple. The gall midge is apparently multivoltine and probably overwinters as an early instar in the plant crown. Host: *E. graminifolia*. Distr.: Maine to Minnesota, south to Virginia.

2b Gall sessile, broadest at base (196f) .... *Rhopalomyia fusiformae* Felt

Galls are monothalamous and green to purple. The biology is probably similar to that of *pedicellata* in 2a. Host: *E. graminifolia*. Distr.: Connecticut and New York.

3a Bud or flower gall 4
3b Leaf or stem gall 7
4a Aborted flower bud (196a) .... *Asphondylia* sp.

c
d
b
e
a
c
c
b
f
g
h

A central cell within the bud encloses the larva or pupa. Host: *E. graminifolia*. Distr.: New Hampshire.

4b Leafy, enlarged bud gall. 5
5a Elongate-globular, spongy gall, enveloped by enlarged outer leaf (196d) . . . . *R. lanceolata* Felt and *Rhopalomyia lobata* Felt

Galls are mono- or polythalamous and light green. The two species, the first from New York, the second from Illinois, may be synonyms. Adults emerge in September so probably overwinter in a gall in the crown of the plant. Host: *E. graminifolia*. Distr.: Maine, New York, and Illinois.

5b Gall not spongy, composed of several imbricated leaves. 6
6a Gall composed of many imbricated leaves resulting from a foreshortened stem, the larval cell lined with white fungal mycelium (196b) . . . . . . *Asphondylia* sp.

Galls are variable in size and mono- or polythalamous. Host: *E. graminifolia*. Distr.: Massachusetts to Maryland.

6b Gall composed of two or more leaves stuck together at their edges to form a loose pouch. *D. carbonaria* Felt and *Dasineura flavicornis* Felt

Galls are green to purple and inhabited by gregarious yellow larvae. Adults of both nominal species were reared in July, so they evidently are at least bivoltine. The two species are probably synonymous. *Camptoneuromyia flavescens* (Felt) has also been reared from such galls and is probably an inquiline. Host: *E. graminifolia*. Distr.: New York.

7a Tapered swelling at base of stem . . . . *Lasioptera cylindrigallae* Felt

This gall has occasionally been attributed to *L. solidaginis*, which forms a similar gall on *Solidago*. The gall midge is probably univoltine. Host: *E. graminifolia*. Distr.: New York and Pennsylvania.

7b Black blister on leaf or upper stem or a 2-leaf gall. 8
8a Black blister on leaf or stem (196e, h). . . . . *Asteromyia euthamiae* Gagné

Galls are occasionally bordered with purple. Leaf galls may dehisce separately from the leaves. *A. euthamiae* has several generations per year and presumably

---

Figure 196. Galls of Asteraceae: *Euthamia graminifolia*: **a,** aborted flower buds, one in 2×: *Asphondylia* sp.; **b,** swollen branch buds: *Asphondylia* sp.; **c,** pediceled, ribbed galls on leaves, stems, and flowers: *Rhopalomyia pedicellata*; **d,** spongy flower head galls: *Rhopalomyia lobata*; **e,** stem blisters: *Asteromyia euthamiae*; **f,** sessile, ribbed galls on stems and leaves (one in 2× with pupal skin in emergence hole): *Rhopalomyia fusiformae*; **g,** two-leaf gall: *Asphondylia* sp.; **h,** leaf blister galls: *Asteromyia euthamiae*.

overwinters as full-grown larvae in the last generation of galls. Hosts: *E. graminifolia*, *E. leptocephala. E. minor*, *E. tenuifolia*. Distr.: Nova Scotia to Minnesota, south to Florida and Louisiana. Ref.: Gagné (1968a).

8b Gall composed of two leaves joined together and forming a green blister where they join (196g) . . . . *Asphondylia* sp.

Galls are similar to those formed on *Solidago* by *Asphondylia solidaginis*. The gall maker is evidently multivoltine. Host: *E. graminifolia*. Distr.: New York.

Flaveria

Tapered stem swellings . . . . *Neolasioptera* sp.

Adults emerged in April and May. Host: *F. linearis*. Distr.: Florida.

Gnaphalium (cudweed)

1a Stem swelling (195) . . . . *Asphondylia* sp.

Plant hairs are longer around the swellings than elsewhere on the stem. Host: *Gnaphalium* sp. Distr.: Oregon.

1b Narrowly ovate, white-woolly bud gall . . . . *Asphondylia* sp.

The gall is about 20 mm long and 10 mm in diameter. Host: *Gnaphalium* sp. Distr.: New Brunswick.

Grindelia (rosin-weed)

1a Leaf blister . . . . *Asteromyia modesta* (Felt)

Galls are 3–5 mm in diameter and have little or no discoloration except for some slight chlorosis. Unlike other species of *Asteromyia*, *A. modesta* does not appear to have an associated fungus. It is probably multivoltine. Hosts: *G. hirsutula*, *G. humilis*, *G. robusta*, and *G. squarrosa*; also *Conyza canadensis*, *Erigeron* spp., and *Solidago* spp. Distr.: Maryland, Minnesota, Texas, and California. Ref.: Gagné (1968a).

1b Bud gall 2

2a Modified, aborted floret (197) . . . . *Rhopalomyia grindeliae* Felt

Galls are hard and capsulelike. Adults emerged from the galls in October. Host: *G.* "*cuneifolia*." Distr.: California.

2b Leafy rosette gall or swollen bud gall 3
3a Leafy rosette gall (198) .... cecidomyiid

Galls are monothalamous, elongate-ovoid, and formed of overlapping bracts. Host: *G. lanceolata*. Distr.: Arkansas. Ref.: Wells (1921b).

3b Globular bud gall (199).................................. cecidomyiid

Galls are firm, green, one- to several-chambered, and usually found in axils of leaves. Host: *G. squarrosa*. Distr.: Kansas.

Gutierrezia

1a Black leaf and stem blister (193) .... *Asteromyia gutierreziae* Felt

Galls are usually monothalamous. The gall midge is probably multivoltine. Hosts: *G. microcephala*, *G. sarothrae*; also *Baccharis angustifolia*, *Gymnosperma glutinosum*, and *Isocoma "palmeri."* Distr.: Utah to New Mexico, west to California, and elsewhere in western North America on other hosts. Ref.: Gagné (1968a)

1b Flower or bud gall 2
2a Modified, aborted floret (230) .... *Rhopalomyia gutierreziae* (Cockerell)

Galls have a hard shell and are slightly longer and wider than normal florets. Adults were reared in October. Host: *G. sarothrae*. Distr.: New Mexico.

2b Bud gall 3
3a Leaves forming gall wider at base than apex (206).......... *Rhopalomyia* sp.

Galls are monothalamous. Host: *Gutierrezia* sp. Distr.: Oregon.

3b Leaves forming galls linear (207) .... *Rhopalomyia* sp.

Galls are monothalamous. Host: *Gutierrezia* sp. Distr.: Oregon.

Gymnosperma

Black leaf and stem blister (as in 193) .... *Asteromyia gutierreziae* Felt

Galls are black, elliptical, slightly convex, and 3–8 mm in diameter. One to several larvae occur in a gall. Hosts: *G. glutinosum*; also *Baccharis angustifolia*, *Gutierrezia* spp., and *Isocoma "palmeri."* Distr.: Texas to Arizona, south to

Mexico, and elsewhere in western North America on other hosts. Ref.: Gagné (1968a).

Haplopappus

Blister gall on leaf (as for 180) .... *Asteromyia chrysothamni* Felt

Galls are blue-gray to black, elongate-elliptical, 5–15 mm, usually only weakly convex, and become hard when mature. Several larvae may live in one gall, and the species is probably multivoltine. Hosts: *H. armerioides*, *H. bloomeri*, *H. linearifolius*; also *Chrysothamnus* spp. Distr.: New York, Alabama, and western North America on other hosts. Ref.: Gagné (1968a).

Helenium (sneezeweed)

1a Larva in pith of stem without apparent deformation.......... *Resseliella* sp.

Host: *Helenium* sp. Distr.: Ontario.

1b Larva in swollen bud or modified, aborted flower...................... 2

2a Swollen, globular, leafy, apical bud gall (200) .... *Asphondylia autumnalis* Beutenmüller

Galls are green, succulent, polythalamous, and variable in shape and size. Adults emerge from the galls in late summer, so the species probably overwinters in another gall in the crown of the plant. Host: *H. autumnale*. Distr.: North Carolina and Missouri.

2b Modified, aborted flower................................ *Asphondylia* sp.

Galls are monothalamous and formed of the enlarged but scarcely thickened corolla of the disk flower. They are about three times the width of a normal

---

Figures 197–205. Galls of Asteraceae. **197,** Infested flower head with enlargement of one gall in 2×, *Grindelia "cuneifolia"*: *Rhopalomyia grindeliae*. **198,** Apical rosette galls, one sectioned to show single larval cell at apex, *Grindelia lanceolata*: cecidomyiid (redrawn from Wells 1921b). **199,** Swollen buds, one sectioned to show larval cells, *Grindelia squarrosa*: cecidomyiid (redrawn from Wells 1921b). **200,** Bud gall, *Helenium autumnale*: *Asphondylia autumnalis* (redrawn from Beutenmüller 1907). **201,** Bud galls, *Helianthus grosseserratus*: *Asphondylia helianthiglobulus*. **202,** Leafy bud gall, *Heterotheca mariana*: *Rhopalomyia chrysopsidis*. **203,** Tubular stem galls, one in section, *Helianthus tuberosus*: *Olpodiplosis helianthi*. **204,** Leaf blister galls, one with pupal skin attached to exit hole on underside of leaf, *Helianthus tuberosus*: *Pilodiplosis helianthibulla*. **205,** Swellings at base of stem, one sectioned to show larval tunnels in pith, *Helianthus divaricatus*: *Neolasioptera* sp.

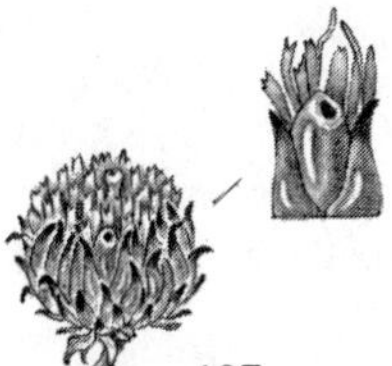
197

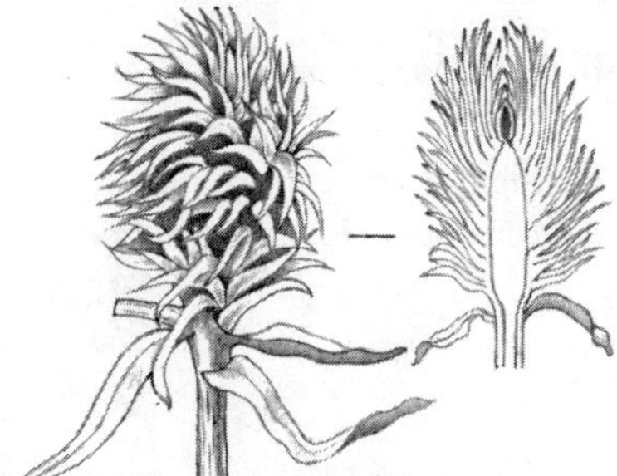
198

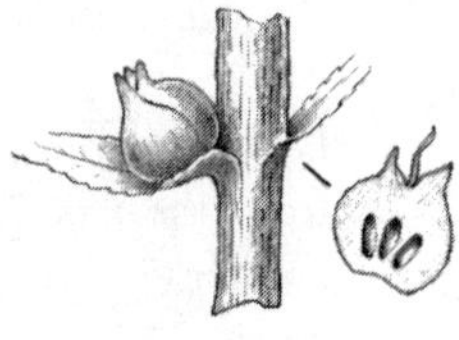
199

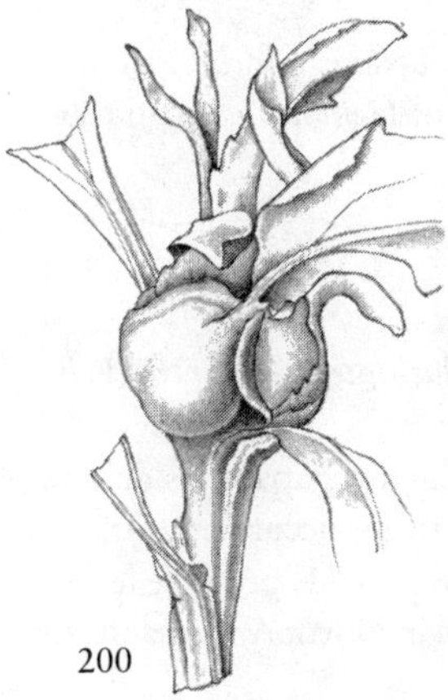
200

201

202

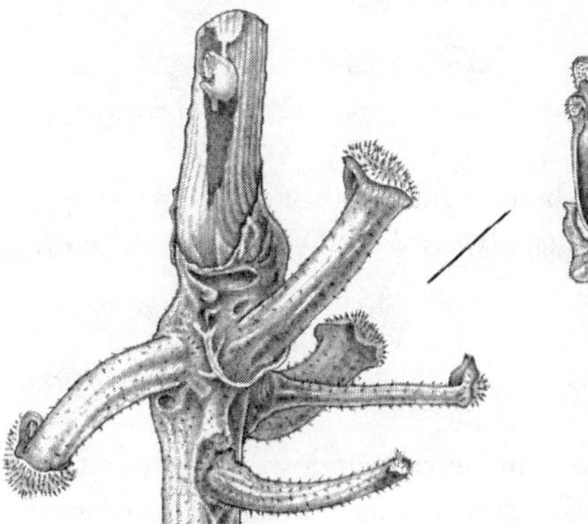
203

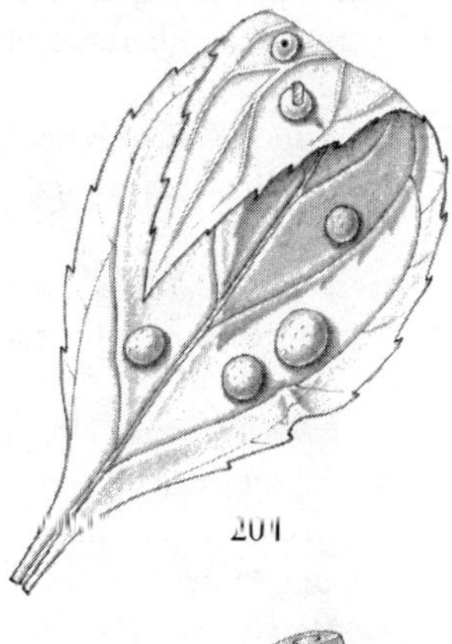
204

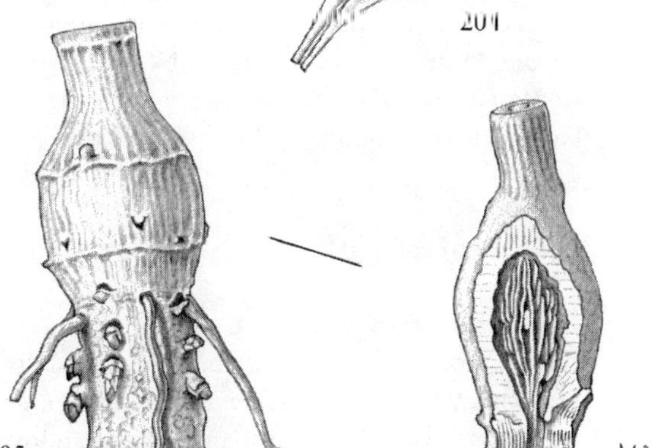
205

flower and protrude slightly above the tops of other disk flowers. Adults emerge from the galls in December. Host: *H. microcephalum.* Distr.: Texas. Ref.: Painter (1935).

Helianthus (sunflower)

In addition to the species keyed below, *Asphondylia helianthiflorae* Felt was reared in late August from unspecified flower galls in heads of *Helianthus* sp. in New York.

1a Larva in or on flower 2
1b Larva in gall on stem, node, bud, or leaf................................ 4
2a Larva in achenes............................ *Neolasioptera helianthi* (Felt)

Infested achenes appear normal or slightly swollen. Larvae pupate inside and emerge through a hole in the crown. At least two generations occur per year, the last overwintering in the achenes. Heavily infested flower heads produce leaves in the center. Hosts: *Helianthus* spp. Distr.: widespread in North America. Ref.: Rogers et al. (1979).

2b Larva free-living in or among florets 3
3a Larva in corolla tubes or floral ovaries .... *Dasineura* sp.

This species is only tentatively referred here. Adults do not fit in any known genus. Hosts: *H. annuus*, *H. grosseserratus*, *H. petiolaris.* Distr.: North Dakota to Texas.

3b Larva free among florets.................................. *Contarinia* spp.

C. schulzi Gagné. This is a severe pest of commercial sunflower. Larvae attack bracts and disc flowers and cause gnarled receptacles, rot, and reduced seed production. The species is multivoltine. Full-grown larvae drop to the soil to pupate. Host: *H. annuus.* Distr.: North Dakota to Texas. Ref.: Schulz (1973). *Contarinia* sp. This species is anatomically distinct from *C. schulzi* but its biology may be similar. Hosts: *H. ciliaris*, *H. petiolaris.* Distr.: Kansas and Texas.

4a Convex leaf gall (204) .... *Pilodiplosis helianthibulla* (Walsh)

Galls are strongly convex on the upper leaf surface, weakly so on the lower surface, from which the adult midges exit. Adults emerge in late summer. Hosts: *H. annuus*, *H. decapetalus*, *H. divaricatus*, *H. grosseserratus*, *H. strumosus*, *H. tuberosus.* Distr.: New York and Ontario to Iowa and Missouri.

4b Gall on bud, node, or stem, rarely leaf 5
5a Tubular gall on all plant parts, usually at nodes (203) ....
.... *Olpodiplosis helianthi* (Brodie)

Galls are mono- or polythalamous and highly variable in shape and length. The open, reflexed ends are filled with fine hairs. Adults emerge through the open end in late spring and in summer, so this species is at least bivoltine. Hosts: *H. annuus*, *H. decapetalus*, *H. divaricatus*, *H. strumosus*, *H. tuberosus*. Distr.: Ontario to Iowa, south to Texas. Ref.: Breland and Schmitt (1948).

5b Globular and succulent or spindleform, woody stem gall.................. 6
6a Gall globular, succulent, along side or at apex of stem (201) ....
.... *Asphondylia helianthiglobulus* Osten Sacken

Galls are variable in size, 5–15 mm in diameter, usually green, occasionally purple. Adults emerge from the galls in summer. The overwintering generation must develop in the crown of the plant. Hosts: *H. annuus*, *H. divaricatus*, *H. grosseserratus*, *H. heterophyllus*, *H. mollis*, *H. neglectus*, *H. niveus*, *H. occidentalis*, *H. petiolaris*, *H. praecox*, *H. strumosus*, *H. tuberosus*. Distr.: widespread in North America.

6b Tapered stem swellings (205).......... *Neolasioptera* spp. and *Resseliella* sp.

N. thurstoni (Brodie). This species is probably univoltine. Hosts: *H. divaricatus* and *H.* spp. Distr.: widespread in North America.
Neolasioptera spp. Two undescribed species other than *N. thurstoni* are known from larvae with distinctly different spatulae. One species was taken in *Helianthus eggertii* in Tennessee and in *H. divaricatus* in Oklahoma, the other was taken in *H. angustifolius* in Texas. Ref.: Rogers et al. (1979).
R. sp. It is possible that the larvae of this species were simply tunneling in pith and were not associated with a swelling. Host: *Helianthus* sp. Distr.: Texas.

Heterotheca

Woolly, leafy bud gall (202) .... *Rhopalomyia chrysopsidis* (Loew)

Galls are polythalamous. Adults emerged in late summer, so the species evidently overwinters as early as first-instar larvae, perhaps in the basal crown. Host: *H. mariana*. Distr.: New York and Maryland.

Hymenoclea

1a Tapered stem swelling .... *Neolasioptera* sp.

Adults emerged from galls in August. Hosts: *H. salsola*, *Hymenoclea* sp. Distr.: Arizona and California.

1b Bud gall 2
2a White-woolly terminal bud gall (211). . . . . . . *Asphondylia* sp. and cecidomyiid

Asphondylia sp. Galls are monothalamous. Adults emerged from the galls in May. Host: *H. salsola*. Distr.: California.
Cecidomyiid. Felt (1940) referred to this species as *Rhopalomyia hymenocleae*, but the species was never formally described and the name is not available. Felt's specimens belong to the Cecidomyiidi but do not fit any known genus. The elongate ovipositor indicates that this species is a plant-feeder. This species is possibly responsible for a different woolly bud gall. Host: *H. salsola*. Distr.: California.

2b Green, smooth, lateral bud gall 3
3a Bud gall pediceled, with leaves and sometimes branches protruding from the apex (210) . . . . *Asphondylia* sp.

Adults were reared in April. Host: *H. salsola*. Distr.: California.

3b Bud gall sessile, terminal (209) . . . . *Asphondylia* sp.

Adults were reared in April. Host: *H. salsola*. Distr.: California.

Hymenopappus

Larva in flower heads . . . . *Contarinia* sp. and *Dasineura* sp.

Larvae of both genera live free among the florets. Host: *H. flavescens*. Distr.: Texas.

Isocoma

Black blister on stem . . . . *Asteromyia gutierreziae* Felt

Swellings are 5–6 mm long. This species is known only from a larva. Host: *I. "palmeri."* Distr.: Texas.

Iva

The USNM collection has a series of adults of an *Asphondylia* reared from an unspecified gall in Mississippi.

Tapered stem swelling . . . . *Neolasioptera* sp.

Swellings are reportedly about 25–30 mm long and 6 mm at the widest diameter. Host: *Iva* sp. Distr.: unspecified.

Lactuca (wild lettuce)

Tapered, woody stem swellings (216) . . . . *Lasioptera lactucae* Felt

Galls are 10–30 mm in length and usually near the base of the stem. Host: *L. canadensis*. Distr.: New Hampshire to Maryland.

Liatris (blazing star)

Irregular, subspherical, bud gall . . . . *Asphondylia lacinariae* Felt

Galls are polythalamous, 20 mm long by 15 mm wide. Adults emerged in early spring. Host: *L. punctata*, *L. pycnostachya*. Distr.: Kansas, Arkansas, and Texas.

Mikania

1a Larva in flower head . . . . *Neolasioptera* sp.

The particular feeding site of this species is unknown. It could be either the fruit or the pedicel. Adults were reared from flowers in early November in southern Florida. Host: *Mikania* sp., probably *batatifolia*. Distr.: Florida.

1b Larva in tapered stem swelling (208). *Neolasioptera* sp.

Larvae overwinter in the stems. Host: *M. scandens*. Distr.: Connecticut and New York.

Palafoxia

A series of *Asphondylia* sp. adults in the USNM was reared from an unspecified gall on *Palafoxia* sp. in Florida.

206

207

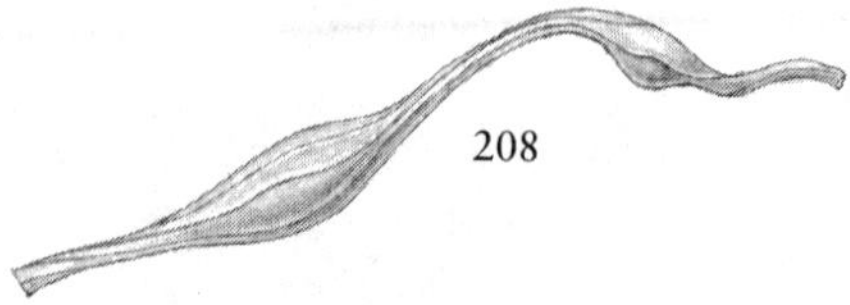
208

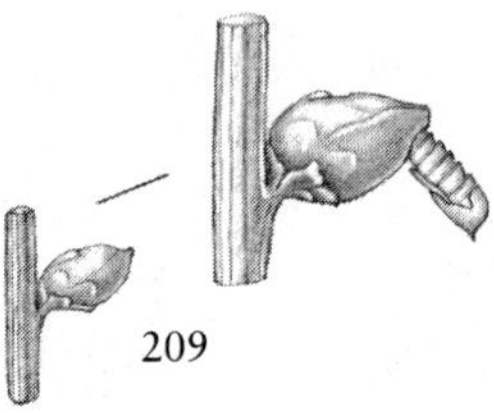
209

210

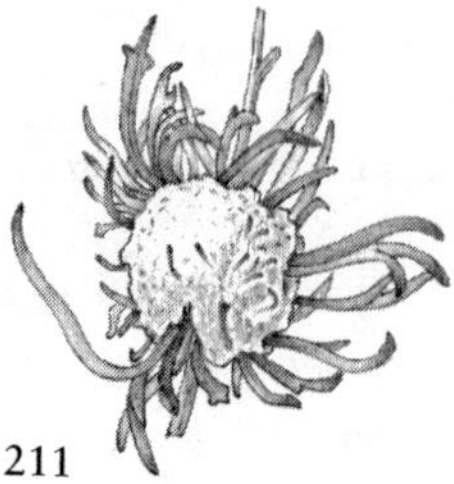
211

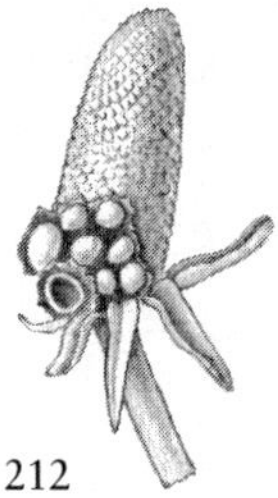
212

214

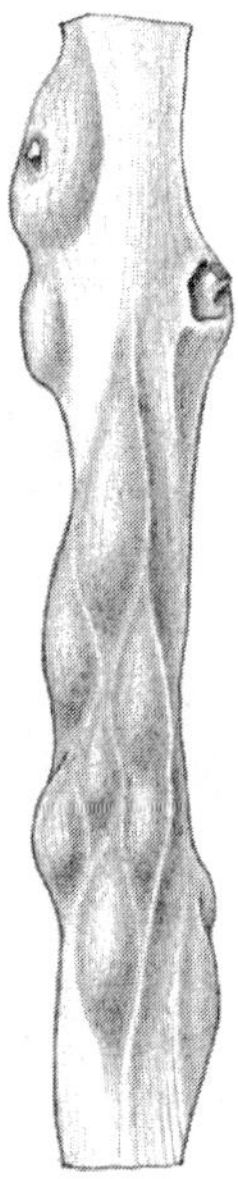
216

213

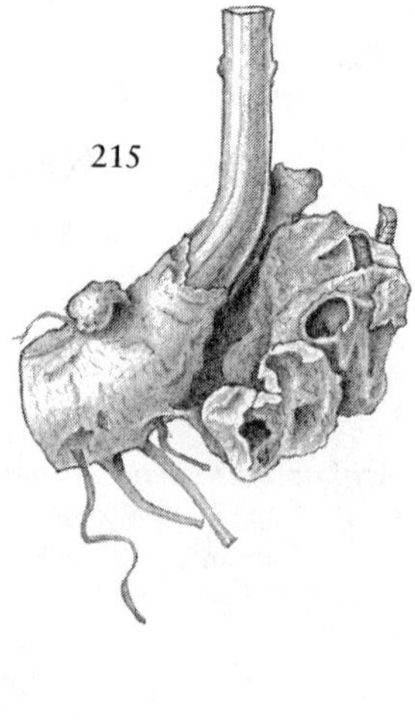
215

DLR

Parthenium

White, woolly galls on axillary bud, young leaf, or inflorescence ....*Contarinia partheniicola* (Cockerell)

Galls are 5–8 mm long, 4–5 mm wide, and occur singly or in small groups. Adults were reared in late summer, winter, and spring, so probably more than one generation occurs per year. Hosts: *P. incanum*; also *Ambrosia* spp. Distr.: New Mexico and California.

Perityle

Larva in flower heads .... *Dasineura* sp.

This species is known only from a series in the USNM. No date or further indication of the damage is known, but the larvae probably range freely among the flowers. Host: *P. coronopifolia*. Distr.: Arizona.

Pluchea

Larva in seed heads .... *Asphondylia* sp.

This species is known only from a series of specimens in the USNM, with no further description of the gall they caused. Adults were reared in April. Host: *P. rosea*. Distr.: Florida.

Prenanthes

Flat, ovoid, purplish leaf blister .... *Cystiphora canadensis* Felt

Galls are approximately 3–5 mm in diameter. This species probably has several generations per year, as do other species of *Cystiphora*. The female inserts her eggs into the developing leaves by means of its pointed, hard ovipositor. Hosts: *P. alba*, *P. altissima*, and *P. serpentaria*. Distr.: Maine and Ontario.

Figures 206–216. Galls of Asteraceae. **206, 207,** Bud galls, *Gutierrezia* sp.: *Rhopalomyia* spp. **208,** Stem swellings, *Mikania scandens*: *Neolasioptera* sp. **209–211,** Bud galls, 209 with specimen on right in 2× and with pupa protruding, *Hymenoclea salsola*: *Asphondylia* sp. **212,** Floret galls, *Ratibida columnifera*: *Asphondylia ratibidae* (after photograph in Felt 1940). **213,** Leafy flower gall, *Rudbeckia hirta*: cecidomyiid. **214, 215,** Bud and root crown galls, respectively, *Rudbeckia laciniata*: *Asphondylia rudbeckiaeconspicua*. **216,** Stem swelling, *Lactuca canadensis*: *Lasioptera lactucae*.

Ratibida (coneflower)

Deformed, aborted florets or swollen apical buds (212) .... *Asphondylia ratibidae* Felt

Galls are monothalamous. This species has at least two generations per year. In summer the galls are in the flower heads. Infested flowers are swollen at their bases, and the florets do not open. From August to October, galls are found in enlarged stem ends or leaf buds. Adults emerging in October must lay eggs on living plant tissue, so this species probably has an early spring bud gall as well. Host: *R. columnifera.* Distr.: Texas. Ref.: Painter (1935)

Rudbeckia (coneflower)

1a Tapered stem swelling..................... *Neolasioptera rudbeckiae* (Felt)

Occasionally one larva lives in the stem pith without causing an enlargement. This species has also been reared from the soft tissue of galls of *Asphondylia rudbeckiaeconspicua*. More than one generation per year may occur because an adult was reared from galls in September. Host: *R. laciniata.* Distr.: New York, Pennsylvania, and Virginia.

1b Bud or flower gall... 2

2a Large, globular, succulent bud gall at base of plant in spring or apex of plant in summer (214, 215) .... *Asphondylia rudbeckiaeconspicua* Osten Sacken

Galls are large and polythalamous. Summer galls begin development in June and July, and have leaves and an occasional flower petal growing from their tips. Adults emerge from these galls in late summer and lay eggs in the next year's buds at the base of the plant. Adults emerge from the resulting galls in the following June. Hosts: *R. laciniata, R. triloba.* Distr.: New York to Maryland. Refs.: Plakidas (1982, 1988).

2b Leafy flower heads (213).................................. ?cecidomyiid

The gall maker is unknown. A predaceous gall midge, *Lestodiplosis rudbeckiae* (Beutenmüller) was reared from infested flower heads. Host: *R. hirta.* Distr.: North Carolina.

Schistocarpha

Two species, an *Asphondylia* and a *Neolasioptera*, were reared in November from flower heads of *Schistocarpha eupatorioides* from Veracruz, Mexico.

Senecio (groundsel)

1a Swollen stem . . . . *Lasioptera arizonensis* Felt

Adults emerged in spring from circular holes in stems. Host: *S. arizonicus*. Distr.: Arizona.

1b Swollen bud or swollen, sterile floret 2
2a Swollen bud . . . . *Asphondylia* sp.

This species is known only from a gall and pupa in the USNM. Host: *S. longilobus*. Distr.: Sonora, Mexico.

2b Swollen, sterile floret . . . . *Asphondylia bea* Felt

Galled florets are swollen basally and remain closed apically. Flower heads with several galls are misshapen. Successive generations emerge during the flowering period in southern California. Host: *S. douglasii*. Distr.: California. Ref.: Needham (1925).

Smallanthus

An *Asphondylia* was reared in November from flower heads of *Smallanthus maculatus* in Chiapas, Mexico.

Solidago (goldenrod)

The grass-leaved goldenrods are now treated under *Euthamia*. The National List of Scientific Plant Names (Anon. 1982) records *Solidago canadensis* and *S. altissima* as separate species. In my own collecting, I have followed Gleason and Cronquist (1963) in considering them the same species. Further, I have allowed *Solidago gigantea* to be subsumed within my concept of *S. canadensis* and thus have not distinguished between those two species in my own work.

In addition to the cecidomyiids keyed below, two others have been reared from galls on goldenrod, but their damage was not described with enough detail to include them in the key. These are *Asphondylia johnsoni* Felt from a goldenrod in Pennsylvania and *Rhopalomyia cruziana* Felt from an unspecified flower gall on a goldenrod in southern California. In addition, a *Camptoneuromyia* sp. has an association with *Asteromyia* leaf blister galls (Weis 1982a). The cocoons in which the *Camptoneuromyia* larvae pupate can sometimes be seen on leaves close by *Asteromyia* galls, and Weis (1982a) reported larvae feeding in the outer layers of those galls.

1a Bud or flower gall 2

1b Stem or leaf gall 11
2a Galls made up of bud leaves, often as rosette on foreshortened stem 3
2b Galls conical, globular, or cylindrical, not formed of leaf rosette 6
3a Closed cluster of several leaves on apex of branch .... *Dasineura radifolii* Felt

Larvae are gregarious among the leaves forming the cluster, which is about 1 cm long. The original series was found among leaves near the base of the plant. This or possibly another *Dasineura* occurs on apical galls on *S. canadensis* and relatives. This species may be multivoltine. Another species, *Dasineura folliculi* (Felt), was reared from a loose pod gall on goldenrod in New Jersey but is probably not a gall maker. Host: *S. juncea* or *S. puberula* and possibly other species of *Solidago*. Distr.: Massachusetts.

3b Rosette galls made up of 25 to hundreds of leaves 4
4a Gall with numerous, imbricated leaves, the inner leaves narrow; on *S. canadensis* and close relatives (218d)............................ *Rhopalomyia* spp.

This gall is common on *Solidago canadensis*. Although superficially similar exteriorly to one on *S. juncea*, the two are structurally different, and an asphondylia is responsible for the one on *S. juncea*. The gall makers live in usually one, but sometimes several, vertical, elongate cells at the very center of the gall. Because adults emerge from the galls in fall, they must lay their eggs at the base of the plant where another bud gall is presumably formed. Five species of *Rhopalomyia* have been described from galls on *S. canadensis*. Larvae of *Clinodiplosis* sp. are sometimes found among the outer leaves of the galls. Although the galls can be found throughout eastern North America, the species reared from them are listed below only with their specific initial collection data:

R. albipennis Felt. Males of this species have white wings and banded legs, so differ from the brown-winged and brown-legged males of the four taxa listed below. Females of *R. albipennis* are superficially indistinguishable from those of the four species listed below. Host: *S. canadensis*. Distr.: New York.
R. capitata Felt. This species may be only a synonym of *R. solidaginis*. Host: *S. canadensis*. Distr.: New York.
R. carolina Felt. This species may be a synonym of *R. solidaginis* also. Host: *S. canadensis*. Distr.: North Carolina.

Figures 217–220. Galls of Asteraceae: *Solidago* spp. **217,** Clustered bud galls, *S. fistulosa*: *Rhopalomyia racemicola*. **218,** Galls on *S. canadensis*: **a,** onion-shaped flower gall: *Schizomyia racemicola*; **b,** vase-shaped flower gall: *Rhopalomyia anthophila*; **c,** cylindrical flower gall: *Rhopalomyia* sp.; **d,** cabbagelike galls, one sectioned to show central larval cell amid modified, linear leaves: *Rhopalomyia solidaginis*. **219,** Galls on *S. canadensis*: **a,** blister galls: *Asteromyia carbonifera*; **b,** two-leaf blister galls with 2× enlargement of the two leaves pulled apart: *Asphondylia solidaginis*; **c,** conical leaf gall: *Rhopalomyia clarkeae*. **220,** Soft, lateral stem gall, *S. canadensis*: *Asteromyia tumifica*.

217
218
a
b
c
d
219
a
b
c
220
DLR

R. inquisitor Felt. This was reportedly reared from the same gall as *R. capitata*, but the two species are distinct. Host: *S. canadensis*. Distr.: New York.

R. solidaginis (Loew). Host: probably *S. canadensis*. Distr.: Maryland. Ref.: Osten Sacken (1862).

4b Gall single or composite, the internal leaves not linear and narrow, merely shortened 5

5a Larva in central cell of rosette gall, the inside lined with white mycelium of symbiotic fungi (221, 222) . . . . *Asphondylia monacha* Osten Sacken

Osten Sacken gave as host of this species *Solidago altissima*, but the type galls appear to be on *S. juncea*. Because this gall is not found in Maryland except on *S. juncea*, Osten Sacken evidently misidentified the host. The gall analagous to this on *S. canadensis* is formed by *Rhopalomyia solidaginis* (see entry 4a of this key). The *A. monacha* gall has an associated fungal symbiont, but the *Rhopalomyia* gall does not. Both species pupate in the gall and emerge as adults in late summer, September 20–25, in Maryland. Orange and active, inquilinous *Youngomyia* larvae have been found, several to an *Asphondylia* cell in late August in Maryland, when the *Asphondylia* larvae were still only in their second instar. When full-grown, *Youngomyia* larvae break out of the gall to drop to the soil. Empty *Asphondylia* larval skins were found in galls vacated by the inquiline. Hosts: *S. juncea*. Distr.: Massachusetts to Virginia west to Minnesota.

5b Larva in bulbletlike gall occasionally present around central cell containing *A. monacha* of 222 . . . . *Rhopalomyia* sp.

The galls are monothalamous, teardrop-shaped, and somewhat shorter than the central cell containing *A. monacha*. Several may surround the central gall. Host: *S. juncea*. Distr.: Maryland.

6a Modified floret 7
6b Bud gall that may be situated among florets 9
7a Gall onion shaped, short-pointed at tip (218a) . . . . *Schizomyia racemicola* (Osten Sacken)

Galls are green to red, smooth, thin-walled, and weakly ribbed. Larvae are orange, very active, and leave the gall to pupate in the soil. Probably only one generation occurs per year. Hosts: *S. canadensis*, *S. leavenworthii*. Distr.: Massachusetts and New York to Florida. Ref.: Gagné (1971b).

7b Gall more or less cylindrical 8
8a Gall widest at base, fuzzy (218b) . . . . *Rhopalomyia anthophila* (Osten Sacken)

Galls are 4–5 mm long, 2 mm wide, usually green-walled but appearing white from the fuzzy covering. Adults were reared from galls in fall. Host: *S. canadensis*. Distr.: Massachusetts, New York, North Carolina, and Missouri.

8b Gall widest near apex, smooth (218c) .... *Rhopalomyia* sp.

This species is known only from its brown galls in the USNM and is only tentatively identified as a *Rhopalomyia*. Host: *S. canadensis*. Distr.: Maryland.

9a Fuzzy, pyramidal gall among flowers (217) ..... *Rhopalomyia racemicola* Felt

This species is only tentatively listed here as causing the galls illustrated, which are from central Florida. *R. racemicola* was reared in autumn in North Carolina from flower heads with galls apparently similar to these. Galls are green, sometimes confluent. In Florida, adults emerged in December. Hosts: *S. canadensis*, *S. fistulosa*. Distr.: North Carolina and Florida.

9b Smooth gall elsewhere on plant 10
10a Smooth, large, apical gall subtended by leaf whorl (223) .... *Rhopalomyia hirtipes* (Osten Sacken)

Galls are succulent and polythalamous. Full-grown larvae are white and pupate in late summer. Adults emerge about two weeks later. Eggs are laid on or near the base of goldenrod stems. These hatch in fall, overwinter as first-instar larvae on the outside of the rootstock, and embed themselves in meristematic tissue the following spring. This species is evidently responsible also for galls on the stolons and crown (226; see entry 10b of this key). Host: *S. juncea*, *S. odora*. Distr.: Massachusetts to Maryland, and Illinois. Ref.: Spence (1969).

10b Bulbous, succulent, smooth node gall, usually near root crown, sometimes at tip of short stunted stem (224, 226)....................... *Rhopalomyia* spp.

Several distinct species have been reared from such galls. They may all have two generations per year.

R. bulbula Felt (224). This was reared from small, bulbous galls similar to small ones attributed to *R. hirtipes*, but adults are different from either *R. hirtipes* or *R. thompsoni*. Host: *Solidago* sp., probably *juncea*. Distr.: Massachusetts.
R. hirtipes (Osten Sacken) (226). See entry 9a of this key.
R. thompsoni Felt. This species was reared from rootstock galls similar to those of *R. hirtipes*, but adults reared from these galls are distinctly different from *hirtipes*. Host: *Solidago* sp., probably *juncea*. Distr.: Massachusetts.

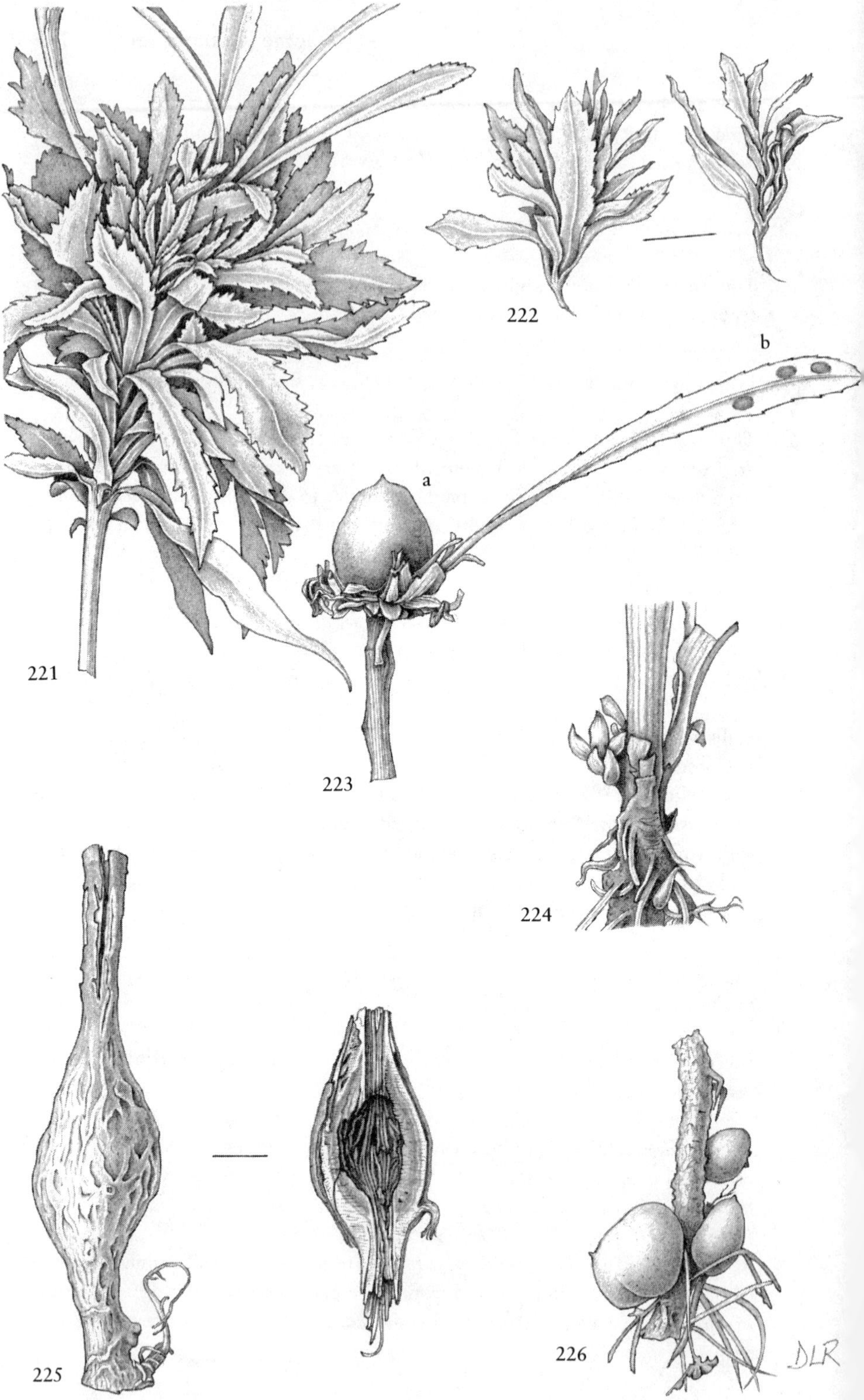
222
b
a
221
223
224
225
226
DLR

11a Stem swelling 12
11b Leaf gall.. 13
12a Swelling woody, evenly tapered; larva in the enlarged pith (225) ....
.... *Lasioptera solidaginis* Osten Sacken

The exterior surface of the gall has the same texture as the rest of the stem. These galls are found near the base of the plant, 5–10 cm above the soil surface. The pith is soft with elongate, vertical larval tunnels. Larvae overwinter in the gall and emerge the following spring. One generation occurs per year. Host: *S. canadensis*, *S. rugosa*. Distr.: Massachusetts to Maryland, and Minnesota.

12b Swelling spongy, an outgrowth usually on one side of the stem (220). ....
.... *Asteromyia tumifica* (Beutenmüller)

Galls are soft, brown, and porous, and are usually found some distance above the base of the stems, higher than galls formed by *Lasioptera solidaginis*. The individual larval cells are oval. This species is the only *Asteromyia* that does not live in leaves or apical stems. It is univoltine. Larvae overwinter in the galls. Host: *S. canadensis*. Distr.: Massachusetts and New York to North Carolina, and Minnesota. Ref.: Gagné (1969c).

13a Gall formed by two leaves appressed together to form a blister gall (219b)
.... *Asphondylia solidaginis* Beutenmüller

Each of the two leaves involved forms one half of the unicellular, ovoid, 4–5 mm diameter, white-fungus-lined gall. This species is multivoltine. Other gall midges have reportedly been reared from these galls: *Trotteria caryae* (Felt) and *Camptoneuromyia adhesa* (Felt). The first is an inquiline and the other two may also be. Host: *S. canadensis*. Distr.: Maine to Virginia, east to Minnesota.

13b Galls involving a single leaf.. 14
14a Blister gall (219a, 223b)............................ *Asteromyia* spp.

A. carbonifera (Osten Sacken). Galls contain a symbiotic fungus that apparently is not eaten by the larva. The fungus turns hard and black and becomes incorporated with the plant tissue as the larva matures. This species is found on most goldenrods and is widely distributed across North America. It is multivoltine. In

Figures 221–226. Galls of Asteraceae: *Solidago* spp. **221, 222,** Leaf cluster galls, *S. juncea*: *Asphondylia monacha*. **221,** Galls in aggregate. **222,** Two separated galls, one with some leaves removed to show central larval cell. **223,** Galls on *S. juncea*: **a,** soft, apical stem gall, all but one of rosette of leaves broken off: *Rhopalomyia hirtipes*; **b,** leaf blister galls: *Asteromyia carbonifera*. **224,** Bulblet galls near stem base, *S. juncea*: *Rhopalomyia bulbula* (after photograph in Felt 1940). **225,** Woody stem galls, one sectioned, *S. canadensis*: *Lasioptera solidaginis*. **226,** Bulblet galls near stem base, *S. juncea*: *Rhopalomyia hirtipes* (after photograph in Felt 1940).

central Illinois it has three generations per year. Development from egg to adult takes 4–5 weeks. Eggs are glued to the underside of the leaf. One to more than 10 larvae occur per gall. As the season progresses, galls develop on leaves progressively farther up the stem. The color of the galls is characteristic of the host species. The size is variable also, 5–15 mm in diameter on *S. canadensis* with one to several larvae, 3–4 mm in diameter on *S. juncea* with one larva. Hosts: most species of *Solidago*. Distr.: coast to coast in North America. Refs.: Gagné (1968a), Weis (1982a).

A. modesta (Felt). This species is less common on goldenrod than *A. carbonifera* and, unlike the latter, does not have a fungus associated with it but lives simply in the mesophyll layer of the leaf and forms only a single convexity on the leaf. Hosts: *S. flexicaulis*, *S. macrophylla*; also *Erigeron* spp., *Conyza* sp., *Grindelia* spp.. Distr.: Minnesota and New Hampshire on *Solidago*. Ref.: Gagné (1968a).

14b Conical, decumbent gall attached to leaf rib (219c) *Rhopalomyia clarkeae* Felt

Galls are 1–2 mm long, inconspicuous, the same color as the leaf and usually on the underside. Hosts: *S. canadensis*, *S. rugosa*. Distr.: New Hampshire and Ontario, south to North Carolina.

Sonchus (sow-thistle)

Leaf blister . . . . *Cystiphora sonchi* (Bremi)

Galls are green or red, their upper surface convex, the lower surface flat. One larva occupies each gall. Fully developed larvae either pupate in the gall or drop to the soil and form a cocoon in which they pupate. There are three generations per year in Canada. Females oviposit into stomatal openings of the lower surface of leaves that are still growing but nearing the end of expansion. This species was introduced into Canada from Europe for the control of its weedy host, a rangeland pest. Host: *S. arvensis*. Distr.: Nova Scotia and Manitoba to Alberta; Europe. Refs.: DeClerck and Steeves (1988), Peschken et al. (in press).

Tetradymia

Woolly, ovoid, lateral bud gall . . . . "*Mayetiola*" *tetradymia* (Felt)

Adults emerged from galls in April. Galls are polythalamous. Host: *T. comosa*. Distr.: California.

Verbesina

1a Aborted, swollen florets (232)....... "*Cecidomyia*" *verbesinae* Beutenmüller

"*Cecidomyia*" *verbesinae* is known only from galls and larvae and cannot now be placed to genus. Galls have one larval cell each, but may be confluent. They terminate in a blunt point and are situated among normal florets. These were found in autumn. The orange-red larvae left the galls to pupate in the ground. Host: *V. alternifolia*. Distr.: North Carolina.

1b Swollen stem, occasionally just under flowers (227) .... *Neolasioptera* sp.

Galls can be found in summer along the whole length of the stem. At that time larvae are in various instars in elongate, fungus-lined tunnels. Larvae are youngest in galls nearest the plant apex. More than one generation may occur per year. Pupation occurs in the galls. The soft parts of the galled stem rot away during winter, the only hard parts left when the larvae pupate being the cylindrical larval tubes made up of the fungal byproducts and a few woody strands. Host: *V. alternifolia*. Distr.: Maryland.

Vernonia (ironweed)

1a Swollen petiole, midrib, or stem ..... *Neolasioptera vernoniae* (Beutenmüller)

Swellings are green to red, 6–10 mm long. Hosts: *V. baldwinii*, *V. noveboracensis*. Distr.: New York to Indiana, south to Virginia, and Nebraska.

1b Enlarged, aborted seed capsule .... *Asphondylia vernoniae* Felt

Galls are globular with longitudinal ribs, dark green, often purple at apex. Adults emerged in autumn. *Youngomyia podophyllae* (Felt), a possible predator, is associated with this gall. Larvae of *Y. podophyllae* leave the galls to pupate in the soil, but larvae of *A. vernoniae* pupate in the galls. Host: *V. noveboracensis* and possibly also *V. blodgettii*. Distr.: Virginia and Florida.

Viguiera

Swollen stem .... *Neolasioptera* sp.

Galls were collected in winter, and adults were reared in spring. Swellings can be slight. Host: *V. annua*. Distr.: Arizona.

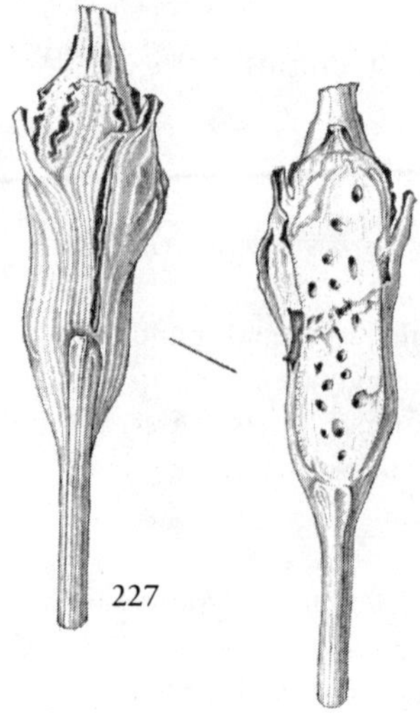

227

228

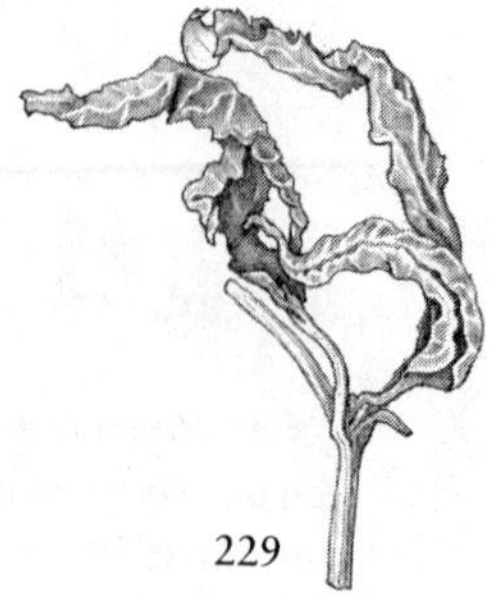

229

230

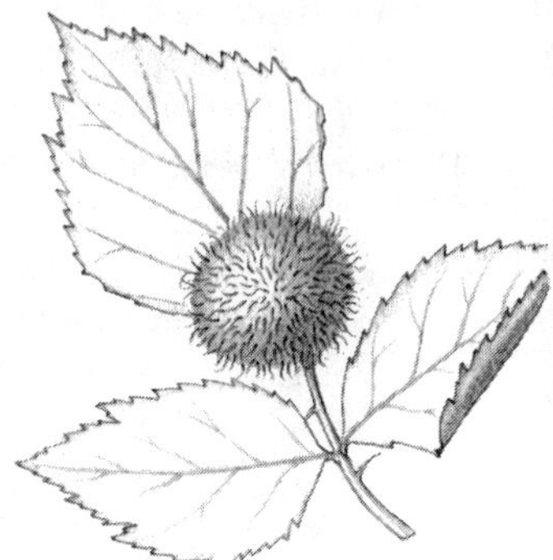

231

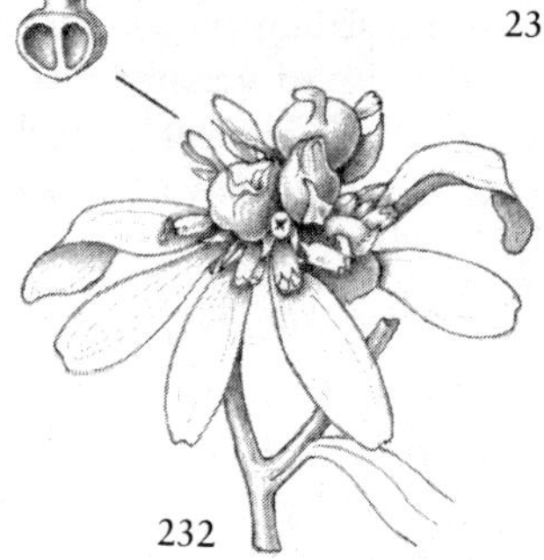

232

233

234

Xanthium (cockleburr)

1a Larva in root. . . . . *Resseliella radicis* (Felt)

Damage or symptoms were not further specified. Host: *Xanthium* sp. Distr.: Alabama.

1b Branch tip galls. . . . . *Asphondylia xanthii* Felt

Galls were not further specified. Adults were reared in June. Host: *X. strumarium*. Distr.: Texas.

Balsaminaceae

Impatiens (jewelweed)

1a Globular, pendant flower bud gall (235a) . . . . *Schizomyia impatientis* (Osten Sacken)

Galls are succulent and 10–15 mm in diameter. The active, red larvae leave the galls in late summer or early fall and drop to the ground to overwinter. Larvae of *Youngomyia podophyllae* (Felt) are occasionally found in these galls as well as those of *Neolasioptera impatientifolia* (below) and those of other *Schizomyia* species on other plants. Whether *Y. podophyllae* is a predator or a benign inquiline is unknown. Hosts: *I. capensis*, *I. pallida*. Distr.: Maine to Virginia.

1b Swollen leaf vein, petiole, or stem 2

2a Swollen midrib or petiole (235b). . . . . *Neolasioptera impatientifolia* (Felt)

Swellings are usually more visible from beneath the leaves, are succulent, partly translucent, light green, and contain one to several larvae and an associated fungus. At least two generations occur per year. Pupation occurs in the galls. Rarely, larvae of *Youngomyia podophyllae* (Felt) can be found in the galls. Hosts: *I. capensis*, *I. pallida*. Distr.: Maine to Virginia.

Figures 227–234. Galls of Asteraceae, Fabaceae, Polemoniaceae, Rosaceae, and Verbenaceae. **227,** Stem swellings, one sectioned to show larval tunnels, *Verbesina alternifolia*: *Neolasioptera* sp. **228,** Swollen, aborted buds, *Desmodium* sp.: *Clinodiplosis meibomiifoliae*. **229,** Rolled leaves, *Verbena hastata*: *Clinodiplosis verbenae*. **230,** Flower bud galls, *Gutierrezia sarothrae*: *Rhopalomyia gutierreziae*. **231,** Spiny, spherical, leaf gall, *Rubus* sp.: "*Cecidomyia*" *muscosa*. **232,** Swollen, aborted buds, one sectioned, *Verbesina alternifolia*: "*Cecidomyia*" *verbesinae* (redrawn from Beutenmüller 1907). **233,** Bud gall, *Aster subulatus*: *Asphondylia* sp. **234,** Healthy and injured buds, some leaves of one bud (enlarged to approximately 3×) removed to show larvae, *Phlox* sp.: *Clinodiplosis phlox* (redrawn from Neiswander 1962).

2b Swollen stem base . . . . *Lasioptera centerensis* Felt

Known only from a larva, this species was reported once, from a "double-celled" gall. Host: *I. pallida.* Distr.: New York.

Betulaceae

Alnus (alder)

1a Enlarged bud (236) . . . . *Dasineura serrulatae* (Osten Sacken)

Affected buds are 6–12 mm in diameter. Each contains several reddish larvae that drop to the soil in fall. Adults emerge in spring. Hosts: *A. crispa, A. rugosa, A. serrulata.* Distr.: Massachusetts and Ontario to North Carolina.

1b Folded leaflet . . . . *Dasineura* sp.

Young leaves are folded in half along the long axis. Larvae spin cocoons in the folds and pupate there. Host: *A. rubra.* Distr.: British Columbia to Oregon.

Betula (birch)

1a Swollen stem, bud, or base of petiole (237) . . . . *Apagodiplosis papyriferae* (Gagné)

Galls appear in midsummer. Full-grown larvae drop to the ground in late September and form cocoons in which they overwinter. Adults emerge in mid-May and lay eggs at bases of leaves on new terminals. Several larvae occur in separate cells in the cortical enlargement. Bud and stem mortality results from girdling. Host: *B. papyrifera.* Distr.: Quebec, Michigan, Washington, and Oregon. Ref.: Wilson (1968d).

1b Swollen bract or ovary of female catkin. . . . . *Semudobia* spp.

Five species of *Semudobia* are known from North America, three native and two originally introduced from Europe in seeds of *Betula pendula.* One of the three native species is Holarctic, the two others are Nearctic only. The biology of all five species is generally similar. Eggs are deposited in spring between the bracts and flowers. Usually only one larva infests a bract or ovary. Pupation occurs in the gall the following spring. Refs.: Roskam (1977), Roskam and Van Uffelen (1981). The five species in our region are:

S. betulae (Winnertz), an immigrant from Europe. Host: *B. pendula.* Distr.: Ontario and Pennsylvania.

S. brevipalpis Roskam, an eastern Nearctic species. Hosts: *B. papyrifera* and *B. populifolia*. Distr.: Quebec and Ontario, south to Maryland.
S. skuhravae Roskam, a Holarctic species. Hosts: *B. alleghaniensis*, *B. occidentalis*, *B. papyrifera*, *B. pendula*, and *B. populifolia*. Distr.: Connecticut and South Dakota to Alberta, south to Colorado.
S. steenisi Roskam, a western Nearctic species. Host: *B. occidentalis*. Distr.: Montana to British Columbia, south to Colorado.
S. tarda Roskam, an immigrant from Europe. Host: *B. pendula*. Distr.: Quebec, Montana, and Idaho.

Carpinus (hornbeam)

Fold along leaf vein (239). .... *Dasineura pudibunda* (Osten Sacken)

Folds usually occur along lateral veins and become hard and brittle while the larvae are still in the first instar. Galls are seen in spring, but larvae develop very slowly and leave the galls only in late summer. One to several larvae live in each gall. Adults have not been reared. Host: *C. caroliniana*. Distr.: Massachusetts and Ontario to Maryland.

Corylus (hazel)

1a Folded leaf. .... cecidomyiid

Folds are reportedly partly covered with red hair. Adults of *Ledomyia coryli* (Felt) were reared from leaf folds at the base of leaves, but this species is probably not a primary plant-feeder. The only available larva associated with this damage is a second instar that cannot be identified. Host: *C. americana*. Distr.: New York.

1b Enlarged bud or deformed catkin. 2
2a Enlarged, aborted bud. .... *Contarinia* sp. and *Dasineura* sp.

The Felt collection contains larvae of each of these genera from bud galls. Adults have not been reared. Hosts: *C. americana*, *C. cornuta*. Distr.: New York.

2b Deformed catkin, its base greatly enlarged (240) ....
.... "*Cecidomyia*" *squamulicola* Stebbins

Galls have been found in September but probably first appear in early spring. Larvae and adults are unknown. Host: *C. americana*. Distr.: Massachusetts and Connecticut.

Ostrya (hop hornbeam)

Folded leaves . . . . cecidomyiid

The larva on which this record (Felt 1940) is based is a second instar and not identifiable. Host: *O. virginiana*. Distr.: New York.

Bignoniaceae

Campsis (trumpet creeper)

1a Larva in curled leaves. . . . . *Contarinia tecomae* (Felt)

Nothing is known of this species beyond the original record. Host.: *C. radicans*. Distr.: New York.

1b Larva in flowers . . . . *Contarinia* sp.

Larvae live near the base of the ovary. Several generations may occur per year. This species is evidently related to another undescribed species inhabiting the flowers of catalpa. Host: *C. radicans*. Distr.: Maryland.

Catalpa

1a Larva in pod, bud, or flower. 2
1b Larva in or on leaf 4
2a Larva in pod. . . . . *Contarinia catalpae* (Comstock)

Infested pods contain many gregarious larvae. Pods often have holes, possibly made by other insects, which may be necessary to allow the gall midge to insert its eggs into the pods. The tissue around the larvae turns brown before the full-grown larvae escape the pods. Host: *C. bignonioides*, *C. speciosa*. Distr.: Maryland to Indiana. Ref.: Gossard (1908).

2b Larva in bud or flower. 3
3a Larva in bud . . . . cecidomyiid

In early summer new shoots become swollen and contain gregarious larvae. Full-grown larvae drop to the ground, after which the shoots turn brown and die. Host: *C. bignonioides*. Distr.: Ohio. Ref.: Gossard (1908).

3b Larva in flower. . . . . *Contarinia* spp.

Two species inhabit the corollas, *Contarinia catalpae* (Comstock) (see entry 2a of this key), which also infests pods, and another, unidentified species of *Contarinia*. The latter is distinct but evidently closely related to the one in flowers of trumpet creeper. Larvae of the latter did not pupate soon after, so this species may have only one generation per year. The similar species on *Campsis* has at least two generations, possibly because *Campsis* flowers bloom throughout the summer. Host: *C. bignonioides*. Distr.: Maryland, Ohio, and Kansas.

4a Discolored, circular leaf spot .... *Neolasioptera* sp.

Spots are about 5 mm in diameter and can be found from May to November. Reportedly they occur in such large numbers as to cause the leaf to wilt. Some larvae drop to the ground to pupate, others pupate in the leaf. Adults were reared in spring from galls collected in November. Host: *C. bignonioides*. Distr.: Pennsylvania, Maryland, and Ohio. Ref.: Gossard (1908).

4b Curled leaf midrib .... cecidomyiid

No further details of this damage were reported (Maze 1949). Host: *C. bignonioides*. Distr.: Pennsylvania.

Chilopsis (desert willow)

Larva in flower buds .... *Contarinia* sp.

Adults were reared in spring. Host: *C. linearis*. Distr.: Texas.

Boraginaceae

Amsinckia

Spherical, soft swellings on inflorescence, stem, or leaf ....
.... *Schizomyia macrofila* (Felt)

Adults emerge in spring and lay eggs directly into plant tissue. Galls grow to 3–15 mm in diameter. One to six larvae develop swiftly in each gall and drop to the soil when full-grown. One generation occurs per year. Hosts: *A. intermedia*, *A. lycopsoides*, *A. spectabilis*. Distr.: California. Refs.: Pantone and Brown (1985), Pantone et al. (1985).

Brassicaceae (Cruciferae)

Cardamine (watercress)

Deformed, irregularly conical flowers .... *Dasineura* sp.

This damage has been noted only in Felt (1940). Host: *Cardamine* sp. Distr.: unspecified.

Erysimum

Larva in flowers .... *Dasineura* sp. and *Prodiplosis* sp.

Adults of both species were reared in spring from unspecified damage. Host: *E. menziesii*. Distr.: California.

Lepidium (peppergrass)

1a Swollen seed capsule .... *Dasineura lepidii* Felt

Infested capsules are darker than normal. Adults have been reared in mid-July. Host: *L. virginicum*. Distr.: Maryland.

1b Stem or root swelling....................................... cecidomyiid

The only reference to this species is in Felt (1940) as *Cecidomyia* sp. Host: *Lepidium* sp. Distr.: unspecified.

Nasturtium

Swollen flower .... *Asphondylia* sp.

Adults were reared in June. Host: *Nasturtium* sp. Distr.: Arizona.

Sisymbrium (hedge mustard)

Swollen bud .... cecidomyiid

Felt (1918a) questionably referred this record to the European *Dasineura sisymbrii* (Schrank). The only available specimen under Felt's note number 1766 has the same correct data for the record but is a *Lestodiplosis*, a predaceous cecidomyiid. No locality was specified; the host was called hedge mustard. All *Sisymbrium* spp. in North America are introduced, so a phytophagous cecidomyiid found on these plants could be an immigrant. Host: *Sisymbrium* sp. Distr.: unspecified.

Stanleya

Swollen flower . . . . *Dasineura stanleyae* (Cockerell)

Sepals of infested flowers are thickened and enlarged. The larvae inside are gregarious. Host: *Stanleya "glauca."* Distr.: Colorado.

Bromeliaceae

Tillandsia

Root nodule . . . . *Neolasioptera* n. sp.

Nodules resemble those on rootlets of poison ivy (147). Mexican plants with such galls are often intercepted at U.S. ports of entry. Hosts: *Tillandsia* spp. Distr.: Mexico and south through tropical South America.

Buxaceae

Buxus (boxwood)

Leaf blister (238) . . . . *Monarthropalpus flavus* (Schrank)

Blisters are slightly convex on the upper side of the leaf, about 5 mm at greatest diameter, and often chlorotic. In each, one or occasionally two or more yellow-green, very active larvae live in a cell surrounded peripherally by a ring of succulent mesophyll tissue. Larvae are in the early last instar at the onset of winter, grow additionally in early spring, pupate there, and emerge in great swarms on a humid morning in spring. Affected leaves die soon after, giving the plant a sickly appearance. Eggs are laid on fresh buds. One generation occurs per year. The species, an immigrant from Europe, was introduced into North America during colonial times with potted boxwood.Hosts: *B. microphylla*, *B. sempervirens*. Distr.: New York to Virginia, Michigan, and California. Refs.: Brewer et al. (1984), Smith and Fisher (1930), Valley (1981).

Cactaceae

Opuntia (prickly pear)

1a Swollen fruit .... *Asphondylia betheli* (Cockerell)

Gall midges emerge from the fruit in early spring and lay eggs on the new, rapidly developing flower buds. Buds begin to develop normally, but several weeks later turn yellowish. Seeds fail to develop, and infested fruit persist on the plant longer than uninfested fruit. Larvae grow slowly at first but more quickly in early winter, and they pupate in the galls in early spring. Infested fruit shrivel and drop off following exit of the adults. Occasionally these fruit become infested with another, gregarious cecidomyiid, *Asynapta cactophila* Samuelson, which feeds in the decaying fruit. Hosts: *O. macrorhiza*, *Opuntia* spp. Distr.: Colorado to California, south to Texas and Oaxaca. Refs.: Ganz (1968), Mann (1969).

1b Swelling or necrotic area on segment (pad)............................. 2

2a Larva from dead tissue at base of spine .... *Contarinia opuntiae* (Felt)

This species was originally found on plants at the New York Botanical Gardens, and since then has been found in Texas and Mexico. Nothing is known of its biology, but the associated necrosis may have developed after the larvae were fully fed. Hosts: *Opuntia* spp. Distr.: Texas and Mexico.

2b Larva in slight swelling, affected area sometimes yellow ....
.... *Neolasioptera* sp. and "*Mayetiola*" spp.

Neolasioptera sp. Many red larvae of this species were found in cells 1 cm in diameter. They spun cocoons when full-grown, pupated in the galls, and adults emerged in March. Hosts: *O. lindheimeri*, *Opuntia* sp. Distr.: Aguascalientes. Ref.: Mann (1969).

"*Mayetiola*" spp. Three separate species have been briefly mentioned and tentatively identified as belonging to this genus: orange larvae on *O. lindheimeri* in Texas; orange larvae on *O. tomentosa* from Cuernavaca, Mexico; and larvae

---

Figures 235–244. Galls of Balsaminaceae, Betulaceae, Buxaceae, and Caprifoliaceae. **235,** Galls of *Impatiens capensis*: **a,** bud gall: *Schizomyia impatientis*; **b,** midrib swelling: *Neolasioptera impatientifolia*. **236,** Swollen bud, *Alnus* sp.: *Dasineura serrulatae*. **237,** Swollen bud and base of petiole, *Betula* sp.: *Apagodiplosis papyriferae*. **238,** Leaf blisters, one leaf with pupal skins attached to exit holes, *Buxus* sp.: *Monarthropalpus flavus*. **239,** Vein folds, *Carpinus caroliniana*: *Dasineura pudibunda*. **240,** Normal and infested catkin, *Corylus* sp.: "*Cecidomyia*" *squamulicola*. **241,** Bud galls, one in section, *Lonicera* sp.: *Rhopalomyia lonicera*. **242,** Swollen flowers, *Sambucus canadensis*: *Schizomyia umbellicola*. **243,** Leaf galls, **a,** vein swellings, **b,** blisters, *Viburnum dentatum*: *Sackenomyia* sp. **244,** Swollen stem, *Viburnum dentatum*: *Neolasioptera viburnicola*.

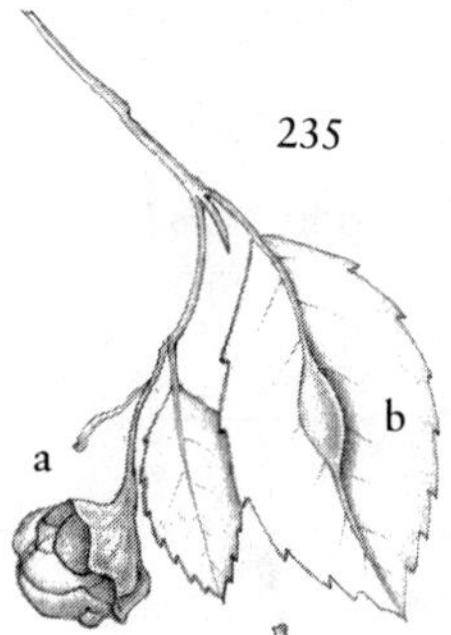

235

236

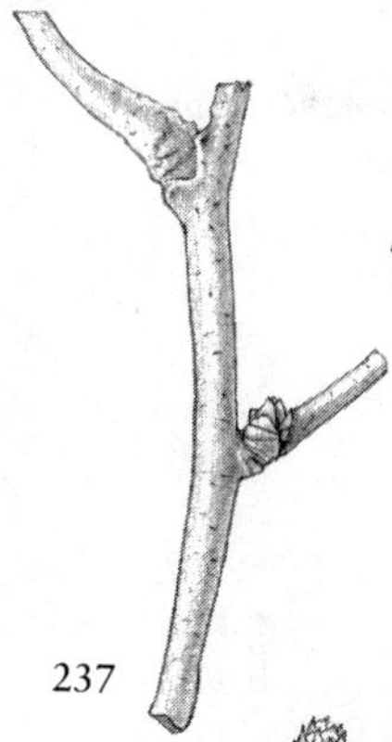

237

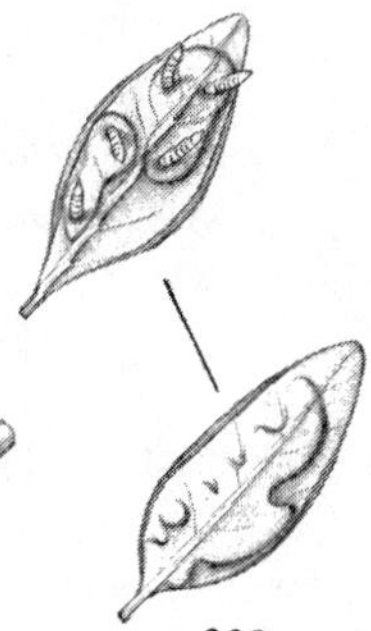

238

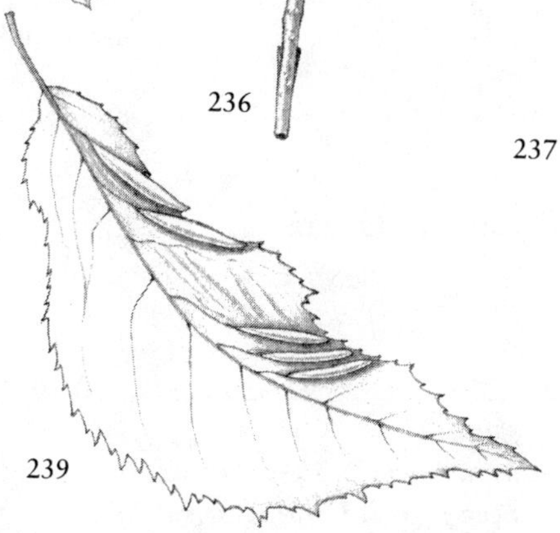

239

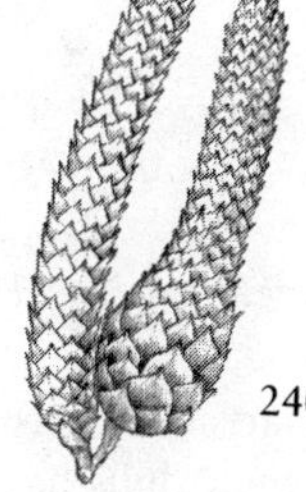

240

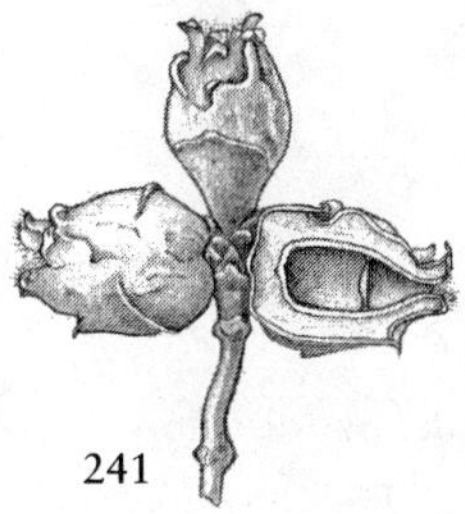

241

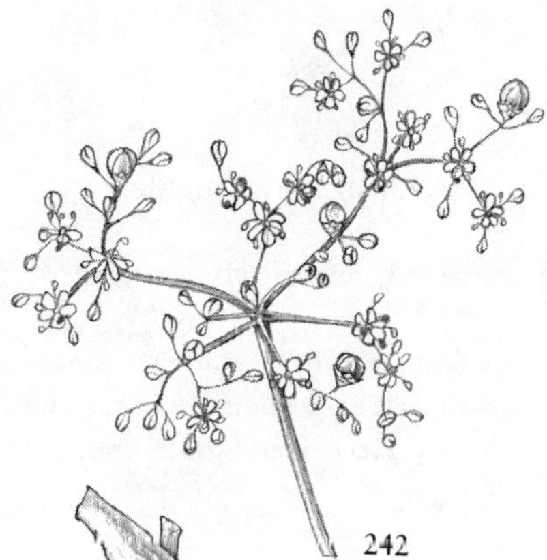

242

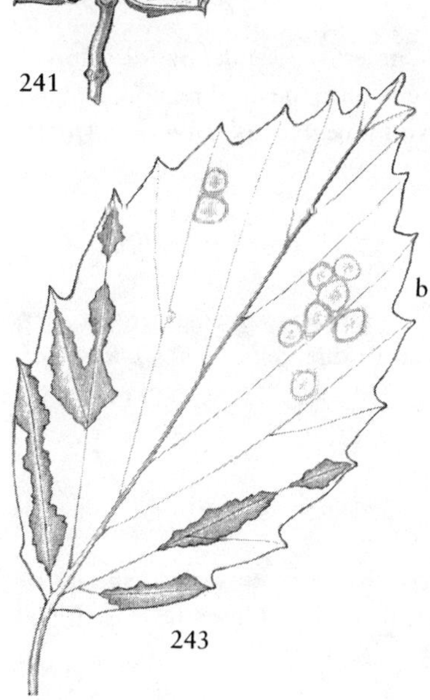

243

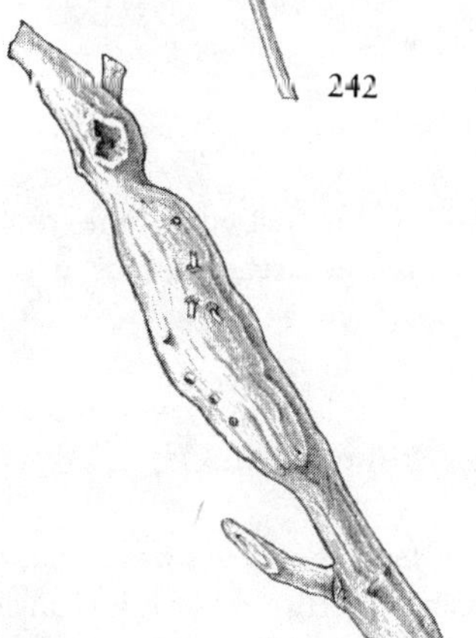

244

from O. *tomentosa* near Mexico City. The specimens appear to be lost. Ref.: Mann (1969).

Cannabaceae

Humulus (hops)

Swollen stem .... *Lasioptera humulicaulis* Felt

Larvae are orange and overwinter in the stems. Adults emerge in spring. Host: *H. lupulus*. Distr.: New York.

Capparidaceae

Cleome

Enlarged or deformed seed pod .... *Contarinia peritomatis* (Cockerell)

Orange larvae live gregariously in the pods. They probably drop to the soil when full-grown. Adults have not been reared. Host: *C. serrulata*. Distr.: New Mexico.

Caprifoliaceae

Diervilla (bush honeysuckle)

1a Enlarged, deformed bud or fruit................. *Asphondylia diervillae* Felt

Galls are about 9 mm long, 3 mm in diameter, green, the slender tips discolored. Most are larger on one side, causing a bending in the opposite direction, making them sickle-shaped. Adults emerged in May and June. Host: *D. lonicera*. Distr.: New York.

1b Stem galls .... *Neolasioptera caulicola* (Felt)

Adults were reared in June from stems showing no apparent enlargement. The pale yellow larvae lived just under the bark in individual cells. Host: *D. lonicera*. Distr.: New York.

Lonicera (honeysuckle)

Three separate species have been taken from swollen buds on various species of *Lonicera* in North America. It is unclear from their descriptions how the galls differ. The three are:

Contarinia sp. The bud is lobulate, 7–13 mm in diameter, and has short, brownish down. Larvae were collected in July. This species was reported as *Cecidomyia* sp. in Felt (1940), but the larvae taken from the galls belong to *Contarinia*. Host: *Lonicera* sp. Distr.: British Columbia.

Rhopalomyia lonicera Felt (241). Enlarged buds are 1–1.5 cm long and flared apically. One larva occurs per gall. The adult emerged from the bud in February. Host: *L. subspicata*. Distr.: California.

A cecidomyiid. The bud is irregular, 2–2.5 cm long. This is a Felt (1940) record only, for which specimens were not kept. Host: *L. ciliosa*. Distr.: unspecified.

Sambucus (elder)

1a Tapered stem swelling. . . . . *Neolasioptera pierrei* (Gagné)

Swellings vary in size depending on the number of larvae within. Larvae overwinter in the stem, and adults emerge in spring. Host: *S. canadensis*. Distr.: Massachusetts and New York.

1b Bud or flower galls 2

2a Swollen bud . . . . *Asphondylia sambuci* Felt

Galls are reportedly greenish-white, 1.5 cm in diameter, with two thickened, projecting bracts. Host: *S. canadensis*. Distr.: Maryland.

2b Enlarged, unopened flower (242) . . . . . *Schizomyia umbellicola* (Osten Sacken)

Galls are many times larger than normal flower buds. The single larva in each gall is orange and very active. It drops to the ground when full-grown and presumably enters diapause until the next spring. Another cecidomyiid, *Youngomyia podophyllae* (Felt), whose role in the gall is unknown, has been reared from these galls and those of other *Schizomyia* spp. as well. Host: *S. canadensis*. Distr.: eastern United States.

Symphoricarpus (snowberry)

1a Swollen twig . . . . cecidomyiid

Mention of this gall appears only in Felt (1918a, 1940) as a *Cecidomyia* sp. It is probably a *Neolasioptera*. Host: unspecified. Distr.: unspecified.

1b Swollen leaf midrib or margin . . . . *Resseliella californica* (Felt)

Galls first appear in May. Larvae are full-grown in July and drop to the ground to pupate. Adults were reared in August. Host: *S. albus*. Distr.: Oregon and California. Ref.: Larew and Capizzi (1983).

Viburnum

1a Leaf gall 2
1b Bud or stem gall 4
2a Marginal leaf roll . . . . *Contarinia* sp.

This is an undescribed species mentioned only as "*Cecidomyia* sp." in Felt (1918a, 1940). The leaf roll is about 1 cm long. The larva, one in each gall according to Felt's unpublished notes, drops to the ground when full-grown. Host: *V. cassinoides*. Distr.: New York.

2b Circular blister gall or vein swelling 3
3a Circular blister gall (243b) . . . . *Sackenomyia commota* Gagné

Galls are green to white, occasionally tinged with red. One larva, rarely two, inhabits each blister and is very active when disturbed. When full-grown, the larva makes a slit through the leaf epidermis and drops to the ground to pupate. Adults have been reared from galls in May, and galls can be seen through September. This species evidently has several generations per year. It is not known whether the same species causes galls on all of the viburnum hosts. Hosts: *V. acerifolium*, *V. cassinoides*, *V. dentatum*, *V. opulus*. Distr.: northeastern United States.

3b Purplish vein swelling (243a) . . . . *Sackenomyia viburnifolia* Felt

The biology of this species generally resembles that of *S. commota*, except that the larvae are not as active. Adults were reared in May from the previous year's galls, but galls can be seen through the summer. Host: *V. dentatum*. Distr.: northeastern United States.

4a Enlarged blossom or bud . . . . *Schizomyia viburni* Felt

This species is known from a female "about to oviposit in flowers," but Felt (1918a, 1940) associates it with an unspecified bud gall. An enlarged flower gall is made by another *Schizomyia* on *Sambucus*, so *S. viburni* may come from enlarged flowers if not buds as well. Host: *Viburnum* sp. Distr.: New York.

4b Swollen stems (244) . . . . *Neolasioptera viburnicola* (Beutenmüller)

Galls are gnarled, 3–8 cm long, and soft-pithy inside. The orange larvae live in channels and overwinter in the galls. Host: *V. dentatum*. Distr.: New York, Connecticut, and Virginia.

Celastraceae

Celastrus (bittersweet)

Irregularly swollen root or stem .... *Neolasioptera fontagrensis* (Gagné)

Galls can be as large as 2.5 cm long and 1 cm in diameter. Larvae overwinter in the gall. Adults have not been reared. Host: *C. scandens*. Distr.: Massachusetts and New York.

Paxistima

Folded, red, terminal leaves .... cecidomyiid

The gall maker was reported only as *Cecidomyia* sp. in Felt (1940). Host: *P. myrsinites*. Distr.: unspecified; western United States.

Chenopodiaceae

Atriplex (saltbush)

1a Galls on *Atriplex canescens*... 2
1b Galls on *Atriplex polycarpa* 7
2a Simple nonleafy enlargements of stem.................................. 3
2b Bud, fruit, or leaf gall, or detachable outgrowths of stem................. 4

3a Swelling gradual, tapered or cylindrical, woody (245) ....
.... *Lasioptera tibialis* Felt and *Neolasioptera willistoni* (Cockerell)

L. tibialis. This species has been reared only once from an apparent *N. willistoni* gall on *A. canescens* in California. It is otherwise known only from *Keckiella* (Scrophulariaceae) (q.v.), and may be an inquiline. Distr.: California. Ref.: Gagné and Hawkins (1983).

N. willistoni. Galls are either terminal, hard, ridged, and situated at tips of fruit-bearing branches (245b), or smooth, softer, and on vegetative stems (245a). Usually only one larva occupies a gall. Adults can be reared almost any time of the year, so the species is apparently multivoltine. Pupation takes place in the galls. Hosts: *A. canescens*, *A. polycarpa*. Distr.: New Mexico to California. Ref.: Gagné and Hawkins (1983).

3b Swelling spherical, soft, and succulent (246a, color plate 2) ....
.... *Asphondylia atriplicis* (Townsend)

Galls are green to red, monothalamous or polythalamous, often in tandem, occasionally in a series as long as 10 cm. All stages of the gall maker can be found all year. Host: *A. canescens*. Distr.: New Mexico to California. Ref.: Hawkins et al. (1986).

4a Leaf blister (246b)................... *Asphondylia atriplicicola* (Cockerell)

Galls are monothalamous, about 3 mm in diameter. All stages of the gall maker can be found all year round. Host: *A. canescens*. Distr.: New Mexico and California. Ref.: Hawkins et al. (1986).

4b Bud, fruit, or stem gall... 5
5a Stem gall (246c)............................ *Asphondylia caudicis* Gagné

Galls are spherical, green, with very short pile, usually monothalamous, and about 6 mm in diameter. Host: *A. canescens*. Distr.: New Mexico and California. Ref.: Hawkins et al. (1986).

5b Axillary bud or fruit gall... 6
6a Densely woolly, spherical gall (247a)... *Asphondylia neomexicana* (Cockerell)

Galls are white, about l cm in diameter including the trichome covering, monothalamous, and occur in summer and fall. Host: *A. canescens*. Distr.: New Mexico and California. Ref.: Hawkins et al. (1986).

6b Short-pilose, oval or spherical gall (247b)....... *Asphondylia gemmae* Gagné

Galls are ovoid, monothalamous, about 3 mm diameter, and occur in summer and fall on buds, flowers, or fruit. Host: *A. canescens*. Distr.: California. Ref.: Hawkins et al. (1986).

7a Large, woolly stem gall not covered with small leaf clusters (249) .... *Asphondylia floccosa* Gagné

Masses of galls are 0.5–1.5 cm in diameter on apices of plants in the summer and are 1–10 cm in diameter and farther down on the plant in winter. They are monothalamous or polythalamous, white, sometimes pink basally. Host: *A. polycarpa*. Distr.: California. Ref.: Hawkins et al. (1986).

7b Smooth, leaf cluster-covered or rough stem gall.......................... 8
8a Foreshortened stem covered with leaf clusters ..... *Asphondylia foliata* Gagné

Beneath the leaf clusters the gall is clublike, 2–4 mm long, and monothalamous. The smooth galls are at the tips of axillary buds of foreshortened stems and are seen only in spring. Host: *A. polycarpa*. Distr.: California. Ref.: Hawkins et al. (1986).

8b Tough-textured stem swelling .... *Asphondylia nodula* Gagné

Galls occur in clusters, are about 4 mm in diameter. They are found in late winter and early spring, are restricted in distribution but may be locally abundant. Host: *A. polycarpa*. Distr.: California. Ref.: Hawkins et al. (1986).

Beta (sugarbeet)

Larva in buds .... *Asphondylia* sp.

Adults in the USNM collection do not bear further information concerning the damage to the plant by the species. They were collected in early spring. An undescribed species of *Asphondylia* is known from Japan (Yukawa 1980). It was intercepted in soybeans once at a U.S. port. Host: *B. vulgaris*. Distr.: Arizona.

Chenopodium (goosefoot)

Stem swelling .... ?*Lasioptera* sp.

This gall is reportedly 1–2 cm long. It is known only from the record in Felt (1940). Host: *C. ambrosioides*. Distr.: unspecified.

Eurotia

1a Irregular bud swelling .... cecidomyiid

This damage is known only from the record in Felt (1940). Host: *Eurotia* sp. Distr.: western United States.

1b Tapered stem swelling .... ?*Neolasioptera* sp.

The swelling is 15 mm long and 5 mm at its widest diameter. The larvae, the only specimens known, probably belong to *Neolasioptera*. Host: *E. lanata*. Distr.: California.

Sarcobatus (greasewood)

1a Swollen bud (250) .... *Halodiplosis sarcobati* (Felt)

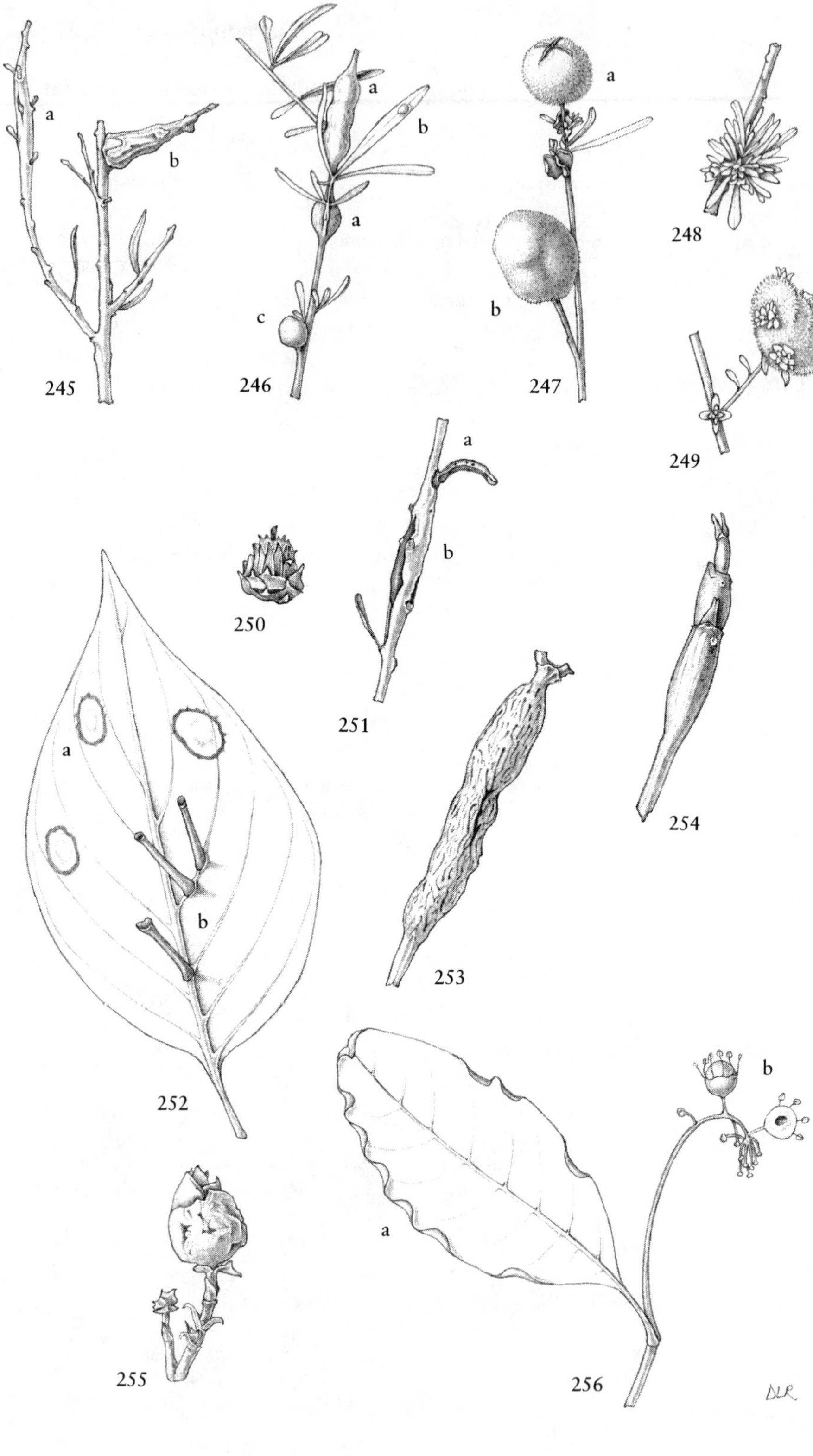
245
a
b
246
a
b
a
c
247
a
b
248
249
250
251
a
b
252
a
b
253
254
255
256
a
b

Enlarged buds are about 1 cm long and 6 mm in diameter and are made up of thickened, appressed bud scales. Pupation occurs inside the buds, and adults emerge in spring. Host: *S. vermiculatus*. Distr.: Idaho and Utah.

1b Swollen leaf or stem 2

2a Swollen, occasionally discolored leaf (251a) ..... *Protaplonyx sarcobati* (Felt)

Adults were reared in spring and summer, so there are probably several generations per year. Adults have an unusual ovipositor modified for piercing plant tissue. Host: *S. vermiculatus*. Distr.: Colorado and Utah. Ref.: Gagné (1969c).

2b Swollen stem (251b)................................. *Neolasioptera* sp.

Adults of this species have not been reared. Host: *Sarcobatus* sp. Distr.: Oregon and Utah.

Suaeda

Globular, leafy, black bud gall .... *Asphondylia dondiae* Felt

The gall is approximately 5 mm in diameter. The one known specimen of the gall midge was reared in April. Host: *S. "multiflora."* Distr.: California.

Convolvulaceae

Calystegia (bindweed)

Tapered stem swelling .... *Neolasioptera convolvuli* (Felt)

Swellings are woody, monothalamous, and 1–2 cm long. Usually a single, orange larva lives in a channel along the pith. The larva overwinters in the gall.

Figures 245–256. Galls of Chenopodiaceae and Cornaceae. **245,** Stem swellings, **a,** smooth, and **b,** rough, *Atriplex canescens*: *Neolasioptera willistoni*. **246,** galls of *Atriplex canescens*, **a,** tumor stem galls: *Asphondylia atriplicis*; **b,** blister leaf gall: *Asphondylia atriplicicola*; **c,** oval twig gall: *Asphondylia caudicis*. **247,** Galls of *Atriplex canescens*, **a,** woolly bud gall: *Asphondylia neomexicana*; **b,** oval bud gall: *Asphondylia gemmae*. **248,** Bud gall, *Atriplex polycarpa*: *Ophiomyia atriplicis* (Diptera: Agromyzidae). **249,** Woolly stem gall, *Atriplex polycarpa*: *Asphondylia floccosa*. **250,** Swollen bud, *Sarcobatus vermiculatus*: *Halodiplosis sarcobati*. **251,** Galls of *Sarcobatus vermiculatus*, **a,** swollen leaf: *Protaplonyx sarcobati*; **b,** swollen stem: *Neolasioptera* sp. **252,** Galls of *Cornus stolonifera*, **a,** leaf spots: *Parallelodiplosis subtruncata*; **b,** leaf tube galls: *Craneiobia tuba*. **253,** Stem swelling, *Cornus racemosa*: *Neolasioptera cornicola*. **254,** Club gall, *Cornus florida*: *Resseliella clavula*. **255,** Swollen bud, *Garrya buxifolia*: *Asphondylia garryae*. **256,** Galls of *Nyssa sylvatica*, **a,** marginal leaf folds: *Phylloxera nyssae* (Phylloxeridae) and cecidomyiid predator *Cartodiplosis nyssaecola*; **b,** Swollen staminate flowers, one in section: *Contarinia* sp.

The adult emerges from the galls in spring. Hosts: *C. sepium*, *C.* sp. Distr.: Massachusetts, New York, Maryland, and Nebraska.

Ipomoea (morning glory)

1a Tapered stem swelling .... *Neolasioptera* sp.

Host: *Ipomoea* sp. Distr.: Florida.

1b Swollen flower .... *Asphondylia* sp.

The particular site of infestation is unknown. The USNM has specimens labelled as reared from flowers and seeds. Host. *I. triloba*. Distr.: Florida.

Cornaceae

Cornus (dogwood)

1a Leaf galls 2
1b Stem or bud galls 3
2a Circular or oval, usually reddish, blister gall (252a) ....
.... *Lasioptera corni* Felt and *Parallelodiplosis subtruncata* (Felt)

L. corni. This species was reported once from these galls. It is a distinctive species and not known from any other host. It could be an inquiline, or it was possibly a contaminant in the container from which it was reared. I have not found *Lasioptera* larvae in the galls that I have seen. Distr.: New York.
P. subtruncata. This species is in every active blister gall I have seen, so is evidently the gall maker. The galls are about 4 mm in diameter and are usually on the leaf laminas but may also occur on the veins and petiole. Since galls are present all summer, perhaps several generations occur each year. Hosts: *C. florida*, *C. racemosa*, *C. stolonifera*. Distr.: Massachusetts to Ontario, south to Maryland, and Minnesota.

2b Tubular leaf gall (252b) .... *Craneiobia tuba* (Stebbins)

These striking, green to pinkish, monothalamous galls are hemispherical on one side of the leaf and tubular on the other without regard to particular leaf surface. The orange larva is in a spherical cell at the base of the gall. Adults have not been reared. Hosts: *C. amomum*, *C. racemosa*. Distr.: Massachusetts, New York, and Illinois.

3a Smooth, clublike enlargement just below bud in apical branches (254) .... *Resseliella clavula* (Beutenmüller)

Eggs are deposited among terminal leaves in spring. Young larvae apparently push their way into the buds and one to several develop in a central cell. The twigs usually do not elongate beyond the galls and there is often a pair of wilted leaves beyond. The swellings are 2–3 cm long, green to red, and especially conspicuous in the fall after the orange larvae have dropped to the ground for the winter. This species has one generation per year. Host: *C. florida*. Distr.: Massachusetts to Ontario and Indiana, south to Virginia. Ref.: Schread (1964).

3b Irregularly tapered, ridged stem swelling (253) .... *Neolasioptera cornicola* (Beutenmüller)

Swellings are woody, one-half to several centimeters long, and contain several yellow larvae that overwinter inside and emerge in spring. One generation occurs per year. Host: *C. stolonifera*. Distr.: New York and Maryland.

Garrya

Greatly swollen, soft, discolored buds (255) .... *Asphondylia garryae* Felt

Adults were reared in spring. Hosts: *G. buxifolia*, *G. fremontii*. Distr.: Oregon and California.

Nyssa (black or sour gum)

1a Swollen, staminate flower (256b)........................ *Contarinia* sp.

Swellings are fruitlike, the pedicellate stamens in a circle around the periphery. One to several larvae occur in a central cavity. Full-grown larvae drop to the soil, where they remain until the following spring. A *Clinodiplosis* sp. was reared from the galls as well. These emerged as adults the same spring as the galls developed, so are presumably secondary, general feeders. Host: *N. sylvatica*. Distr.: Maryland.

1b Leaf or twig galls 2

2a Marginal leaf folds (256a) .... *Cartodiplosis nyssaecola* (Beutenmüller)

This species was originally believed to cause this gall, which is actually formed by *Phylloxera nyssae* Pergande (Homoptera: Phylloxeridae). *C. nyssaecola* is a predator on the phylloxerids. Host: *N. sylvatica.* Distr.: New York and Maryland.

2b Swollen twigs .... cecidomyiid

Swellings are reportedly about 6 cm long. Larvae associated with this damage were apparently not preserved. Host: *N. sylvatica.* Distr.: Virginia.

Cucurbitaceae

Citrullus (watermelons), *Cucumis* (melons), *Cucurbita* (squashes)

1a Curled leaf buds and rolled leaves .... *Prodiplosis citrulli* (Felt)

Larvae are white, gregarious, and develop rapidly. Infested leaves die after the full-grown larvae drop to the soil. Several generations probably occur per year. Two other gall midges were reportedly reared together from curled, slightly enlarged tips of melon (Lintner 1897). One, *Contarinia setigera* (Lintner), known from a single specimen, is very similar to the very distinctive *Contarinia pyrivora*, the pear midge. If *setigera* is *pyrivora*, it might have been a contaminant in the rearing container. The other gall midge was an aphid predator, *Aphidoletes aphidimyza*, so it is possible that the curled tips of Lintner's melons were caused by aphids. Hosts: cultivated *Citrullus*, *Cucumis*, and *Cucurbita.* Distr.: Arizona. Ref.: Wehrle (1946).

1b Swollen flower or bud .... *Asphondylia* sp.

A series of an asphondylia in the USNM is labelled as reared from melon, but its damage is unspecified. Host: cultivated *Cucumis* sp. Distr.: Arizona.

Cupressaceae

Chamaecyparis

1a Larva in cone (258) .... *Janetiella siskiyou* Felt

Infested cones may appear normal to slightly deformed. Larvae live among the scales. They spin a cocoon in summer and pupate there in spring. Adults emerge shortly after. Probably only one generation occurs per year. Host: *C. lawsoniana.* Distr.: Oregon; also Europe, where it is now well established. Ref.: Meijere (1935), Nijveldt (1984).

1b Larva in swollen branchlet tip (257) .... *Contarinia* sp.

Branchlet tips are swollen. Larvae of this undescribed species pupate in the galls in spring and emerge shortly after. Two generations may occur per year. Host: *C. nootkatensis*. Distr.: British Columbia and Washington.

Cupressus (cypress)

1a Greatly enlarged bud (260).................... *Walshomyia cupressi* Gagné

Adults emerge from the buds in April. Apparently only one generation occurs per year. Hosts: *C. pygmaea*, *C. sargentii*. Distr.: California.

1b Bud of normal size, becoming brown after larva exits (259)... *Contarinia* spp.

C. rugosa Gagné (259). The infestation is noticeable in spring, when the buds turn first red then brown after the full-grown larvae drop to the soil. Host: *C. arizonica*. Distr.: Californica and Hawaii, where it was introduced with nursery stock. Ref.: Gagné (1986a).
Contarinia sp. The damage is probably similar to that of *Contarinia rugosa*, but adult specimens are distinctly different. Host: *C. macrocarpa*. Distr.: California. Ref.: Gagné (1986a).

Juniperus (juniper)

1a Slight swellings at base of needle (261)........... *Contarinia juniperina* Felt

Infestation can cause extensive dieback and unsightliness after the larvae leave. Eggs are laid in spring. Larvae live in the cambial layer in mines up to 25 mm long. One mine may contain one to several larvae, which, in Pennsylvania, overwinter in the twigs as second-instar larvae. Full-grown larvae escape in spring and pupate in the soil. Adults emerge a short time later. Hosts: *J. chinensis*, *J. communis*, *J. horizontalis*, *J. sabina*, *J. squamata*, *J. virginiana*. Distr.: Ontario to Pennsylvania, west to Nebraska and Kansas. Refs.: Steinhauer (1983), Wheeler et al. (1981).

1b Gall on bud tip or fruit 2
2a Branch tip gall consisting of imbricated needles, enlarged or not 3
2b Enlarged deformed fruit.. 5
3a Gall surrounded by two to several needles, barely larger in diameter than normal twig (262, 269) .... *Oligotrophus* spp. and *Walshomyia insignis* Felt

O. betheli Felt (269). The needles covering a gall are green at first and turn red or black only when the larva inside approaches full growth. Two to several generations occur per year. The last overwinters in the galls as full-grown larvae, pupates there in spring, and emerges as adults shortly after. Eggs are laid in groups on the buds, but each branch tip has only one gall and each gall one cecidomyiid. Affected buds die after the gall midges emerge. Heavily infested plants have a light brown cast. Hosts: *J. ashei*, *J. chinensis*, *J. horizontalis*, *J. scopulorum*, *J.* "*utahensis*," *J. virginiana*. Distr.: New Jersey to British Columbia and California. Ref.: Appleby and Neiswander (1965), Wheeler et al. (1981).

Oligotrophus sp. (262). This species, probably an *Oligotrophus*, is known only from galls and pupae. The galls differ from those of *O. betheli* only by having the tips of the outer leaf scales recurved. Host: *J. scopulorum*. Distr.: Colorado.

W. insignis. Galls were apparently only collected once and were reported as yellowish bud galls, hardly distinguishable from normal buds. That short description could also fit the description of *O. betheli* galls, but the one known specimen, a female, of *W. insignis* is distinct from *O. betheli*. Host: *J. scopulorum*. Distr.: Texas.

3b Gall much larger than normal twig diameter, consisting of many enlarged needles 4

4a Gall burrlike, the leaves short, pointed (268) . . . . . *Oligotrophus juniperi* (Felt)

This species is known from a female and the gall illustrated in figure 268. Host: *Juniperus* sp. Distr.: Colorado.

4b Leaves forming gall appressed, although tips may be reflexed (265, 266, 270) . . . . *Oligotrophus pattersoni* White and *Walshomyia* spp.

O. pattersoni (266). Galls are 10–15 mm long, 8–10 mm in widest diameter. Host: *J. ashei*. Distr.: Texas. Ref.: White (1950).

Walshomyia spp. (265, 270). Two kinds of galls from the USNM collection are shown, to give an idea of the variety of imbricate bud galls that can be found on junipers. These are presumed to be caused by *Walshomyia* spp. The gall in figure 265 is from *J. occidentalis* in California; that shown in figure 270 is from *J.* "*utahensis*" in Colorado.

5a Gall elongate-cylindrical, brown (263) . . . . *Walshomyia* sp.

These galls from the USNM collection are presumed to be formed by a *Walshomyia*. Host: *J.* "*utahensis*." Distr.: Colorado.

5b Gall ovoid, globular, pink to dark purple, sometimes split at apex when mature (264) . . . . *Walshomyia* spp.

Galls of all three species are much alike, pink to purple, 5–10 mm long, 4–8 mm at widest diameter. Each contains one larva in a central cell. The larva overwinters in the gall and pupates in spring. Adults of the three species below are distinct from one another.

W. juniperina Felt. Host: *J. californica*. Distr.: California.

W. sabinae (Patterson). Patterson's species was recorded from *J.* "*sabinoides*" in Texas and based only on the gall. Adults of an alleged synonym of this gall midge, *W. sabinae* Felt, were reared from *J. monosperma* in Colorado and from *J.* "*utahensis*" in Utah. Specimens have not been reared from galls of *W. sabinae* (Patterson), which will have to be done to determine whether the two names are distinct. Patterson's species may be synonymous with *W. texana*, which was reared from *J.* "*sabinoides*."

W. texana Felt. Host: *J.* "*sabinoides*." Distr.: Texas.

Libocedrus (incense cedar)

Large bud gall . . . . *Walshomyia* sp.

Galls are reportedly tapered and about 2 cm long and 0.5 cm in diameter (Felt 1940). They were evidently not kept, but adults in the USNM reared from typical damage belong to *Walshomyia*. Host: *L. decurrens*. Distr.: California.

Thuja (arborvitae)

Larva in cone (267) . . . . "*Mayetiola*" *thujae* (Hedlin)

Eggs are laid in early spring on the inner surface of cone scales. Newly hatched larvae migrate to the fertile portion of the cone and feed on the surface of developing seeds and the larger scales. Larvae do not become embedded in the tissue. Damaged cones may be smaller than normal but not otherwise conspicuous. Seeds rarely survive an infestation. Full-grown larvae spin cocoons in the cones in late summer or early fall. Adults emerge the following spring. Host: *T. plicata*, western red cedar. Distr.: British Columbia to Oregon. Ref.: Hedlin (1964b).

Cyperaceae

Carex (sedge)

1a Swelling or tuber-shaped outgrowth at base of plant (271, 272) . . . . *Planetella* spp.

These species may have but one generation per year. Adults emerge in spring. Hosts: many species of *Carex* including *C. vesicaria*. Distr.: at least throughout northern Holarctic Region. Ref.: Frohne (1939).

257

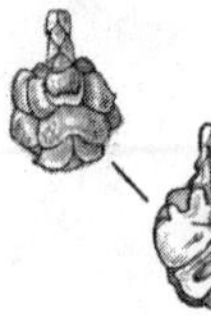
258

259

260

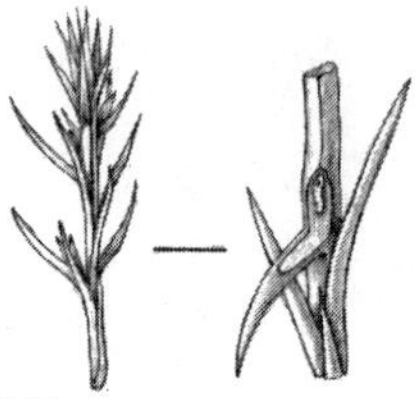
261

262

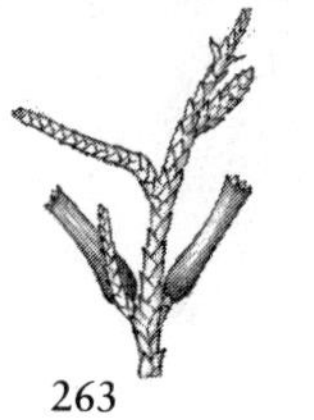
263

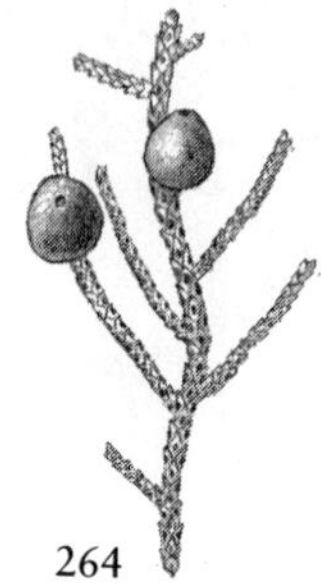
264

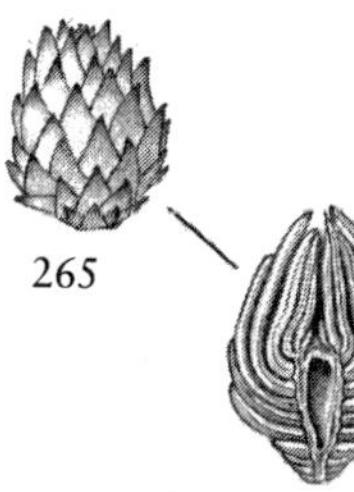
265

266

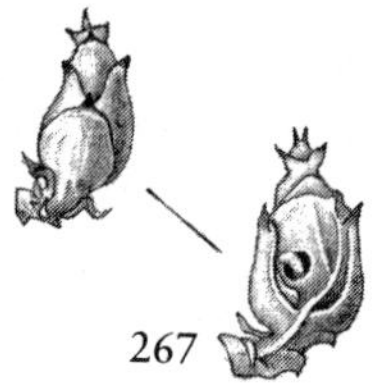
267

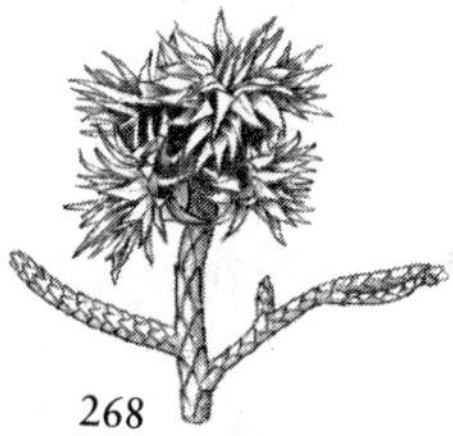
268

270

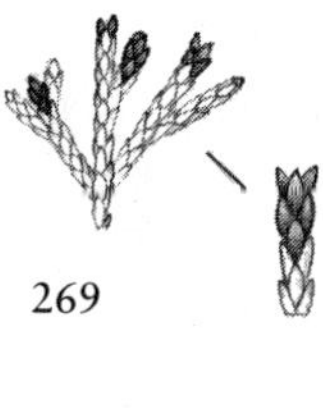
269

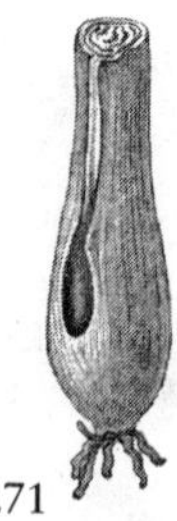
271

272

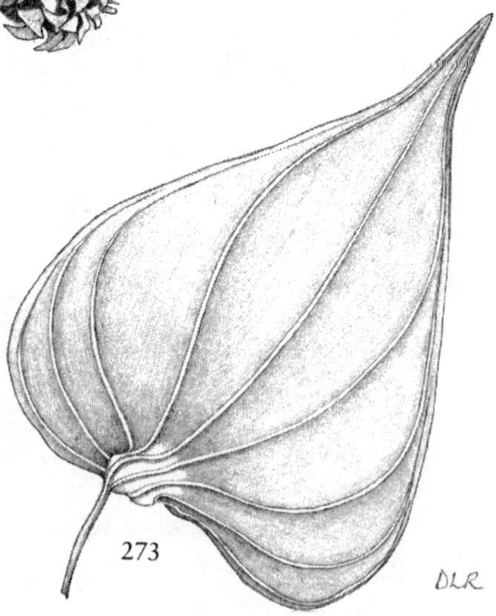

273

1b Deformed fruit .... cecidomyiid

Full-grown larvae were found in deformed perigynes in June. Adults have not been reared. Host: *C. praegracilis*. Distr.: Nebraska.

Cyperus

From unspecified shoot gall at base of plant .... *Planetella caudata* (Felt)

Host: The plant host was only tentatively identified as a *Cyperus*. Distr.: Tennessee.

Dulichium

Larvae in fruit .... *Ametrodiplosis dulichii* (Felt)

This species was reportedly reared from fruit. Host: *D. arundinaceum*. Distr.: Massachusetts.

Dioscoreaceae

Dioscorea (yam)

Irregular, woody swelling at base of leaf blade (273) .... *Neolasioptera* sp.

Adults of this species have not been reared. Larvae evidently overwinter in the gall. Host: *D. villosa*. Distr.: New York.

Figures 257–273. Galls of Cupressaceae, Cyperaceae, and Dioscoreaceae. **257,** Swollen branchlet tips, *Chamaecyparis nootkatensis*: *Contarinia* sp. **258,** Cones, one with scales removed to show larval chambers, *Chamaecyparis lawsoniana*: "*Janetiella*" *siskiyou*. **259,** Branchlet tips, *Cupressus arizonica*: *Contarinia rugosa*. **260,** Enlarged bud, *Cupressus sargentii*: *Walshomyia cupressi*. **261,** Branch swellings, one in 2× showing larval cell (adapted with permission from Haseman and McLane 1940, *Annals of the Entomological Society of America* 33: 615, © 1940, Entomological Society of America), *Juniperus communis*: *Contarinia juniperina*. **262,** Swollen branchlet tips, *Juniperus scopulorum*: *Oligotrophus* sp. **263,** Deformed fruit, *Juniperus* "*utahensis*": *Walshomyia* sp. **264,** Deformed fruit, *Juniperus occidentalis*: *Walshomyia* sp. **265,** Enlarged buds, one sectioned to show larval chamber, *Juniperus occidentalis*: *Walshomyia* sp. **266,** Enlarged bud (redrawn from White 1950, courtesy of University of Texas Press), *Juniperus ashei*: *Oligotrophus pattersoni*. **267,** Cones, one with some scales removed to show larva in cell, *Thuja plicata*: "*Mayetiola*" *thujae*. **268,** Enlarged burrlike bud, *Juniperus* sp.: *Oligotrophus juniperi*. **269,** Swollen, discolored branchlet tips, *Juniperus virginiana*: *Oligotrophus betheli*. **270,** Enlarged buds with recurved bracts, *Juniperus* "*utahensis*": *Walshomyia* sp. **271,** Swollen stem base, partially cut open to reveal a larval cell, *Carex* sp.: *Planetella* sp. **272,** Ovoid swellings at base of stem, *Carex* sp.: *Planetella* sp. **273,** Leaf base swelling, *Dioscorea villosa*: *Neolasioptera* sp.

Eleagnaceae

Shepherdia (buffalo-berry)

Enlarged bud (274) .... *Asphondylia shepherdiae* Felt

Infested buds are thin-walled, yellowish-green to red, and 4–5 mm in diameter. Adults were reared in late June. Host: *S. canadensis*. Distr.: Colorado.

Ericaceae

Gaylussacia (see *Vaccinium,* entry 3a)

Rhododendron (azalea, rhododendron)

1a Enlarged bud (277) .... *Asphondylia azaleae* Felt

Infested buds are about 8 mm long and monothalamous. Adults emerge from the buds in spring when the plant begins to bloom. Hosts: *R. nudiflorum*, *R. roseum*. Distr.: New York, Maryland, and Virginia.

1b Rolled, distorted, young leaf .... *Clinodiplosis rhododendri* (Felt)

Leaves attacked in the bud stage may be killed, and young leaves may be dwarfed or deformed; all develop an in-rolling of the margin. Affected tissue dies after full-grown larvae drop to the soil. Larvae develop rapidly and are full-grown in several days. The species is multivoltine. Hosts: *R. catawbiense*, *R. maximum*. Distr.: New York to North Carolina. Ref.: Specker and Johnson (1988).

Vaccinium (blueberry, cranberry)

1a Swollen berry ... *Dasineura* sp. (color plate 3C) and *Lasioptera fructuaria* Felt

Dasineura sp. Infested fruit enlarge sooner than normal ones and become bright red instead of blue. Full-grown larvae leave the berries, which then die. Host: *V. stamineum*. Distr.: Georgia.
Lasioptera fructuaria. One to several reddish larvae are found in a berry, which grows deformed. Larvae leave the berries in late July and drop to the ground. Adults emerge the following June. Hosts: *V. lamarckii*, *V. myrtilloides*. Distr.: Maine. Ref.: Barnes (1948a).

1b Bud or leaf gall 2

2a Swollen bud with curled leaves (276) .... *Dasineura cyanococci* Felt, *Dasineura oxycoccana* (Johnson), and *Prodiplosis vaccinii* (Felt)

These three species reportedly cause similar damage. First the terminal buds and, later in the season, the lateral buds become swollen and their outer leaves cupped. Larvae live on the inside surface of the curled leaves. They may pupate in the buds (*D. oxycoccana*, reportedly) or drop to the ground when full-grown and the bud dies, turning black and brittle. It is not known whether the two dasineuras are the same species: one is on cranberry, the other on blueberry. Larvae of the *Prodiplosis* and a *Dasineura* were collected together from the same buds on "wild lowbush blueberries" in Pennsylvania.

D. cyanococci. This species was reared from an apical bud gall in September and resembles *D. oxycoccana*. Larvae taken from commercial blueberry bud galls in New Jersey could be this species. Adults emerged 10–15 days after full-grown larvae dropped to the soil. Hosts: *V. ashei*, *V.* "*canadense*," and possibly commercial blueberries. Distr.: Massachusetts, New Jersey, and Georgia.
D. oxycoccana. The yellow to red larvae either pupate in the buds or drop to the ground. There may be several generations per year, from early May to September. The later brood is the most injurious because injured tips rarely recover in time to set fruit buds for the following season. Host: *Vaccinium* sp. (cranberry). Distr.: Massachusetts, New Jersey, Michigan, Wisconsin, and Washington. Ref.: Barnes (1948a).
P. vaccinii. Several generations of this species may occur per year. Eggs are laid in unfolding leaf buds. Full-grown larvae drop to the soil where, reportedly, they form cocoons. Host: *Vaccinium* sp. (commercial blueberries). Distr.: New Jersey and Pennsylvania. Ref.: Driggers (1926).

2b Leaf gall or spot 3
3a Succulent, kidney-shaped midrib leaf gall (278) .... "*Cecidomyia*" *vaccinii* Osten Sacken

Galls are red, about 10 mm long, 5 mm high, and grow on the upper surface of the leaf. When larvae are fully developed in late summer, galls open along the scalloped upper margin and the one to two reddish larvae per gall drop to the ground. Adults are unknown, and the larvae cannot be placed in any known genus. Hosts: *V. stamineum*, *V.* spp., and possibly *Gaylussacia*, according to Osten Sacken (1862). Distr.: Maine to Georgia.

3b Leaf spot (275) .... *Contarinia* sp.

Larvae live in a sticky substance in a slight depression on the underside of the leaf. The upper side of the leaf is slightly dimpled. Usually one larva, but as many as three, live on one spot. The spots are oval, about 4 mm wide, 10 mm long, and reddish or a lighter green than the leaf when viewed from above. Larvae were found in early August at high altitudes. Full-grown larvae drop to the soil. Hosts: *Vaccinium* spp., both red- and blue-berried. Distr.: Washington.

Euphorbiaceae

Acalypha (three-seeded mercury)

Flower buds with swollen, imbricated leaves (295) .... *Resseliella* sp.

The original record (Felt, 1940) was based on a larva that appears to belong to this genus, although the pair of enlarged, terminal papillae have an unusual shape for *Resseliella*. Hosts: *A. rhomboidea*, *A. virginica*. Distr.: New York and Illinois.

Euphorbia (spurge)

Swollen bud (137) .... *Dasineura* sp. near *capitigena* (Bremi)

This species was purposely introduced into Montana in 1985 to help control leafy spurge, an introduced weedy plant poisonous to cattle. Heavily attacked plants do not produce flowers. Host: *E. esula*. Distr.: North Dakota and Montana; Europe. Refs.: Pecora et al. (in prep.), Solinas and Pecora (1984).

Fabaceae (Leguminosae)

Three gall midge genera, *Asphondylia*, *Dasineura*, and *Neolasioptera*, predominate on legumes. *Asphondylia* spp. form many flower and pod galls, particularly in southwestern North America. Those on the subfamily Mimosoideae appear distinct

Figures 274–288. Galls of Eleagnaceae, Ericaceae, and Fabaceae. **274,** Enlarged buds, *Shepherdia canadensis*: *Asphondylia shepherdiae*. **275,** Leaf spots, **a** from above, **b** from below showing larva, *Vaccinium* sp.: *Contarinia* sp. **276,** Rolled leaf, *Vaccinium* sp.: *Dasineura cyanococci*. **277,** Enlarged buds, one in section, *Rhododendron nudiflorum*: *Asphondylia azaleae*. **278,** Succulent midrib fold gall, *Vaccinium* sp.: "*Cecidomyia*" *vaccinii*. **279,** Leaf cone galls, *Amorpha canescens*: *Rhopalomyia* sp. (redrawn from Wells 1921b). **280,** Marginal leaf roll, *Amorpha canescens*: cecidomyiid (redrawn from Wells 1921b). **281,** Pod galls, *Cyamopsis tetragonoloba*: *Asphondylia websteri*. **282,** Joined, swollen leaflets, *Acacia greggii*: *Contarinia* sp. **283,** Swollen buds, *Desmodium* sp.: *Dasineura meibomiae*. **284,** Marginal leaf rolls, *Robinia pseudoacacia*: *Obolodiplosis robiniae*. **285,** Persistent bud galls, *Prosopis glandulosa*: *Asphondylia* sp. **286,** Flower galls, *Prosopis glandulosa*: *Asphondylia prosopidis*. **287,** Leaflet galls, *Gleditsia triacanthos*: **a,** leaf blisters: *Meunieriella aquilonia*; **b,** leaf folds: *Dasineura gleditchiae*. **288,** Swollen stem, *Gleditsia triacanthos*: *Neolasioptera brevis*.

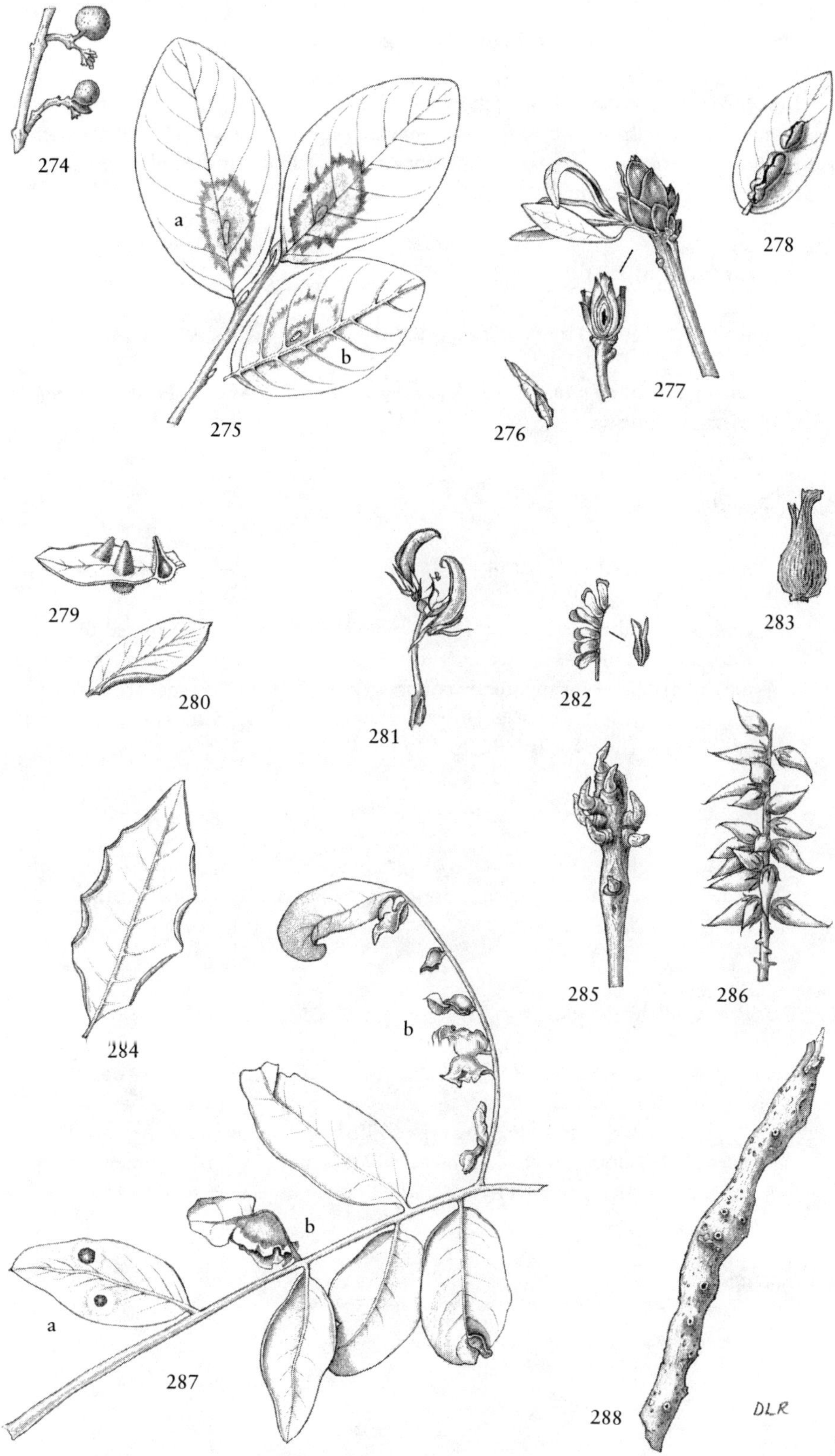
274
275
a
b
276
277
278
279
280
281
282
283
284
285
286
287
a
b
b
288
DLR

from those on the subfamilies Papilionoideae and Caesalpinioideae, and each of these groups is distinct from Old World asphondylias on Fabaceae (Gagné and Wuensche 1986). *Dasineura* spp. are responsible for many leaf folds, and *Neolasioptera* spp. chiefly form stem swellings.

Acacia (acacia)

Swollen bases of joined pairs of leaflets (282) .... *Contarinia* sp.

Each gall contains one larva. Adults were reared in March. Host: *A. greggii*. Distr.: California.

Amorpha

1a Elongate-conical gall on leaflet, rachis, or flower stalk (279)... *Rhopalomyia* sp.

Galls are reddish and about 4 mm long. They have been attributed to either *Rhopalomyia* or *Diarthronomyia* (the second now a subgenus of the first), but the gall makers were apparently not preserved. *Rhopalomyia* spp. are associated principally with Asteraceae. Host: *A. canescens*. Distr.: Minnesota, Nebraska, and Kansas. Ref.: Wells (1921b).

1b Marginal leaf roll (280)...................................... cecidomyiid

Rolled margins are reportedly 3–5 mm long and yellowish. Specimens of the gallmaker were apparently not preserved. Host: *A. canescens*. Distr.: Kansas. Ref.: Wells (1921b).

Amphicarpa (hog peanut)

Stem swelling .... *Neolasioptera* sp.

Originally referred to *Lasioptera* (Felt 1940) based on the gall alone, this species more likely belongs to *Neolasioptera*, which is much richer in numbers of stem gall-making species. Host: *A. bracteata*. Distr.: unspecified.

Astragalus

Petiole swelling .... *Dasineura* sp.

Swellings are irregular, warty, 5–10 mm long, and 5–8 mm in diameter. Referred to *Cecidomyia* by Felt (1940), larval specimens in the Felt collection belong to *Dasineura* in the broad sense. The adult has not been reared. Host: *Astragalus* sp. Distr.: New Mexico.

Baptisia

Slightly swollen stem .... *Neolasioptera* sp.

The gall maker is unknown and only tentatively referred to *Neolasioptera*. Host: *Baptisia* sp. Distr.: New York.

Cassia

Swollen twig .... *Neolasioptera cassiae* (Felt)

Swellings are irregular, elongate, about 5 cm long and 5–6 mm in diameter. Adults emerge in spring. Host: *C. nictitans*. Distr.: Arizona.

Crotalaria (rattlebox)

1a Larva in tapered stem swelling .... *Neolasioptera crotalariae* (Stebbins)

Host: *C. sagittalis*. Distr.: Massachusetts.

1b Larva in pod .... *Asphondylia* sp.

Adults were reared from pods, but the damage was not further specified. Host: *C. pumila*. Distr.: Florida.

Cyamopsis (guar or clusterbean)

1a Closed, hard, swollen buds and sterile, deformed pod (281) .... *Asphondylia websteri* Felt

This native American species has crossed over to guar, an introduced plant from India. There are several generations per year in Arizona. The cecidomyiid probably overwinters as an egg or first instar in dormant buds of its native host plant or alfalfa and not guar, which overwinters only as seed. Hosts: *C. tetragonoloba*; also *Medicago*, *Parkinsonia*, and *Simmondsia* (Simmondsaceae). Distr.: Oklahoma to California. Refs.: Rogers (1972a, b), Gagné and Wuensche (1986).

1b Open, swollen, discolored bud .... *Contarinia texana* (Felt)

Females deposit eggs in flower buds. Several larvae in each bud cause swelling and discoloration. Buds abort after the full-grown larvae leave to pupate in the ground. There may be several generations per year. Up to 90% of the buds can be aborted in some fields. Host: *C. tetragonoloba*; also *Parkinsonia*. Distr.: Oklahoma and Texas to Arizona. Ref.: Rogers (1972a, b, 1974).

Cytisus

Closed, swollen flower .... *Asphondylia* sp.

This damage was attributed to *Asphondylia cytisi* Frauenfeld, but the gall makers were not preserved. The host for this record, exotic to North America, was growing in the New York Botanical Gardens. No other reports are known of the European *A. cytisi* in North America. Host: *Cytisus* sp.. Distr.: New York. Ref.: Batra and Batra (1967).

Dalea

Larva in open flower bud .... *Contarinia* sp.

Adults were reared in June. Host: *D. enneandra*. Distr.: Texas.

Desmanthus

Larva in open flower bud .... *Contarinia* sp.

Adults were reared in June. Host: *D. illinoensis*. Distr.: Texas.

Desmodium (tick-trefoil)

1a Swollen stem............................ *Neolasioptera desmodii* (Felt)

Swellings may be irregular and are as long as 25 mm. Larvae overwinter in the stems, and adults emerge in spring. Felt (1940) attributed another stem gall, from *Desmodium grandiflorum*, to *Neolasioptera hamata* (Felt), a species originally described from a female caught in flight. The larvae he collected from that gall resemble in every respect those of *N. desmodii*, so the two species may be synonyms. Hosts: *D. acuminatum*, *D. canadense*, *D. cuspidatum*, *D. grandiflorum*. Distr.: New York and New Jersey.

1b Bud gall 2
2a Swollen, aborted bud (228)...... *Clinodiplosis meibomiifoliae* (Beutenmüller)

Several larvae in a bud arrest the growth of leaves and can kill the buds. Full-grown larvae drop to the soil to pupate. Adults were reared in July from larvae collected two weeks earlier. Several generations may occur per year. Host: *D. canadense*. Distr.: New Jersey.

2b Hard, elongate-oval, hollow bud gall (283) .... *Dasineura meibomiae* (Beutenmüller)

Galls are monothalamous, 8–22 mm long, 4–8 mm wide, and found in late summer in clusters of three or more. Hosts: *D. laevigatum*, *D. pauciflorum*, *D. rigidum*. Distr.: New York and North Carolina.

Genista

Folded, apical leaflet .... *Clinodiplosis* sp.

This damage was reported as possibly caused by the European *Dasineura genistamtorquens* (Kieffer) (Felt 1911b, 1940), but the available specimens from that damage in the Felt collection comprise a larva and a female of a *Clinodiplosis*. The damage does resemble that usually made by *Dasineura* on Fabaceae. Host: *G. tinctoria*. Distr.: Massachusetts.

Gleditsia (honeylocust)

1a Swollen stem (288)........................... *Neolasioptera brevis* Gagné

In spring eggs are laid in rows on new shoots, where larvae develop in tunnels. Adults emerge from the galls the following spring. Infested twigs cease growing and eventually die. Sprouting may occur from the base of the swellings, changing the growth form of the tree. Host: *G. triacanthos*. Distr.: Pennsylvania to

Michigan and Illinois. Refs.: Gagné and Valley (1984), Wilson and Heaton (1987).

1b Leaflet gall 2
2a Leaflet folded in half, swollen (287b) . . . *Dasineura gleditchiae* (Osten Sacken)

Eggs are laid singly or in clusters on the upper surface of developing leaflets. The outer edges of the leaflets wrap around the gregarious larvae to produce the podlike galls. There may be five to seven generations per year. All but the last pupates in the gall; the last probably drops to the soil to pupate. After larvae have finished feeding, the leaflets die. Repeated defoliation can cause the death of small branches. Host: *G. triacanthos*. Distr.: widespread in North America, probably spread to the west coast with nursery stock; introduced also into Europe. Ref.: Schread (1959).

2b Flat, ovoid blister (287a) . . . . *Meunieriella aquilonia* Gagné

Eggs are laid singly on either surface of developing leaflets, usually the lower. Each blister contains one larva, which breaks out and drops to the ground when full-grown. The spot is inconspicuous, showing only a little chlorosis until the larva escapes, when it turns brown. Two or more generations occur per year. Host: *G. triacanthos*. Distr.: Pennsylvania. Ref.: Gagné and Valley (1984).

Lathyrus (vetchling, wild pea)

Tightly rolled leaflet . . . . *Dasineura maritima* Felt

Adults were reared in spring from gregarious larvae collected in damaged leaflets the previous fall. Host: *L. maritimus*. Distr.: Massachusetts.

Lespedeza (bush-clover)

1a Swollen stem . . . . *Neolasioptera* sp.

Figures 289–300. Galls of Euphorbiaceae and Fabaceae. **289,** *Parkinsonia aculeata*: *Asphondylia websteri*: **a,** normal flower; **b,** injured bud with pupa emerging. **290,** Injured, sectioned flower buds (2×), *Parkinsonia aculeata*: *Asphondylia websteri*: **a,** with pupal cell at top; **b,** cell opened to show pupa; **c,** with two spherical cells of *Tetrastichus* (Hymenoptera), an inquiline; **d,** with two larval *Asphondylia* cells. **291–293,** *Lupinus* sp. **291,** Rolled leaves: *Dasineura* sp. **292,** Swollen petiole and stem: *Neolasioptera lupini*. **293,** Swollen buds: *Dasineura lupini* (redrawn from Russo 1979, courtesy of The Boxwood Press). **294,** *Medicago sativa*: *Asphondylia websteri*: **a,** infested pods; **b,** normal, curled pods. **295,** Bud galls, *Acalypha rhomboidea*: *Resseliella* sp. **296,** Bud galls, *Lespedeza violacea*: cecidomyiid. **297, 298,** Pod galls, *Rhynchosia texana*: *Asphondylia* sp. **299, 300,** *Trifolium repens*. **299,** Damaged flower heads with full-grown larvae of *Dasineura leguminicola* (after photograph in Barnes 1946b). **300,** *Dasineura trifolii* (redrawn from Forbes 1909): **a,** folded leaf gall, **b,** one pulled open to show larvae.

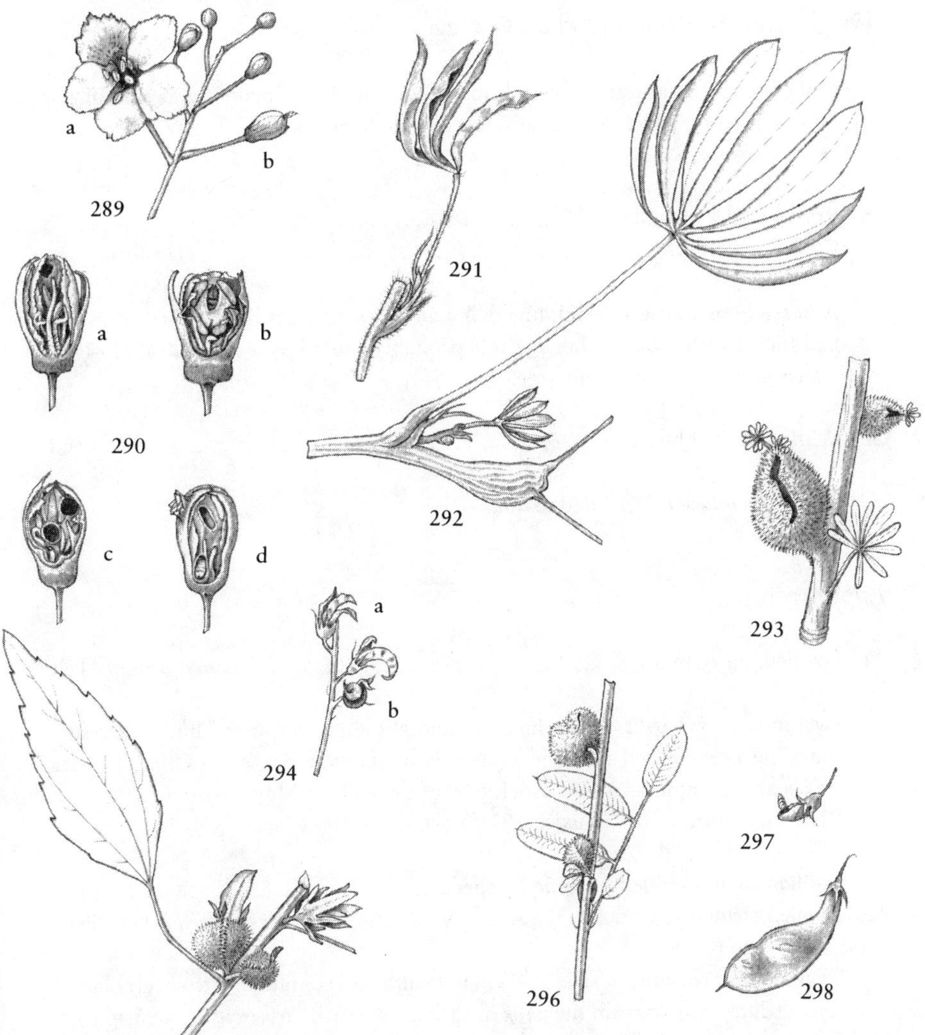
289
a
b
290
a
b
c
d
291
292
293
294
a
b
295
296
297
298
299
300
a
b
DLR

Only the gall is known. The gall maker is tentatively identified as one of the neolasiopteras, which are usually responsible for stem galls. Host: *L. violacea.* Distr.: unspecified.

1b Rolled leaflet or bud gall 2
2a Rolled leaflet . . . . ?*Dasineura* sp.

A possibly predaceous cecidomyiid, *Youngomyia* sp., has been reared from this gall (Felt 1940), but the damage was probably caused by a dasineura. Host: *L. "decumbens."* Distr.: Tennessee.

2b Whitish-haired bud gall (296) . . . . cecidomyiid

Host: *L. violacea.* Distr.: Alabama

Lupinus (lupine)

1a Swollen, hairy bud gall (293) . . . . *Dasineura lupini* Felt

Swellings are up to 2 cm in diameter and made of imbricated bud scales surrounding a central cell containing one to several larvae. Sometimes the galls bear leaves at their apices. Adults emerge from the galls in May. Host: *L. arboreus.* Distr.: California. Ref.: Russo (1979).

1b Swollen stem or rolled or folded leaflet 2
2a Swollen stem (292) . . . . *Neolasioptera lupini* (Felt)

Swellings are variable in size, 1–3 cm in length and contain one to several larval cells. Adults emerge from the galls in spring. Hosts: *L. arboreus*, *L. densiflorus*, *L. varicolor.* Distr.: Washington to California.

2b Tightly rolled leaflet or vein fold (291) . . . . *Dasineura* sp.

Rolled leaflets are usually young and less than 2 cm long; midvein folds are found on full-grown leaves. The vein folds are barely swollen and contain one or two larvae only, while the rolls contain several larvae. Adults have not been reared, so it is not known whether both kinds of damage are made by the same dasineura. Hosts: *L. albifrons*, *L. arboreus*, *L. chamissonis*, *L. varicolor.* Distr.: Washington to California.

Medicago (alfalfa)

Pod gall (294) . . . . *Asphondylia websteri* Felt

Infested pods swell and do not curl as normal pods do. The gall maker is multivoltine and pupates in the pod. This species probably overwinters as an egg or first instar in dormant buds of this or its native hosts. It is a native American species that has evidently crossed over to this exotic plant. A different, readily distinguishable species of *Asphondylia* forms the same damage on alfalfa in Europe. Hosts: *M. sativa*; also *Cyamopsis tetragonoloba*, *Parkinsonia* spp., and *Simmondsia* (Simmondsiaceae). Distr.: Texas to California. Refs.: Gagné and Wuensche (1986), Webster (1912).

Mimosa

1a Flower bud gall . . . . *Asphondylia mimosae* Felt

Found in clusters on flower peduncles, this gall is analogous to that of *Asphondylia prosopidis* on *Prosopis*. It is thin-walled, smooth, greenish-pink, and about 7 mm long by 3 mm in diameter. Pupation occurs inside the gall. The gall may fall off the tree before the adult gall midge emerges. Several generations may occur per year. Host: *M.* "*fragrans*." Distr.: Texas. Ref.: Painter (1935).

1b Cone-shaped bud, thorn, or petiole gall. . . . . *Asphondylia* sp.

Galls are brown and about 9 mm long. This gall was also attributed to *Asphondylia mimosae*, but, if the situation on *Mimosa* is analagous to that on *Prosopis* (q.v.), the two different types of galls are made by different species of *Asphondylia*. Host: *M.* "*fragrans*." Distr.: Texas. Ref.: Painter (1935).

Parkinsonia

Swollen, unopened, aborted flower (289, 290) or distorted pods . . . . *Asphondylia websteri* Felt and *Contarinia texana* Felt

A. websteri. Male and female flower parts are incorporated to form a smooth larval cell. Adults pupate in the aborted flowers. There are several generations per year. *Parkinsonia* is one of the native American hosts and reservoirs for this pest of alfalfa and guar, although it belongs to a different subfamily of Fabaceae. Hosts: *P. aculeata*, *P. florida*, and *P. macrophyllum*; also *Cyamopsis tetragonoloba*, *Medicago sativa*, and *Simmondsia* (Simmondsiaceae). Ref.: Gagné and Woods (1988).

C. texana. Larvae are gregarious and can be found throughout the year in flowers and pods. Adults drop to the ground to pupate. This species has several generations per year. Hosts: *P. aculeata*, *P. macrophyllum*; also *Cyamopsis tetragonoloba*. Distr.: Oklahoma to Mexico, west to Arizona. Ref.: Rogers (1972b).

Phaseolus (bean)

Swollen, deformed pod .... *Asphondylia* sp.

This species has been repeatedly reared from beans intercepted at U.S.-Mexican ports of entry. Host: *Phaseolus* sp. (green beans). Distr.: Guatemala and Mexico.

Pithecellobium

Atrophied seed pods .... *Asphondylia* sp.

Adults emerged in May. Host: *P. keyense*. Distr.: Florida.

Prosopis (mesquite)

1a Smooth, dehiscent flower bud gall (286)..... *Asphondylia prosopidis* Cockerell

Six to eight generations of these galls develop between May and September. The galls consist of the aborted and enlarged ovary of the flower, are smooth and pink to green, monothalamous, thin-walled, conical, and 10 mm long by about 3 mm in diameter. Each generation lasts 2–3 weeks. The galls dehisce at about the time the adults are ready to leave the galls. Host: *P. glandulosa*. Distr.: Texas and New Mexico. Ref.: Rogers (1973).

1b Rough, persistent flower bud gall (285).................... *Asphondylia* sp.

This is a brown, scaly, spindle-shaped gall, 6 mm long by 2 mm wide, and formed from the enlarged ovary and style of the flower atop a thickened peduncle. These galls develop in late April to early May and remain on the tree all year. Eggs hatch in spring and the larvae remain in an early instar through the winter. The main growth occurs the following spring. This species, unlike that in the abortive bud gall, is univoltine. Host: *P. glandulosa*. Distr.: Texas and New Mexico. Ref.: Rogers (1973).

Rhynchosia

Atrophied or deformed pod (297, 298) .... *Asphondylia* sp.

Adults were reared in September from pods in various stages of development. Host: *R. texana*. Distr.: Texas.

Robinia (black locust)

1a Tightly folded, swollen leaflet .... *Dasineura pseudacaciae* (Fitch)

Larvae are gregarious and pupate in the podlike swellings. The species is multivoltine. Host: *R. pseudoacacia*. Distr.: New York and New Jersey.

1b Rolled margin of leaflet (284) .... *Obolodiplosis robiniae* (Haldemann)

Larvae are gregarious, multivoltine, and pupate in the rolled margins. Host: *R. pseudoacacia*. Distr.: Maine to Maryland.

Thermopsis

Thickened, podlike terminal leaflet .... *Dasineura* sp.

Host: *T. rhombifolia*. Distr.: Colorado.

Trifolium (clover)

1a Leaflet folded and swollen along midvein (300) ..... *Dasineura trifolii* (Loew)

Eggs are laid on developing leaflets. Larvae feed gregariously along the midrib covered by the expanding leaflet halves. Full-grown larvae spin a cocoon either in the swelling or in the soil. Several generations occur per year. This species originated in Europe and was recognized in North America in 1879. Hosts: *T. pratense*, *T. repens*. Distr.: Massachusetts to Maryland, west to Nebraska; Europe. Ref.: Barnes (1946b).

1b Aborted florets (299) .... *Dasineura* spp.

The two following species from clover flowers are host specific and anatomically distinct. They cause a considerable reduction in seed production. Both were first discovered in the United States although they are immigrants from Europe. Eggs are laid on the flower heads before the florets open. Larvae enter the developing florets and destroy the ovules. Full-grown larvae drop to the soil where they spin a cocoon. Several generations occur per year.
Dasineura gentneri Pritchard. Hosts: *T. repens*, *T. hybridum*. Distr.: widespread in North America; Europe. Ref.: Bishop (1954).
Dasineura leguminicola (Lintner). Host: *T. pratense*. Distr.: widespread in North America; Europe. Ref.: Barnes (1946b).

Fagaceae

Castanea (American chestnut, chinquapin)

1a Irregular swelling on burr (301) ..... "*Cecidomyia*" *chinquapin* Beutenmüller

Each swelling contains a single, orange-red larva that drops to the ground at the end of summer. Adults have not been reared but larvae resemble those of "*Rhopalomyia*" *castaneae* of entry 2a of this key. Host: *C. pumila*. Distr.: Virginia and North Carolina.

1b Bud or leaf galls 2
2a Enlarged, clustered buds (302)............... "*Rhopalomyia*" *castaneae* Felt

Reddish larvae were taken from galls in July to September. The only known adult was reared the following June. Larvae resemble those of "*Cecidomyia*" *chinquapin* of entry 1a of this key in that they also have only two papillae on each of the two lateral groups of thoracic papillae and three pairs of long setae on the terminal segment. The adult specimen, a female placed by Felt in *Rhopalomyia*, generally resembles females of that genus, but all other known rhopalomyias pupate in their galls, unlike the present species. Possibly that female was a contaminant in the rearing container. Host: *C. dentata*. Distr.: New York, West Virginia, and North Carolina.

2b Leaf gall.. 3
3a Swollen petiole .... cecidomyiid

All that is known of this species is that it forms an irregular gall about 12 mm in length (Felt 1940). Host: *C. pumila*. Distr.: unspecified.

3b Growth or fold on leaf blade 4
4a Flat, circular, warty growth, usually between veins (303c) ....
.... *Dasineura dentatae* (Stebbins)

Galls are small, roughened blisters, 2–3 mm in diameter, that are evident on both sides of the leaf. The surface of the gall beneath the leaf dehisces as a cap to reveal the larva in a smooth, craterlike depression. A single larva lies between both gall halves and presumably drops to the ground to pupate. The gall is

Figures 301–304. Galls of Fagaceae: *Castanea* spp. **301,** Husk swellings, one in 2×, *C. pumila*: "*Cecidomyia*" *chinquapin*. **302,** Swollen, clustered buds, *C. dentata*: "*Rhopalomyia*" *castaneae*. **303, 304,** Leaf galls, *C. dentata*. **303,** Lower surface of leaf, **304,** upper surface; **a,** vein folds, with enlarged cross section in side view: *Macrodiplosis castaneae*; **b,** leaf blade folds: *Contarinia* sp.; **c,** capsule galls, two in 2×, one with cap partially separated: *Dasineura dentatae*; **d,** ribbed vein folds, *Macrodiplosis* sp.

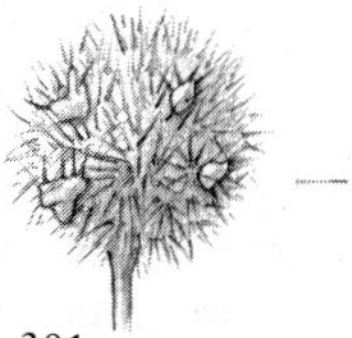

301

302

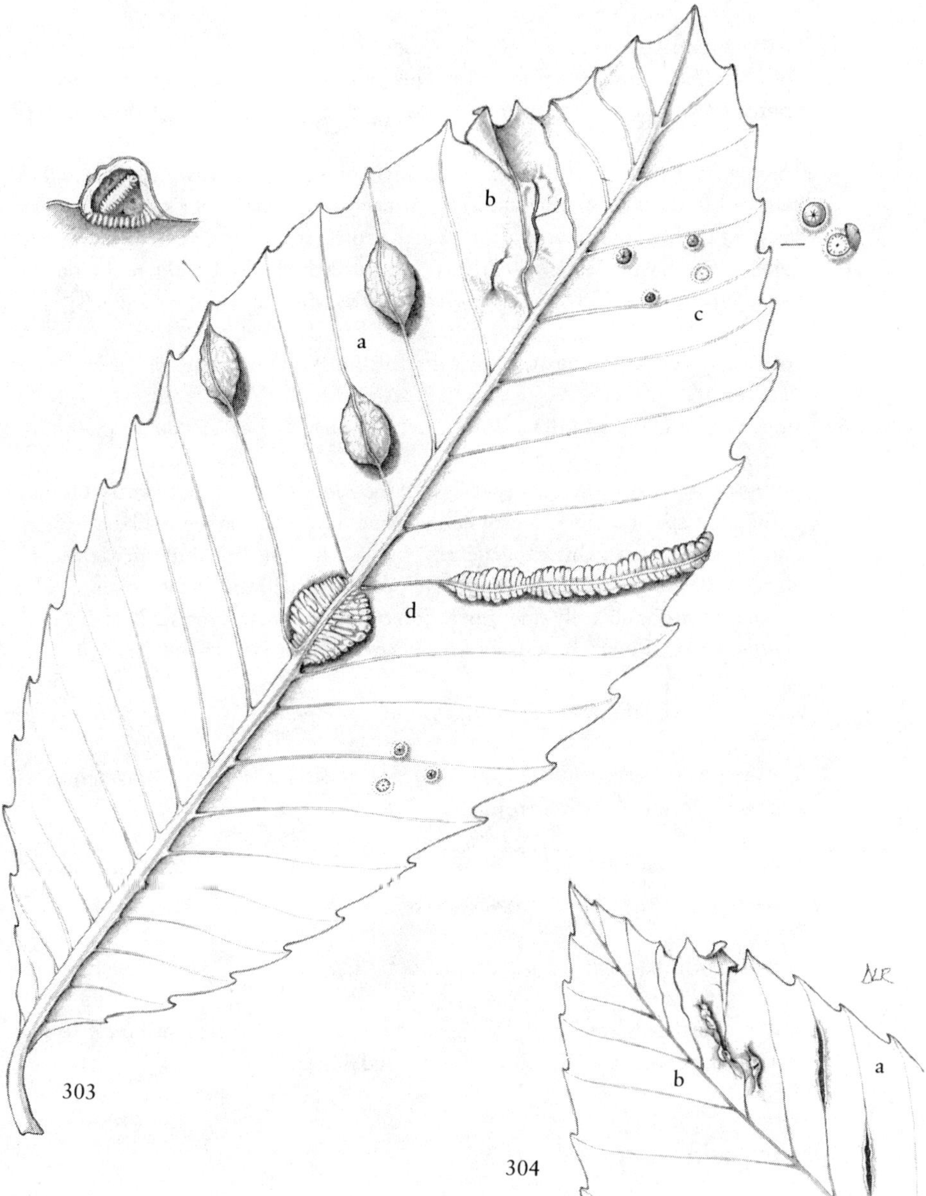

303

304

apparently the same as that attributed to an eriophyid mite (Felt 1918a, 1940, Jarvis 1909, Stebbins 1910) and a cynipid (Felt 1940). This cecidomyiid, tentatively placed in *Dasineura* because only larvae are available, has only four terminal papillae and one group of two lateral papillae each on either side of the spatula. Host: C. *dentata*. Distr: Massachusetts, Ontario, and Ohio.

4b Leaf or vein fold 5
5a Fold between and sometimes across veins, without conspicuous invagination on opposite side of leaf (303b, 304b) .... *Contarinia* sp.

These galls distort unfolding leaves in early spring. The narrow, longitudinal or diagonal folds, 1 to several cm long, project out of the upper side of the leaf, contrary to those of *Macrodiplosis*, which project from the lower side of the leaf. Full-grown larvae escape in early spring through the longitudinal slit on the lower side of the leaf. Host: C. *dentata*. Distr.: Maryland and Georgia.

5b Fold along vein with conspicuous cylindrical or bulbous enlargement on lower side of leaf 6
6a Smooth, globular fold (303a, 304a)....... *Macrodiplosis castaneae* (Stebbins)

Galls develop in spring. This species and the *Macrodiplosis* sp. of entry 6b may be the same species, but adults of neither kind have yet been reared. Larvae from smooth and rough galls are generally similar. In June and July larvae escape through the elongate opening that develops along the vein on the upper surface of the leaf. Probably only one generation occurs per year. Hosts: C. *dentata*, C. *pumila*. Distr.: Massachusetts and New York to North Carolina.

6b Rough, ribbed, usually elongate fold (303d, color plate 3D) .. *Macrodiplosis* sp.

See remarks under entry 6a. Hosts: C. *dentata*, C. *pumila*. Distr.: Massachusetts and New York to North Carolina.

Castanopsis (chinkapin)

Vein swelling or rolled leaf .... cecidomyiid

Nothing further is known of this species beyond the original record (Felt 1940). Host. *Castanopsis* sp. Distr.: western United States.

Quercus (oak)

Most gall midges of oaks belong to one of two genera, *Macrodiplosis* and *Polystepha*. *Macrodiplosis* spp. cause extensive leaf and vein fold galls on white and red oaks in eastern North America. Galls mature in spring and often completely envelop the new leaves (307). The often gregarious, white larvae develop quickly and drop to the soil, where they remain until the following year. Galls of *Polystepha* are one-celled spot or blister galls on red and black oaks in eastern North America. Galls are evident in late summer or early fall. Several galls may occasionally coalesce, but larvae remain in separate cells. Galls develop slowly, and the usually orange larvae reach their full size in fall. Most larvae break out of the galls in autumn, but some remain inside the fallen leaves through the winter. *Macrodiplosis* and *Polystepha* species appear to be univoltine. Some kinds of galls can be seen on many species of oak, but it is not known whether they are all made by the same species of gall midge. Further, the damage made by some species has not always been clearly described. For these reasons, the following key will list for most gall midge species only the type host and type locality, even though the galls may be widespread in eastern North America.

Also on oaks are several species of *Lasioptera*, mainly in twig and petiole galls that are probably initiated by cynipid wasps, and a few species of the ubiquitous genera *Contarinia* and *Dasineura*. *Dasineura* larvae can sometimes be found in galls formed by *Macrodiplosis* spp., and the gall of *Macrodiplosis qoruca* was once attributed to *Parallelodiplosis acernea* (Felt) (as *florida*). The latter species may have entered the gall only after the gall maker had left. In addition, *Dasineura florida* Felt, originally described as reared from an "oak-like gall" resembling that of a *Polystepha*, was subsequently listed as responsible for a gall on oak (Felt 1940). Notes relating to that species in the USNM indicate that the host was unknown, so *D. florida* is not considered further here. On several oaks in western North America, one or more species of an unknown genus near *Dasineura* forms a peculiar invagination of the leaf margin.

The greatest majority of galls on oak are formed by cynipid wasps. These are not treated here, but Felt (1940) is still an adequate guide for their identification. All complex galls of oak are formed by cynipids. Gall midges form simple galls on oak. Some cynipid galls, however, are simple swellings of stems and petioles and as such superficially resemble galls of some cecidomyiids. In such cases, some examples should be opened to determine whether the galls are caused by cynipids or cecidomyiids.

1a Gall on fruit or flower 2
1b Gall on bud, twig, or leaf 4
2a Larva in acorn, between seed coats . . . . *Dasineura glandis* Felt

No further details are known of the biology of this species. Host: *Quercus* sp. Distr.: Missouri.

2b Larva in catkin or flower 3
3a Larva on male catkin. . . . . *Contarinia* n. sp.

The specific damage to the catkins made by this undescribed species is unknown, but in spring the full-grown yellow larvae drop in such great numbers as to cover

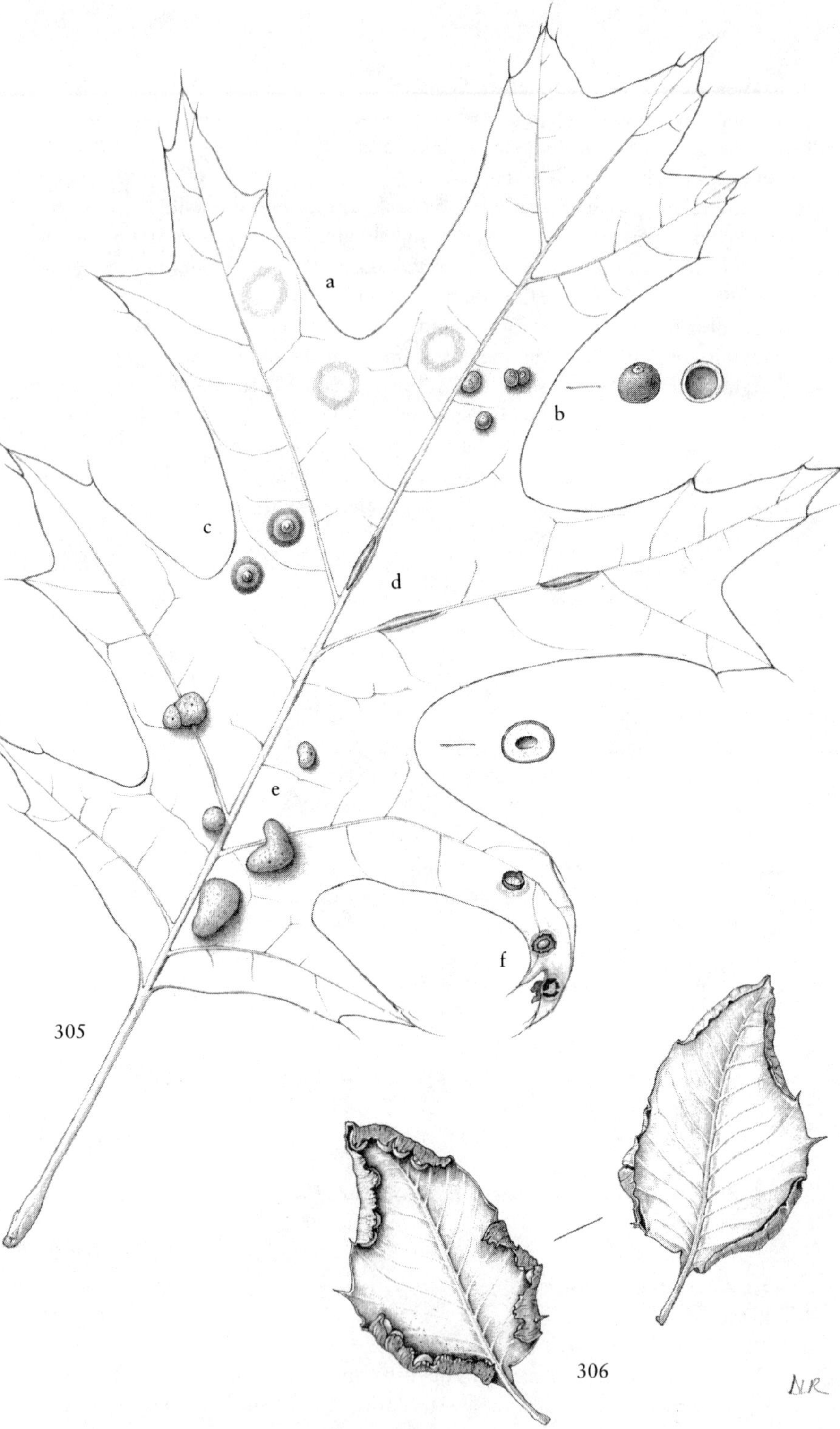

305

306

the surface of the ground until they are able to dig into the soil. The larvae presumably remain there until the following spring. Host: *Q. palustris*. Distr.: Maryland.

3b Larva from unspecified oak blossom gall. . . . . *Lasioptera querciflorae* Felt

This species has been recorded also from various cynipid galls. See entry 5a for further remarks. Hosts: *Quercus* spp. Distr.: Ontario to Oregon, south to Maryland and Arizona.

4a Aborted bud . . . . *Contarinia* sp.

Larvae were taken from between imbricated bud scales in early June. Host: *Q. rubra*. Distr.: New York.

4b Woody twig swelling or leaf gall 5

5a Tapered, woody swellings on twig, petiole, or midvein, some of them caused by cynipid wasps . . . . *Lasioptera* spp.

Four species of *Lasioptera* have been described from oak. Their antennae have an unusually large number of flagellomeres for the genus. These gall midges are evidently not restricted to particular species of oak, and at least *L. querciflorae* Felt appears to be an inquiline in galls formed by Cynipidae. The most commonly collected species, *L. querciperda* Felt, has been reared from leaf midrib and twig galls of white oak and from petiole and midrib galls of bur oak and red oak; *L. querciflorae* was reared from unspecified blossom galls on an oak in Arizona, twig galls on white oak in Ontario, cynipid twig galls on *Q. phellos* in Maryland, and empty leaf galls of the cynipid *Bassettia ligni* Kinsey on *Q. garryana* in Oregon; *L. howardi* Felt was reared from twig galls in California; and *L. quercirami* Felt from twig galls in Utah. Larvae attributed to *L. querciflorae* from Oregon and Maryland and to *L. querciperda* from New York are atypical for *Lasioptera* for their long setae. In addition, those larvae are red and their integument is pebbled throughout. Pupation takes place inside the galls.

5b Leaf fold, spot, or blister 6

6a Circular, flat or hemispherical, closed gall (as in 305b, c) . . . *Polystepha* spp. 7

6b Leaf fold, always with a slitlike opening (as in 306–309) 12

Figures 305, 306. Galls of Fagaceae: *Quercus* spp. **305,** Leaf galls, *Q. rubra*, upper leaf surface: **a,** simple leaf spots: *Polystepha quercifolia*; **b,** thin-walled, spherical galls, with examples in 2×, one sectioned: *Polystepha globosa*; **c,** nippled blister galls: *Polystepha ?sobrina*; **d,** vein swellings: *Polystepha americana*; **e,** hard, woody galls with enlarged section in 2×: *Polystepha pilulae*; **f,** hard, woody gall with dehiscent outer skin, visible on both sides of leaf: *Polystepha symmetrica*. **306,** Rolled leaf margins, lower surface on left, upper on right: *Q. chrysolepis*: "*Dasineura*"-*silvestrii*.

7a Thin-walled, elongate blister on leaf vein (305d) ....
.... *Polystepha americana* (Felt) and *Polystepha podagrae* (Felt)

P. americana. This species was taken originally from slight swellings, nearly the same color as the leaf, on lateral veins. Host: *Q. rubra*. Distr.: New York.
P. podagrae. This species was reared from purplish blisters, about 8 mm long, on the underside of mid or lateral veins. Host: *Q. velutina*. Distr.: Massachusetts.

7b Circular, flat or convex cell between veins 8
8a Rough, convex, thick-walled, brown gall, several often coalesced (305e, f)... 9
8b Smooth, flat or convex, thin-walled cell (305a, c) 10
9a Gall equally prominent on both sides of leaf, outer surface peeling away at end of season (305f) .... *Polystepha symmetrica* (Osten Sacken)

This gall is uncommon. It is possible the species that forms it is responsible also for the gall in 305b. Host: *Q. falcata*. Distr.: Maryland.

9b Gall evident only on upper side of leaf, outer surface not dehiscent (305e)
.... *Polystepha pilulae* (Beutenmüller)

This is one of the most conspicuous *Polystepha* galls and is commonly seen in autumn on many kinds of oaks. Hosts: *Quercus* spp. Distr.: New York.

10a Spherical gall (305b)............................ *Polystepha globosa* (Felt)

This gall is usually found on the lower side of leaves. Host: *Q. velutina*. Distr.: Massachusetts.

10b Flat and inconspicuous or slightly raised blisters....................... 11
11a Gall not raised above surrounding leaf tissue (305a) ....
.... *Polystepha quercifolia* (Felt)

Often accompanied by reddish discoloration, the gall is otherwise inconspicuous. Host: *Q. rubra*. Distr.: Maryland.

11b Gall protruding on both surfaces of leaf (305c) ....
Polystepha pustulata (Felt), *P. pustuloides* (Beutenmüller), *P. serrata* (Felt), *P. simpla* (Felt), and *P. sobrina* (Felt)

Figures 307–309. Galls of Fagaceae: *Quercus* spp. 307, Vein fold galls, lower surface on left, upper surface, opened to show larvae, on right, *Q. alba*: *Macrodiplosis niveipila*. **308,** Marginal leaf rolls, lower leaf surface on left, upper on right, *Q. vaccinifolia*: *Contarinia* sp. **309,** Leaf galls, *Q. rubra*, lower leaf surface: **a**, hairy vein pocket: *Macrodiplosis ?niveipila*; **b**, smooth, thin-walled pocket with lateral cross section in 2× showing larva: *Macrodiplosis majalis*; **c**, short, striate vein pocket: *Contarinia* sp.; **d**, marginal leaf roll: *Macrodiplosis erubescens*; **e**, smooth vein pocket: *Macrodiplosis qoruca*.

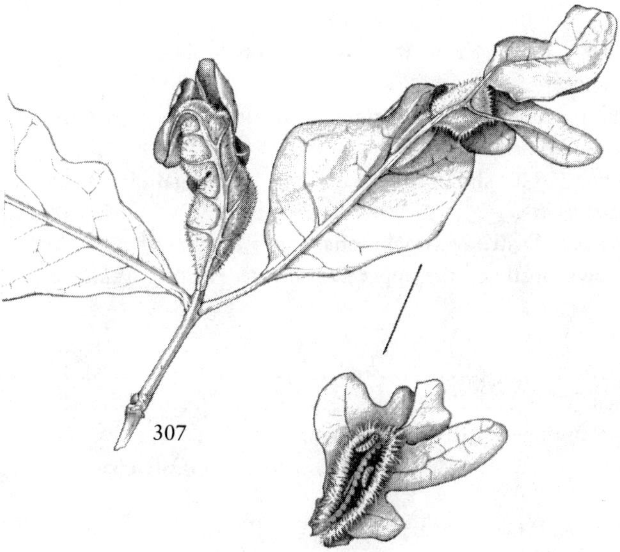

307

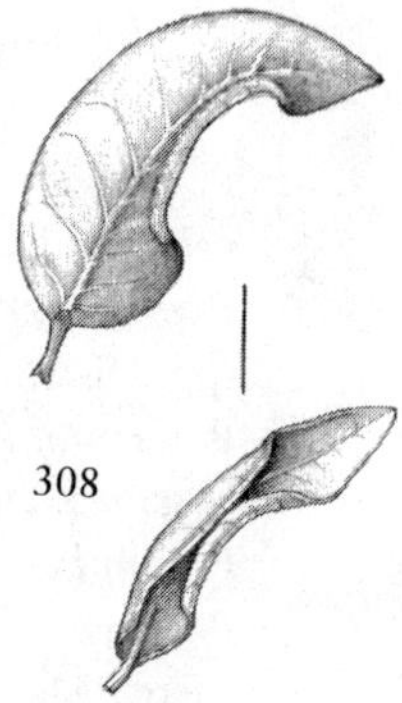

308

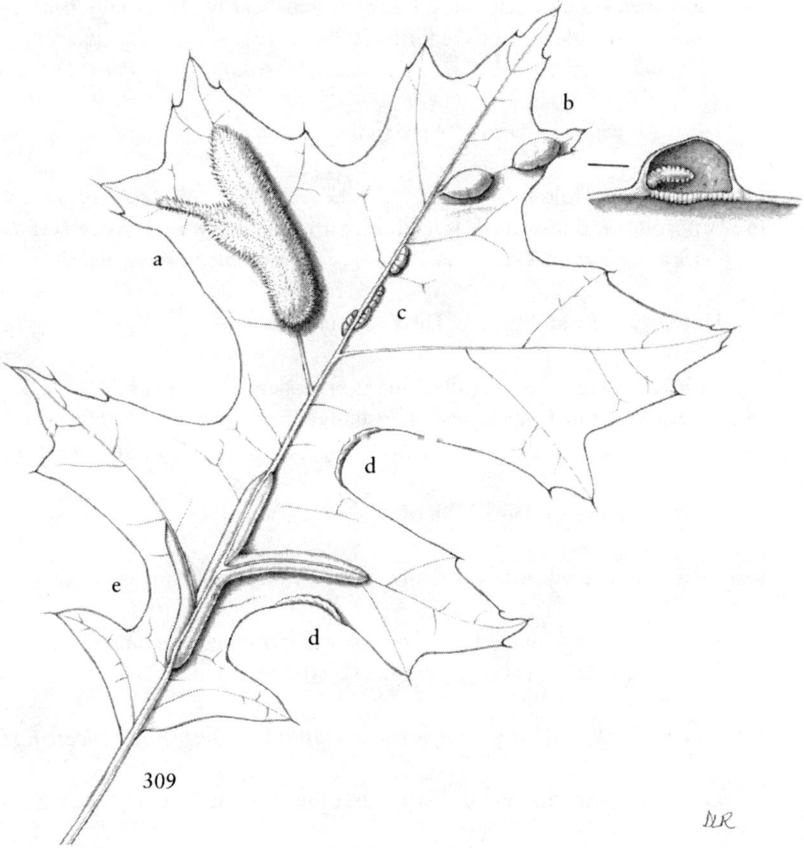

309

P. pustulata. The typical gall shows on both surfaces of the leaf. Host: *Q. velutina*. Distr.: Massachusetts.
P. pustuloides. This gall protudes on both surfaces of the leaf. Hosts: *Quercus* spp. Distr.: New Jersey and North Carolina.
P. serrata. A photo in Felt (1940) shows a central nipple on this gall. Host: *Q. coccinea*. Distr.: Massachusetts.
P. simpla. Host: *Q. velutina*. Distr.: Massachusetts.
P. sobrina. This gall shows mostly on the upper leaf surface. Host: *Q. velutina*. Distr.: Massachusetts.

12a East of the Great Plains 13
12b West of the Great Plains 17
13a Folded sinus, usually green tinged with red (309d) .... *Macrodiplosis erubescens* (Osten Sacken)

Host: *Quercus* sp., "red oak." Distr.: Maryland.

13b Leaf folded along or between veins 14
14a Short, pea-sized pouch gall on underside of leaf blade, usually near leaf vein, thin-walled, pale green to reddish (309b) .... *Macrodiplosis majalis* (Osten Sacken)

Host: *Q. palustris*. Distr.: Maryland.

14b Elongate vein fold.. 15
15a Amorphous, thin-walled, large fold with white down on inside surface, often deforming leaf (307) .... *Macrodiplosis niveipila* (Osten Sacken)

Hosts: *Q. alba* and *Q.* spp. Distr.: Maryland.

15b Regular, narrow, thick-walled, more or less annulate vein fold............ 16
16a Vein fold parallel sided, smooth, usually longer than 1 cm (309e) .... *Macrodiplosis qoruca* (Walsh)

Host: *Quercus* sp. Distr.: Illinois.

16b Vein fold tapered, striate, usually about 5 mm long (309c)..... *Contarinia* sp.

Larvae drop to the soil in summer and presumably remain there until the following spring. Host: *Q. palustris*. Distr.: Maryland.

17a Folded, thickened lobe along leaf edge (306) .. "*Dasineura*" *silvestrii* (Trotter)

Only larvae are known of this species; they do not fit in any known genus but belong to the Oligotrophini. Hosts: *Q. chrysolepis*, *Q. garryana*, *Q. vaccinifolia*. Distr.: Oregon and California.

17b Elongate, marginal leaf roll (308) .... *Contarinia* sp.

Larvae apparently drop to the soil in summer. Host: *Q. vaccinifolia*. Distr.: Oregon.

Geraniaceae

Stem blisters on *Geranium* have been attributed to a cecidomyiid (Felt 1940), but the larvae associated with the damage belonged to *Mycodiplosis*. These feed on rust spores. The gall possibly was formed by a rust.

Gnetaceae

Ephedra

1a Subglobular, nonresinous stem swelling (312)... *Lasioptera ephedrae* Cockerell

The monothalamous swellings are dimpled, the same color as the normal stem, and about 10–15 mm long and 5 mm in diameter at midlength. The larva feeds along the pith. Pupation occurs in the gall, and the pupa exits in spring through a depression on one side where the wall is thinner. Hosts: *E. antisyphilitica*, *E. nevadensis*, *E. trifurca*. Distr.: New Mexico to California.

1b Irregular, resinous stem swelling (313) .... *Lasioptera ephedricola* Cockerell and *Neolasioptera* sp.

L. ephedricola. Swellings are rugose, usually darker than the normal stem, 1–3 cm long, and usually polythalamous. Occasionally a gall with only one larva inside will not show a conspicuous swelling. Larvae live in fungus-lined tunnels, the fungus turning black when mature. Adults emerge in February to March. Hosts: *E. californica*, *E. nevadensis*, *E. trifurca*. Distr.: New Mexico to California.

Neolasioptera sp. A few specimens of an unidentified species were reared with *L. ephedricola* from galls on *Ephedra* sp. in Arizona.

Hamamelidaceae

Hamamelis (witch-hazel)

In addition to the gall midges listed in the following key, several cecidomyiids, the predators *Aphidoletes aphidimyza* and a *Lestodiplosis* sp. and a presumably fungus-feeding *Clinodiplosis* sp., have been reared from leaf-spine galls formed by *Hormaphis hamamelidis* Fitch (Aphididae).

1a Circular leaf blister (315a) .... *Contarinia* sp.

Each spot contains a white larva that leaves when full-grown through a slit in the leaf surface. Adults have not been reared. Host: *H. virginiana*. Distr.: Massachusetts and Maryland.

1b Vein fold (315b).. cecidomyiid

Folds are reportedly 1–3 cm long and found in spring. In addition to the specimen referred to as a "*Cecidomyia*" and which is apparently lost, another cecidomyiid was reared from "wrinkled, discolored, intervenal folds" from New York (Felt 1940). That specimen, a female, belongs to the predaceous genus *Lestodiplosis*. Host: *H. virginiana*. Distr.: Massachusetts and Connecticut. Ref.: Wells (1914).

Hippocrataceae

Hippocratea

Irregular root swelling (314) .... *Neolasioptera* sp.

Swellings may contain several larvae. Adults have not been reared. Host: *H. volubilis*. Distr.: Florida.

Hypericaceae

Hypericum (St. John's-wort)

1a Enlarged seed capsule.............................. *Dasineura toweri* Felt

Affected seed capsules are about 5 mm long and 3 mm in diameter. Adults were reared in early fall. Host: *H. mutilum*. Distr.: Massachusetts.

1b Swollen bud (317).......................... *Zeuxidiplosis giardi* (Kieffer)

Buds are ovoid, slightly pointed at the tip; the terminal leaves are incurved, swollen, and usually red. Larvae may be gregarious, and pupation occurs in the bud. The species is multivoltine. This cecidomyiid was introduced from Europe into California for the control of the alien *Hypericum perforatum*. The gall midge has apparently crossed over to a native American *Hypericum*. Hosts: The alien *H. perforatum* and the native *H. concinnum*. Distr.: California. Refs.: Barnes (1952), Gordon and Neser (1986), Harris (1966).

Triadenum (marsh St. John's-wort)

Irregular stem swelling (316) .... *Neolasioptera triadenii* (Beutenmüller)

Swellings are about 8–12 mm long and monothalamous. The orange larvae overwinter in the stems and pupate there the following spring. Host: *T. virginicum*. Distr.: Massachusetts, New York, and New Jersey.

Juglandaceae

Carya (hickory)

Seventeen named species of cecidomyiids are known from hickories. Twelve belong to *Caryomyia*, a North American genus restricted to complex galls on hickory leaflets. Only the galls with known *Caryomyia* species are illustrated here. Many other kinds of *Caryomyia* galls are known, but their gall makers have not yet been described. The majority of those were illustrated by Wells (1915). The remaining five cecidomyiids on *Carya* are: *Parallelodiplosis caryae*, an inquiline in *Caryomyia* galls; "*Cecidomyia*" *glutinosa*, which lives in sticky spots on the underside of leaflets; and three *Contarinia* species, one forming a nut husk gall, another a leaflet blister gall, and the third living as an inquiline in a gall formed by a phylloxerid (Homoptera). Phylloxeridae (Homoptera) also form galls on hickory and pecan leaflets. These can be distinguished externally from most cecidomyiid galls by the conspicuous opening on one side of the leaflet (310m).

1a Swelling on surface of husk (311) .... *Contarinia nucicola* (Osten Sacken)

Swellings cover a deep-seated, monothalamous, larval cell. Full-grown larvae leave through a cylindrical escape tube lined with hairs. This species was originally collected in August in New York, but in Maryland it can be seen in early July, and in Georgia in June, when husks have just reached full size. Adults have not yet been reared. This species is probably univoltine. Hosts: *C. glabra*, *C. tomentosa*. Distr.: New York, Maryland, and Georgia.

1b Gall or sticky spot on leaflet 2

2a Yellowish leaf spot on underside of leaflet (310g) ....
.... "*Cecidomyia*" *glutinosa* Osten Sacken

Spots are covered with a sticky substance. One larva lives on each spot and drops to the soil when full-grown. The spot then dries and turns brown. The yellowish-

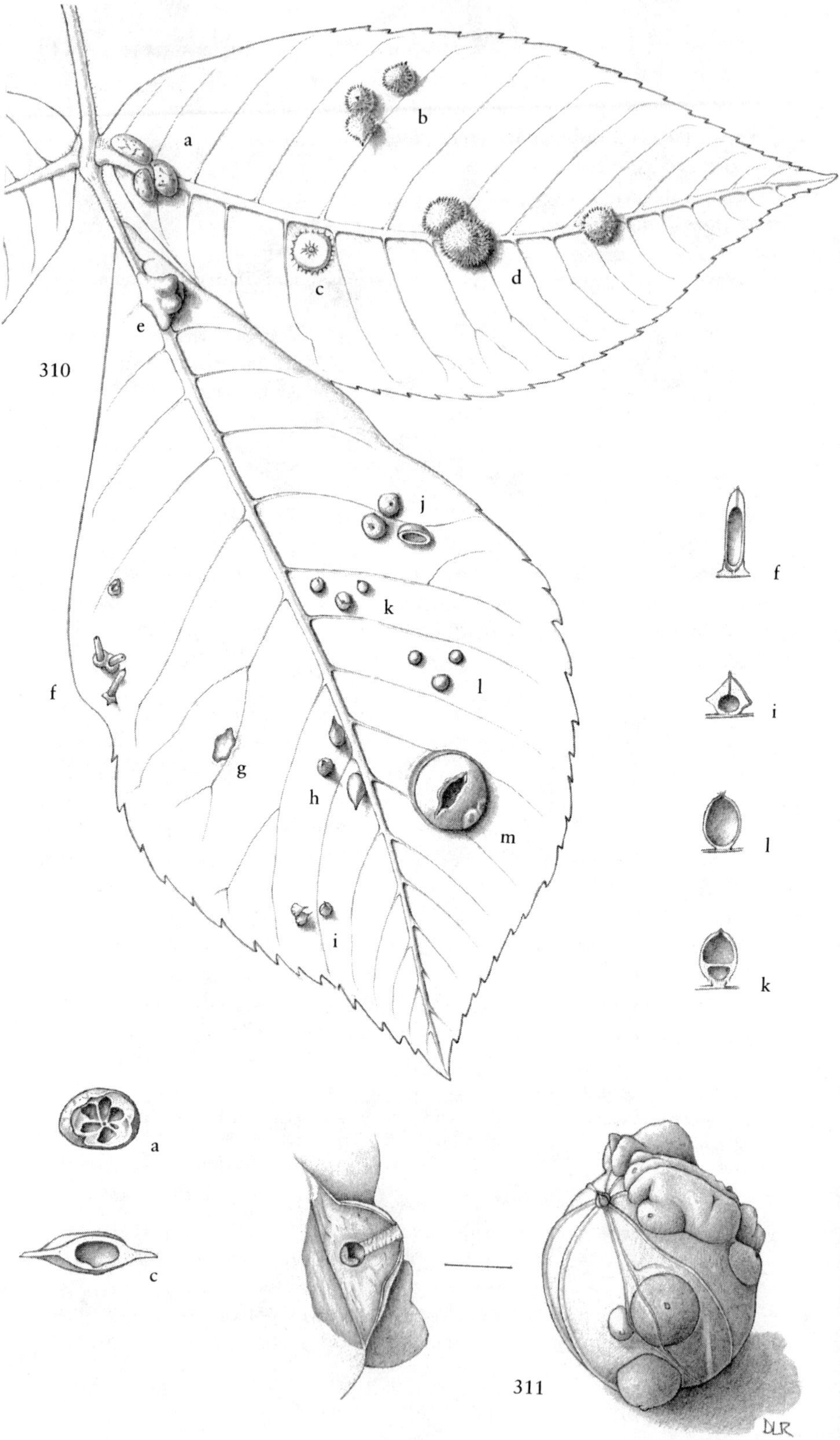

310

311

orange larvae are notable for their conspicuous, long, flattened dorsal lobes. These lobes are similar to those of some pine resin midges of the genus *Cecidomyia*, but the hindmost pair of larval spiracles are not greatly enlarged in "C." *glutinosa* larvae, unlike in *Cecidomyia* spp. (figs. 87, 88), nor are they at the posterior end of the body. This species is known only from the larval stage and cannot be placed in any known genus, although it belongs to the Oligotrophidi. Hosts: *C. illinoensis*, *Carya* spp. Distr.: Connecticut to Georgia and west to Missouri, Louisiana, and Texas.

2b Blister, fold, or complex gall 3

3a Simple, nondehiscent, blister gall or midrib fold, or in pouch gall formed by phylloxerans 13

3b Complex, tubular, conical, or globular gall on leaflets; usually dehiscent, often dropping to ground before leaves . . . . *Caryomyia* spp. and *Parallelodiplosis caryae* (Felt) 4

Caryomyia spp. Several kinds of galls may be found on one leaflet and most can be found on any species of hickory (color plate 4C, D). All these galls are monothalamous, except for that of *Caryomyia cynipsea*, which is sometimes polythalamous. Galls contain a single, white, short larva. The spatula varies with the species but is never clove-shaped, with its anterior end very wide or considerably narrowed. Larvae remain in the galls through winter, and adults emerge in spring. Only the galls with named caryomyias definitely associated with them are illustrated here. At least 30 other kinds are known. Particular hosts and geographic range are not given for these species because they appear to be common on many hickories in North America. Interestingly, none has been reported from pecan, *Carya illinoensis*.

P. caryae. This species has been reared several times from at least two kinds of *Caryomyia* galls, in which it is presumably an inquiline. The pink or orange, longer and more tapered larvae, and their more active habits when disturbed, distinguish them from the short, flat, white, quiescent *Caryomyia* larvae.

4a Gall elongate-tubular, green to brown, set in socket on leaflet and readily detachable (310f) . . . . *Caryomyia tubicola* (Osten Sacken)

Figures 310, 311. Galls of Juglandaceae: *Carya* spp. **310,** Galls on lower side of leaflets, examples of some drawn also in 2× section: **a,** smooth, spherical galls: *Caryomyia cynipsea*; **b,** hairy, onion-shaped galls: *Caryomyia holotricha*; **c,** flat, dimpled blister gall: *Contarinia* sp.; **d,** downy, spherical galls: *Caryomyia persicoides*; **e,** vein swelling of Phylloxeridae containing inquilinous *Contarinia* sp.; **f,** cylindrical, detachable galls: *Caryomyia tubicola*; **g,** leaf spot of "*Cecidomyia*" *glutinosa*; **h,** smooth, onion-shaped galls: *Caryomyia caryaecola*; **i,** smooth, pyramidal galls: *Caryomyia sanguinolenta*; **j,** smooth, depressed-spherical galls, one in section: *Caryomyia thompsoni*; **k,** smooth, spherical galls: *Caryomyia inanis*; **l,** spherical, nippled galls: *Caryomyia caryae*; **m,** leaf blade gall: *Phylloxera* sp. (Phylloxeridae). **311,** Husk swellings, one in 2× section to show larval cell and escape passage: *Contarinia nucicola*.

4b Gall conical or globular 5
5a Gall conical, widest at base, tapered to pointed apex, smooth 6
5b Gall globular; if apex pointed, gall greatly constricted at base, with or without hairs 7
6a Gall angular at base, usually ridged, often red or purple, sometimes green (310i) *Caryomyia sanguinolenta* (Osten Sacken)
6b Gall smoothly rounded, tapering to pointed apex, light brown (310h) . . . . *Caryomyia caryaecola* (Osten Sacken)
7a Gall smooth 8
7b Gall covered with sparse to dense hairs 10
8a Gall sessile, irregularly spherical, usually polythalamous (310a) . . . . *Caryomyia cynipsea* (Osten Sacken)
8b Gall stalked, spherical, monothalamous 9
9a Gall spherical, yellowish, nippled, thin-walled, with single cavity (310l) . . . . *Caryomyia arcuaria* (Felt), *C. caryae* (Osten Sacken), and *C. similis* Felt

These three species are listed together here because the differences reported in the thickness of the gall wall and whether there is an internal projection in the larval cavity are difficult to determine from the galls' exteriors (Felt 1921). Whether differences among the adults exist will have to await a revision of these species.

9b Gall irregularly spherical, green or brownish, with empty cavity above larval chamber (310k) . . . . *Caryomyia inanis* Felt
10a Sparse hairs on gall not obscuring the gall surface 11
10b Dense hairs on gall completely obscuring the surface 12
11a Gall yellowish-green or brown, spherical, thick-walled, with short hairs . . . . *Caryomyia antennata* Felt
11b Gall green to black, depressed-spherical, indented at apex, thin-walled, with sparse, long hairs (310j) . . . . *Caryomyia thompsoni* Felt

The gall illustrated here for this species is the kind Felt (1908) originally described for *C. thompsoni*, although a different kind was subsequently attributed to this gall midge (Felt 1940).

12a Onion-shaped gall, thin-walled, bulbous at base, pointed apically, the hair long, reddish (310b) . . . . *Caryomyia holotricha* (Osten Sacken)
12b Irregularly spherical, thick-walled, occasionally nippled, the hair short-downy, tan to brown (310d) . . . . *Caryomyia persicoides* (Osten Sacken)
13a Hard blister gall, equally apparent on both surfaces of leaf (310c) . . . . *Contarinia* sp.

Blisters are monothalamous. Host: *Carya* spp. Distr.: New York and Maryland.

13b Soft midrib swellings visible from one side of leaf and formed by phylloxerids (310e) . . . . *Contarinia* sp.

This species is evidently an inquiline. Larvae were found in early June; adults have not been reared. Host: *C. illinoensis*. Distr.: Louisiana.

Juglans (black walnut, butternut)

Folded, swollen leaflets .... *Contarinia* sp. and *Dasineura* sp.

Leaflets were evidently attacked when growing rapidly in early spring. Larvae in the Felt collection that were taken from *J. cinerea* in New York belong to *Contarinia*, and larvae in the USNM that were taken from *J. nigra* in Virginia belong to *Dasineura*.

Juncaceae

Juncus (rush)

Swollen stems .... *Procystiphora* spp.

P. coloradensis Felt. Adults were reared in July. Host: "possibly *Carex*"; *Carex* might well be the host, but because other known *Procystiphora* spp. come from *Juncus*, it is likely that *P. coloradensis* does also. Distr.: Colorado.

P. junci Felt. Adults were reared in October. Host: *J. dudleyi*. Distr.: Illinois.

Lamiaceae (Labiatae)

Collinsonia (horse-balm)

1a Swollen, unopened flower .... *Schizomyia* sp.

The greenish-yellow larva of this unnamed species is very active when disturbed. In late September it leaves the gall and drops to the ground to pupate. Host: *C. canadensis*. Distr.: Maryland.

1b Leaf gall 2

2a Onion-shaped gall (318a) .... *Dasineura collinsoniae* (Beutenmüller)

Galls are green, sparsely haired, thin-walled, about 5–7 mm high and 3–5 mm in basal diameter, and usually on the underside of the leaf. The single white larva in each gall breaks out to drop to the soil in early fall. Larval characters place this species in the Oligotrophidi. It is tentatively placed in *Dasineura* until adults are reared. Host: *C. canadensis*. Distr.: New York and Maryland.

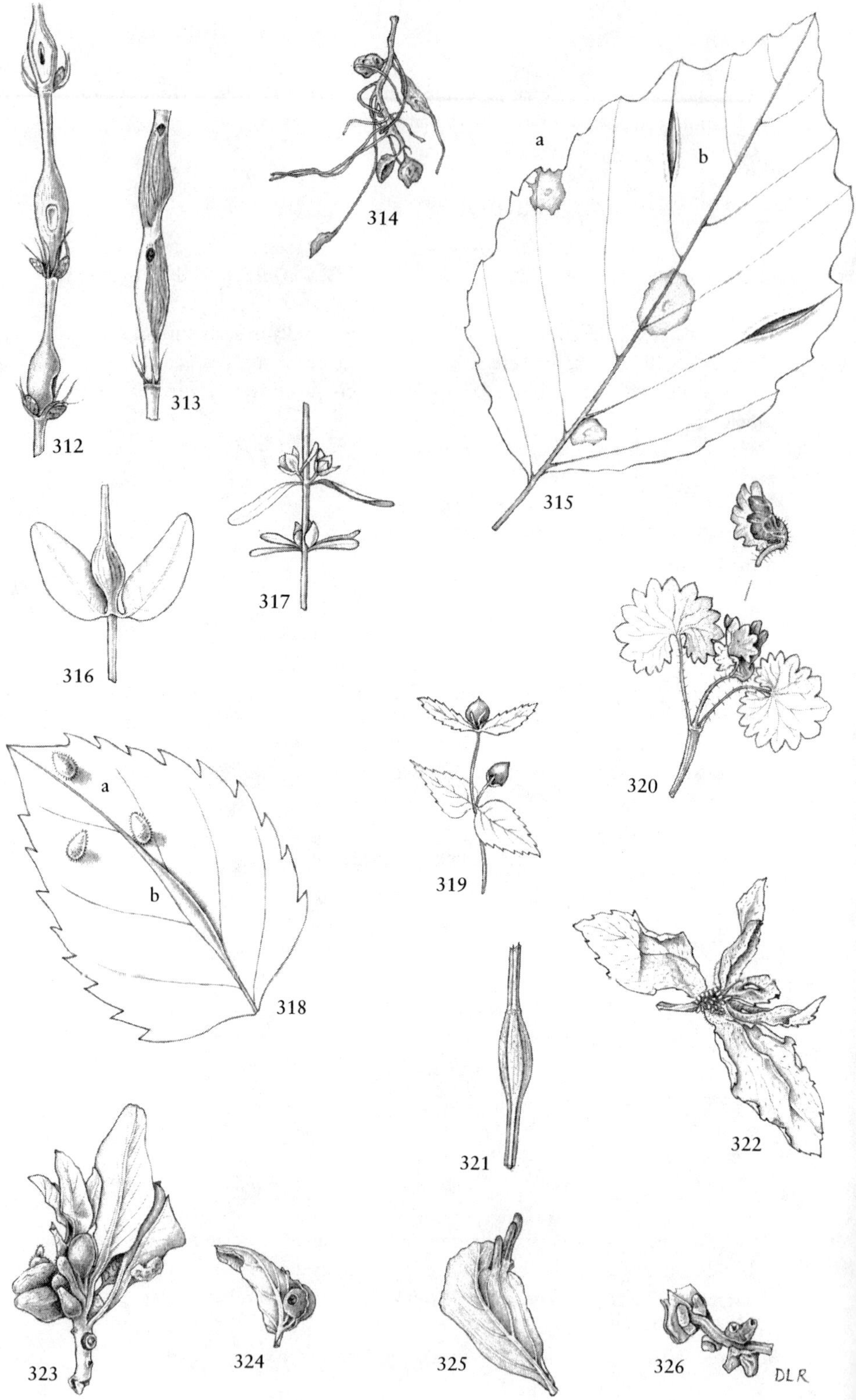
312
313
314
315
a
b
316
317
318
a
b
319
320
321
322
323
324
325
326
DLR

2b Swollen midrib or lateral vein (318b). . . *Lasioptera collinsonifolia* (Beutenmüller)

Galls are green, tapered, and 1–2 cm long. The one to several white larvae in the central tunnel of each gall overwinter in the gall and pupate the following spring. Host: *C. canadensis*. Distr.: New York and Maryland.

Cunila

Swollen bud gall (319) . . . . cecidomyiid

Galls are 2–4 mm in diameter, green at first, becoming bluish-purple. The branch tips are killed. Host: *C. origanoides*. Distr.: Indiana. Ref.: Wells (1921b).

Galeopsis (hemp nettle)

Irregular stem swelling . . . . *Neolasioptera galeopsidis* (Felt)

Galls are spongy, polythalamous, and 1–2 cm long. The orange larvae live in elongate silken chambers. Host: *G. tetrahit*. Distr.: Massachusetts.

Glecoma (ground-ivy)

Folded, swollen leaf base (320) . . . . *Dasineura glechomae* (Kieffer)

One or more white larvae live in an inpocketing at the base of a leaf. Host: *G. hederacea*. Distr.: Maryland; immigrant from Europe.

Hedeoma

Swollen flower bud . . . . cecidomyiid

Figures 312–326. Galls of Gnetaceae, Hamamelidaceae, Hippocrataceae, Hypericaceae, and Lamiaceae. **312,** Smooth, dimpled stem swellings, *Ephedra* sp.: *Lasioptera ephedrae*. **313,** Rough, resinous, stem swellings, *Ephedra* sp.: *Lasioptera ephedricola*. **314,** Root nodules, *Hippocratea volubilis*: *Neolasioptera* sp. **315,** Leaf galls, *Hamamelis virginiana*: **a,** blister gall: *Contarinia* sp.; **b,** vein fold: cecidomyiid (redrawn from Wells 1914). **316,** Stem swelling, *Triadenum virginicum*: *Neolasioptera triadenii*. **317,** Swollen buds, *Hypericum perforatum*: *Zeuxidiplosis giardi* (after photograph in Given and Woods 1964). **318,** Leaf galls, *Collinsonia canadensis*: **a,** onion-shaped galls: *Dasineura collinsoniae*; **b,** swollen vein: *Lasioptera collinsonifolia*. **319,** Swollen buds (redrawn from Wells 1921b), *Cunila origanoides*: cecidomyiid. **320,** Vein fold, *Glecoma hederacea*: *Dasineura glechomae*. **321,** Stem swelling, *Monarda fistulosa*: *Neolasioptera monardi*. **322,** Vein folds, *Mentha arvensis*: *Dasineura aromaticae*. **323–326,** Tubular leaf and stem galls, *Salvia* spp.: *Rhopalomyia* spp. **323,** *S. apiana*: *R. audibertiae*. **324,** *S.* "*nivea*": *R.* sp. **325,** *S.* "*californica*": *R. salviae*. **326,** *S. apiana*: *R. audibertiae*.

The calyx is swollen and nutlets do not develop. Host: *H. pulegioides*. Distr.: Indiana. Ref.: Wells (1921b).

Lycopus (bugleweed)

Swollen stem .... *Neolasioptera* spp.

Neolasioptera lycopi (Felt). Swellings are monothalamous, subglobular, hard, and about 1 cm long. The pale orange larva lives along the central axis and overwinters in the gall. Hosts: *L. "communis," L. uniflorus*. Distr.: New York and Maryland.
Neolasioptera mitchellae (Felt). Swellings are polythalamous and spongier and more elongate than those of *N. lycopi*. Larvae overwinter in the gall. Host: *L. virginicus*. Distr.: Maryland.

Mentha (mint)

1a Swollen stem .... *Neolasioptera menthae* Felt

Galls are polythalamous, pithy, and 8–15 mm long. Larvae are orange and overwinter in the gall. Host: *Mentha* sp. Distr.: Illinois.

1b Swollen bud or crumpled, folded leaf 2
2a Swollen bud .... *Dasineura piperitae* Felt

Galls are reportedly loose and about 5 mm long. Larvae are pale green. Adults were reared in early August. *Giardomyia piperitae* (Felt) also was reared from this kind of gall, but it may have been feeding on fungus. Hosts: *M. gentilis, M. piperita*. Distr.: New York.

2b Crumpled, folded, young leaf (322) .... *Dasineura aromaticae* Felt

Larvae are gregarious within the leaf folds. Infested leaves are usually very hairy. The larvae form cocoons and pupate among the distorted leaflets. *Giardomyia menthae* Felt has also been reared from these galls but may have been feeding on fungi. Hosts: *M. arvensis, M. spicata*. Distr.: Massachusetts and New York.

Monarda (horsemint)

1a Elongate, tapered stem swelling (321) .... *Neolasioptera monardi* (Brodie)

Swellings are 1–3 cm long. Host: *M. fistulosa*. Distr.: Ontario.

1b Ovate outgrowth of root stalk .... cecidomyiid

The gall was reportedly about 2 cm in diameter. Specimens of the gall maker and gall are not available, but the latter has been illustrated elsewhere (Felt 1940). Host: *Monarda* sp. Distr.: Ontario.

Nepeta (catnip)

Stem swelling .... cecidomyiid

Swellings were reportedly about 1 cm long. They are probably formed by a *Neolasioptera*. Host: *N. cataria*. Distr.: unspecified.

Prunella (self-heal)

Leaf pocket or leaf roll .... cecidomyiid

This damage was reported by Maze (1949) with no further details. Host: *P. vulgaris*. Distr.: Pennsylvania.

Pycnanthemum (mountain mint)

Root gall .... cecidomyiid

No further details were given of this damage (Felt 1940). Host: *Pycnanthemum* sp. Distr.: unspecified.

Salvia (sage)

Tubular outgrowth of leaf (323–326) .... *Rhopalomyia* spp.

Sages in western North America evidently support many kinds of galls formed by *Rhopalomyia* spp., and several kinds are represented in the USNM. Galls are monothalamous even though they may form close aggregates. Larvae pupate in the galls, and the adults emerge through a hole at the apex. Two species of gall midges have been described from *Salvia*.

R. audibertiae Felt. Galls are conical, narrowing from base, and the same color as the leaf (323, 326). Adults were reared from the galls in summer. Host: *S. apiana*. Distr.: Baja California and Mexico.

R. salviae Felt. Galls are cylindrical and reddish (325). Host: *S. "californica."* Distr.: California.

Scutellaria (skullcap)

Oval stem swelling .... cecidomyiid

Galls are reportedly about 1 cm long (Felt 1940). They are probably an elongate stem swelling and caused by a *Neolasioptera.* Host: *Scutellaria* sp. Distr.: unspecified.

Teucrium

Curled and deformed leaf .... cecidomyiid

This damage was reported without further details (Maze 1949). Host.: *T. canadense.* Distr.: Pennsylvania.

Lauraceae

Lindera (spice-bush)

1a Rolled leaf or leaf margin................ *Contarinia* sp. and *Dasineura* sp.

Larvae of both genera have been found in the rolled leaves. Galls are noticeable as soon as the leaves are fully expanded. Larvae drop to the ground when full-grown. Infested leaves turn black and drop off shortly afterward. Host: *L. benzoin.* Distr.: New York and Maryland.

1b Irregular stem swelling (330).......... *Neolasioptera linderae* (Beutenmüller)

Galls contain several orange larvae, each in a long chamber. Adults emerge in spring from galls of the previous year. Host: *L. benzoin.* Distr.: New York and Virginia.

Sassafras (sassafras)

1a Curled leaf or leaf margin .... *Dasineura sassafras* Felt

This species is known only from a female reared in August from larvae collected earlier the same year. Host: *S. albidum.* Distr.: Ontario.

1b Slightly convex leaf spots (329) .... *Contarinia* sp.

Spots are about 3 mm in diameter and the same color as the leaf when the larva is present. After the full-grown larvae escape in late spring, the spots turn black. Host: *S. albidum*. Distr.: Pennsylvania and Maryland.

Liliaceae

A number of American Liliaceae have simple leaf spot galls about 8 mm in diameter (327, 328a) made by unidentified cecidomyiids for which the adult stage has not been reared. Each spot contains a single larva that leaves the gall when full-grown. At that time the spot becomes translucent and more noticeable. The gall makers are listed below only as "cecidomyiid." Two other species, *Ledomyia crispata* (Felt) and *Camptoneuromyia rubifolia* Felt, have been indirectly associated with this damage, but they are probably not primary plant feeders. Larvae taken from the galls resemble Alycaulini and Lasiopterini in that they have only four lateral papillae on each side of the thoracic segments, but that is a character that could have been derived separately in other Lasiopteridi.

Allium (onion)

Aborted seed pod .... cecidomyiid

This damage was reported without further details (Maze 1949). Host: *A. canadense*. Distr.: Pennsylvania.

Brodiaea

Swollen stem .... *Lasioptera* sp.

Adults emerged in spring from galls collected the previous fall. Host: *Brodiaea* sp. Distr.: California.

Lilium (lily)

Swollen to normal-appearing stem .... *Neolasioptera* sp.

Orange larvae were found in great numbers in the pith of the upper parts of the stalks and flower peduncles. The infestation caused bud-drop and leaf necrosis of the flowering spike and distorted, ragged flowers below, resulting in heavy losses in some plantings. Adults were reared in August, indicating that this species has more than one generation per year. Felt (1940) identified this as *Neolasioptera hibisci*, which typically comes from stem galls on *Hibiscus* (Malvaceae). The specimens reared from *Lilium* are poorly mounted and not comparable to those from *Hibiscus*, but future rearings should resolve this identification. Hosts: *L. auratum*, *L. canadense*, *L. regale*, *L. speciosum*, *L. tigrinum*. Distr.: New York and Wisconsin. Ref.: Imle and Hartzell (1939).

Maianthemum (lily-of-the-valley)

1a Swollen root. . . . . *Dasineura torontoensis* Felt

Swellings are gradual, tapering at each end, monothalamous, and about 8 mm long. Adults were reared in May, probably from galls held over the winter. Host: *M. canadense*. Distr.: Ontario.

1b Leaf spot . . . . cecidomyiid

See note under Liliaceae. Host: *M. canadense*. Distr.: Massachusetts.

Polygonatum (Solomon's seal)

Leaf spot . . . . cecidomyiid

See note under Liliaceae. Host: *P. biflorum*. Distr.: Massachusetts and Maryland.

Smilacina (false Solomon's seal)

1a Deformed berry . . . . *Asphondylia smilacinae* Felt

Adults emerged from berries in late September to early October. Host: *S. racemosa*. Distr.: Maryland and Virginia.

1b Swollen root or leaf spot 2

2a Swollen root. . . . . *Dasineura smilacinae* Bishop

Swellings are reportedly about 8 mm in diameter. Larvae apparently overwinter in the galls. Host: *S. racemosa*. Distr.: New York, Ontario.

2b Leaf spot (327) . . . . cecidomyiid

See note under Liliaceae. Host: *S. racemosa*. Distr.: Maryland.

Smilax (greenbrier)

1a Leaf spot (328a) .... cecidomyiid

See note under Liliaceae. The leaf spot on this plant has been attributed to *Camptoneuromyia rubifolia* Felt (Felt 1918a), but that species is probably only an inquiline. Larvae taken from galls have fewer papillae than found on *Camptoneuromyia* larvae. Hosts: *Smilax* spp. Distr.: Massachusetts to Maryland.

1b Reddened, marginal leaf roll (328b) .... *Dasineura smilacifolia* Felt

Rolled margins are commonly seen in early spring. Each contains one to several larvae. Hosts: *Smilax* spp. Distr.: Massachusetts and Maryland.

Uvularia (bellwort)

Leaf spot .... cecidomyiid

See note under Liliaceae. This gall was attributed to *Ledomyia crispata* (Felt), which was reared from a container with leaves of this host (Felt 1915a). Larvae normally found in *Uvularia* leaf spots do not belong to *Ledomyia*. Host: *U. sessilifolia*. Distr.: Massachusetts and Maryland.

Yucca (yucca)

Larva in pod .... *Ledomyia yuccae* (Felt)

Although reared from pods, this species is presumably a secondary feeder and is possibly not specific to yucca. It may have been feeding on decaying plant tissue. A predaceous cecidomyiid, *Lestodiplosis yuccae* Felt, was also reared from the pods at the same time. Host: *Y. angustifolia*. Distr.: unspecified, presumably southwestern United States.

Linaceae

Linum (flax)

Swelling on branch of flower head (334) .... *Neolasioptera* sp.

Galls were collected in November with apparently diapausing larvae. Host: *L. striatum*. Distr.: New York.

Loasaceae

Mentzelia

Enlarged flower bud .... *Asphondylia mentzeliae* Cockerell

Adults were reared in August. Host: *M. multiflora*. Distr.: New Mexico.

Lobeliaceae

Lobelia (lobelia)

Elongate swollen stem (333) .... *Neolasioptera* sp.

The gall maker is only tentatively referred to *Neolasioptera*. Host: *L. puberula*. Distr.: North Carolina. Ref.: Wells (1921a).

Loganiaceae

Buddleja

Enlarged flower bud .... *Asphondylia buddleia* Felt

Aborted flower buds are about twice normal size. The corolla fails to open, and one larva occurs in each affected flower. Adults were reared in summer. Host: *B. racemosa*. Distr.: Texas. Ref.: Painter (1935).

Lythraceae

Decodon (water willow)

Thickened, overlapping bud leaves .... cecidomyiid

Infestations involve the 2–3 uppermost bud leaves on both terminal and lateral buds. Swollen buds are 3–5 mm long. Each contains a single larva. Host: *D. verticillatus*. Distr.: Connecticut. Ref.: Wells (1914).

Figures 327–337. Galls of Lauraceae, Liliaceae, Linaceae, Lobeliaceae, Magnoliaceae, Myricaceae, Myrtaceae, and Nyctaginaceae. **327,** Leaf spots, *Smilacina racemosa*: cecidomyiid. **328,** Leaf galls, *Smilax* sp.: **a,** leaf spot: cecidomyiid; **b,** marginal leaf roll: *Dasineura smilacifolia*. **329,** Leaf spots, *Sassafras albidum*: **a,** with active larvae: *Contarinia* sp.; **b,** necrotic spots after larvae have escaped. **330,** Swollen stem, *Lindera benzion*: *Neolasioptera linderae*. **331,** Leaf rolls, *Myrica asplenifolia*: *Janetiella asplenifolia*. **332,** Swollen bud and rolled leaves, *Myrica cerifera*: *Prodiplosis myricae*. **333,** Swollen stem, *Lobelia puberula* (redrawn from Wells 1921a): *Neolasioptera* sp. **334,** Swollen stems, *Linum striatum*: *Neolasioptera* sp. **335,** Leaf galls (lower leaf surface), *Liriodendron tulipifera*: **a,** leaf spots: *Resseliella liriodendri*; **b,** swollen veins: *Resseliella tulipiferae*. **336,** Leaf galls, *Eugenia axillaris*: **a,** fuzzy, cylindrical galls: *Stephomyia eugeniae*; **b,** smooth conical galls: *Stephomyia* sp. **337,** Woody leaf galls, *Pisonia* "*subcordata*": *Bruggmannia pisonifolia*.

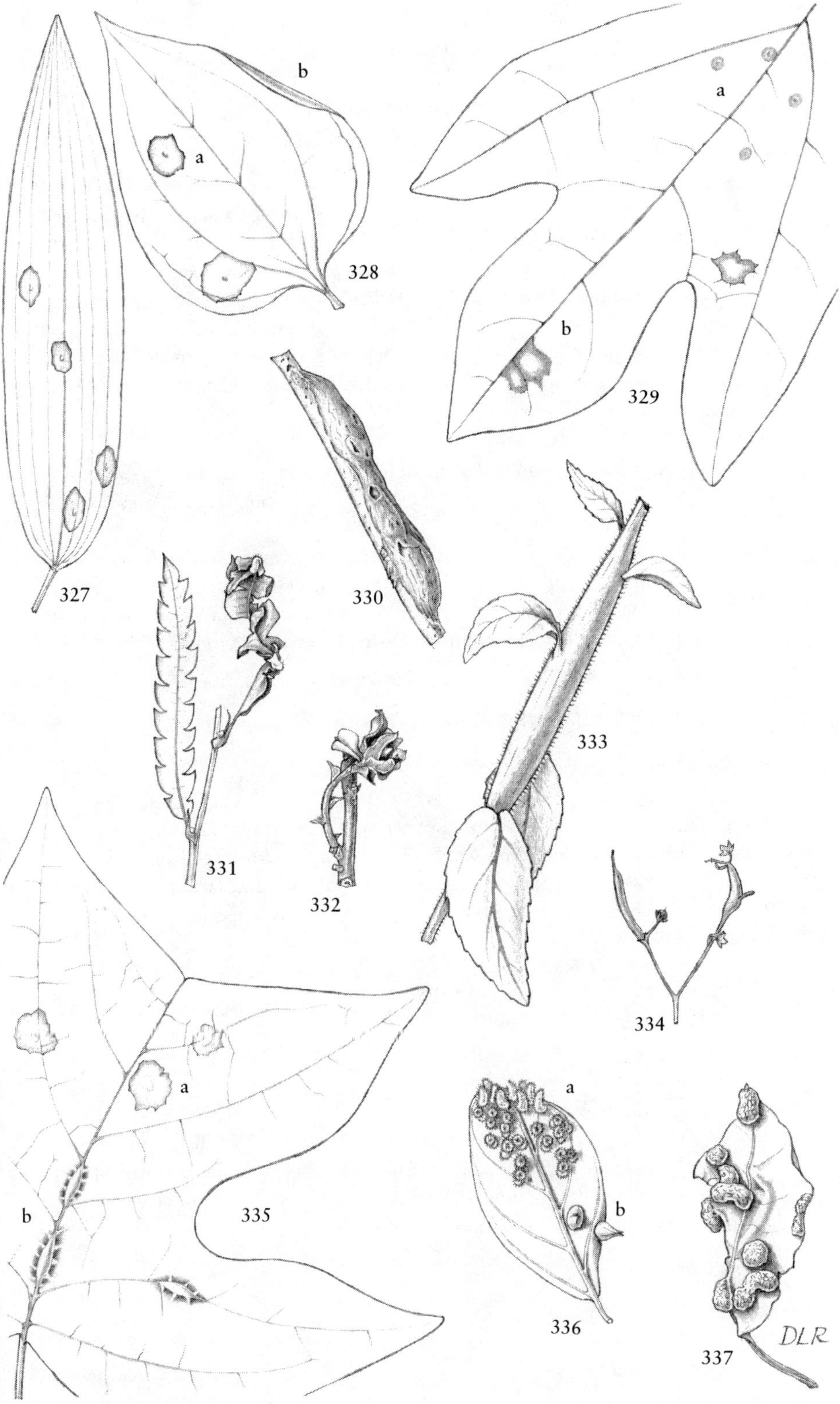
327
328
a
b
329
a
b
330
331
332
333
334
335
a
b
336
a
b
337
DLR

Magnoliaceae

Liriodendron (tulip tree, yellow poplar)

1a Green, yellow, or black leaf spot, sometimes with discoloration beyond gall circumference (335a) .... *Resseliella liriodendri* (Osten Sacken)

Galls are monothalamous. Full-grown larvae leave through a slit along the edge of the spot and drop to the soil. The spots turn black after the larvae leave. Several generations occur per year. Galls appear in late May in Maryland and can be seen on new leaves all summer long. Host: *L. tulipifera*. Distr.: Massachusetts to North Carolina.

1b Swollen midrib or vein on underside of leaf (335b) .... *Resseliella tulipiferae* (Osten Sacken)

Swellings may contain one to several larvae. These can be seen in Maryland by early June and appear through the summer on new leaves. Probably several generations occur per year. Full-grown larvae cut an exit hole and drop to the ground to pupate. Host: *L. tulipifera*. Distr.: Ontario to Virginia.

Magnolia (magnolia)

Leaf spot, 4.5 mm diameter .... *Contarinia* sp.

Only larvae are known of this undescribed species. Leaf spots are similar to those formed on *Liriodendron tulipifera* by *Resseliella liriodendri* and show the same exit slit. In Maryland, larvae are found in mid-July and later, so the species apparently has several generations per year. Host: *M. virginiana*. Distr. Maryland and Florida.

Malvaceae

Abutilon

Elongate, aborted flower .... *Asphondylia abutilon* Felt

Infested flowers are about 7 mm long. They remain closed, and the internal parts do not develop. Galls are monothalamous. Adults emerge from the galls in late summer and fall. Host: *A. incanum*. Distr.: Texas.

Gossypium (cotton)

1a Leaf midrib fold or leaf roll .... cecidomyiid

Larvae are gregarious and deform the young expanding leaves. Specimens of the gall maker are not available, but it is possible they belong to one of the species listed in entry 1b. Host: *G. thurberi*. Distr.: Arizona. Ref.: Pierce and Morrill (1914).

1b Swollen flower bud. . . *Contarinia gossypii* Felt and *Prodiplosis longifila* Gagné

C. gossypii. This species, known otherwise from the West Indies, was reported as very common in Florida (Rainwater 1934), but specimens from that record are not available. Flowers reportedly remain closed and rot follows, but it is possible the Florida damage was caused by *P. longifila*. Host: *Gossypium* sp. Distr.: ?Florida; West Indies.

P. longifila. This species was reared from wild cotton buds in 1967 and may possibly be the same species reported earlier from Florida as *C. gossypii*. It is multivoltine and has extensive geographical and host ranges. Hosts: *Gossypium* spp.; also *Citrus* (q.v.); in South America, various Solanaceae (including potato), Chenopodiaceae, Fabaceae, and Euphorbiaceae. Distr.: Florida; also Colombia and Peru.

Hibiscus (rose-mallow)

Swollen stem . . . . *Neolasioptera hibisci* (Felt)

Galls are irregular, ridged, and 2–3 cm long. Larvae develop in tunnels, mainly in the pith, overwinter in the stalks, and pupate there in spring. Host: *H. moscheutos*. Distr.: New York, North Carolina, and Louisiana.

Kosteletzkya (saltmarsh mallow)

Tapered stem swelling . . . . *Neolasioptera* sp.

Adults were reared in July in Florida, so this species probably has more than one generation per year. Hosts: *K. virginica*, *K.* sp. Distr.: Florida and Texas.

Sphaeralcea

Larva in bud . . . . *Contarinia* sp.

Adults were reared from buds in spring, but the damage was not specified further. Hosts: *S. coccinea*, *S. emoryi*. Distr.: Texas and Arizona.

Moraceae

Ficus (fig)

Enlarged seed capsules . . . . *Ficiomyia* spp.

F. birdi Felt. The monothalamous, enlarged seed capsules are five times larger than normal. Pupation occurs in the fruit. Adults were reared in February. Host: *F. carica*. Distr.: Florida.

F. perarticulata Felt. This species was reared from fruit. No further details about the damage were reported. Host: *F. aurea*. Distr.: Florida.

Myricaceae

Myrica

1a Swollen bud and rolled leaf (332) . . . . *Prodiplosis myricae* (Beutenmüller)

Larvae are gregarious and drop to the ground when full-grown. Adults emerged in early August, so the species may have at least two generations per year. Host: *M. cerifera*. Distr.: New Jersey.

1b Swollen vein, causing severe leaf distortion (331) . . . *Janetiella asplenifolia* (Felt)

Midveins are folded, fleshy, reddish, and cause extreme folding of the leaf. Larvae are gregarious. Host: *M. asplenifolia*. Distr.: Massachusetts.

Myrtaceae

Eugenia

1a Deformed fruit . . . . *Dasineura eugeniae* Felt

Adults were reared in March and August from deformed fruit, 5–7 mm in diameter, bearing separate oval cells. Host: *E. foetida*. Distr.: Florida.

1b Complex leaf gall 2
2a Fuzzy leaf gall (336a) . . . . *Stephomyia eugeniae* (Felt)

Galls are monothalamous, boot-shaped, and covered with white hairs. Similarly shaped, but smooth galls are in the USNM collection; it is possible they are made by a different species of *Stephomyia*. Pupation occurs in the gall, and pupae break out of the toe of the boot. In aggregate the galls sometimes cover the entire surface of a leaf. Adults were reared from the galls in March. Hosts: *E. axillaris*, *E. foetida*. Distr.: Florida.

2b Conical, smooth leaf gall (336b) .... *Stephomyia* sp.

Hosts: *E. axillaris*, *E. foetida*. Distr.: Florida.

Nyctaginaceae

Allionia

Swollen stem .... *Neolasioptera allioniae* (Felt)

Swellings are polythalamous, tapered, 1.5 to 2 cm long, and usually near a node. Larvae overwinter in the stem and pupate in spring. Hosts: *A. nyctaginae* and *Allionia* sp. Distr.: Colorado and Texas. Ref.: Painter (1935).

Boerhavia

Swollen seed .... *Asphondylia boerhaaviae* Möhn

Host: *Boerhavia* sp. Distr.: Mexico (unspecified further) and El Salvador.

Mirabilis

Stem swelling .... *Neolasioptera* sp.

Host: *M. linearis*. Distr.: Arizona.

Pisonia

In addition to the species below, *Bruggmannia mexicana* (Felt), known from a single male specimen, was possibly reared from *Pisonia aculeata* (the type is labeled "*Smilax*") in Tamaulipas, Mexico. The damage was characterized as a "subconic stem gall" (Felt 1911b).

Hard, woody, convex leaf gall (337) .... *Bruggmannia pisonifolia* (Felt)

Larvae pupate in the galls. Hosts: *P.* "*nigricans*," *P.* "*subcordata*." Distr.: Florida and St. Vincent, West Indies.

Oleaceae

Forestiera

Complex, cylindrical, hairy leaf gall .... cecidomyiid

Galls are 2–3 mm high, 1 mm in diameter, monothalamous, and appear in clusters. Adults emerged from the galls in March. This undescribed species belongs to the supertribe Oligotrophidi. Host: *F. segregata*. Distr.: Florida.

Fraxinus (ash)

1a Swollen seed .... *Contarinia* sp.

Larvae are gregarious in the slightly swollen exocarp. Adults were reared the spring following the collection of larvae in mid-July. The damage is similar to that of the European *Contarinia marchali* Kieffer on a European *Fraxinus*. Host: *F. pennsylvanica*. Distr.: Ontario.

1b Leaflet roll or fold... 2

2a Folded leaflets showing only a slight swelling along midrib (339) .... *Dasineura apicata* Felt and *D. fraxinifolia* (Felt)

These are distinct species of *Dasineura*, but larvae have not been definitely associated with one or the other name. Gregarious larvae develop on very young leaflets and drop to the ground when full-grown. Probably only one generation occurs per year. Another cecidomyiid, *Rhizomyia fraxinifolia* Felt, was also reared from similar damage but it is probably not a primary plant-feeder. Host: *Fraxinus* spp. Distr.: New York and Pennsylvania.

2b Succulent midrib or petiole swelling or hemispherical leaflet gall 3

3a Succulent midrib or petiole swelling opening on the upper surface of the leaflet (338a, color plate 1D–F) .... *Contarinia thalictri* (Felt) and *Dasineura tumidosae* Felt

C. thalictri. These larvae are less commonly found in these galls, so may be inquilinous. They are also white and gregarious and have been reared in long series. Males are remarkable in that they have antennae shaped as for the female. Another species of this genus, *Contarinia canadensis* Felt, has been reported from these galls (Felt 1940), but there is no evidence that the specimens *C. canadensis* is based on were reared from ash. Hosts: *Fraxinus* spp. Distr.: Ontario to Maryland.

D. tumidosae. Larvae are gregarious, white, and drop to the ground when full-grown, around the first week of June in Maryland. Leaflet swellings are barely noticeable from the top of the leaflet but are large and pendant below. Galls are sometimes reddish. Larvae overwinter in the soil. Hosts: *Fraxinus* spp. Distr.: New England to Michigan, south to Virginia, and California.

3b Globular or bullet-shaped swelling on upperside of leaflets along veins, opening on lower surface of the leaflet (338b) .... *Dasineura pellex* (Osten Sacken)

Galls may be green, purple, or black. The white larvae drop to the soil in about the second week of June in Maryland and overwinter there. Adults have not been reared. *Neolasioptera fraxinifolia* Felt was recorded once from this gall, but it may have been a contaminant in the rearing container. The larvae in the galls have a full complement of lateral papillae, unlike larvae of *Neolasioptera*. *Neolasioptera* spp. usually are associated with woody twig or petiole swellings. Host: *Fraxinus* sp. Distr.: throughout northeastern North America.

Onagraceae

Epilobium (fireweed)

Rolled leaves of young bud .... *Contarinia zauschneriae* (Felt)

Larvae reportedly live in the center of the young folded or curled leaves. Adults were reared in December. Hosts: *E. canum*. Distr.: California.

Ludwigia (water primrose)

Variably swollen and distorted pod .... *Asphondylia* sp.

This is an undescribed species. Adults emerged from the galls in November and December. Host: *Ludwigia* sp. Distr.: Florida.

Oenothera (evening primrose)

Rolled leaves of young bud .... *Contarinia zauschneriae* (Felt)

The species is based on specimens from California; the New Jersey record is only tentatively placed here. Larvae were found in midsummer at the bases of the leaves of the apical rosette. Hosts: *O. biennis*, *O. californica*. Distr.: New Jersey and California. Ref.: Dickerson and Weiss (1920).

Orchidaceae

Cattleya and Other Tropical Orchids

Root nodules (367) ...... *Neolasioptera* sp. and *Clinodiplosis cattleyae* (Molliard)

Neolasioptera larvae are usually found in orchid root swellings, so are evidently the gall makers. *Clinodiplosis cattleyae* may be specific to orchid because it has been reared from the nodules several times, but it could be an inquiline. Hosts: *Cattleya* spp. and other orchids. Distr.: tropical America; often intercepted at U.S. ports of entry.

Hookera

Flower stalk swelling .... cecidomyiid

Swellings were reportedly about 4 cm long and 1.5 cm in diameter. The host was given as "*Hookera* or *Triteleia* (*brodiaea*)" (Felt 1940). *Triteleia* and/or *Brodiaea* are Liliaceae, so the identification of this host was probably based only on vegetative parts. Host: ?*Hookera*. Distr.: unspecified.

Papaveraceae

Papaver (poppy)

Larva in base of stem .... *Karshomyia caulicola* (Coquillett)

This species probably feeds in decaying plant tissue. It has been found also in several localities in Europe, and in The Netherlands it feeds on molds growing in the stems of various commodities (Nijveldt 1985). It is the only species of the large genus *Karshomyia* known to occur in both North America and Europe, so it could be an immigrant. Host: *Papaver* sp. Distr.: New Hampshire; Europe.

Figures 338–345. Galls of Oleaceae and Pinaceae. **338,** Leaf galls, *Fraxinus americana*: **a,** swollen midribs, one opened to show larvae: *Dasineura tumidosae*; **b,** pellet galls: *Dasineura pellex*. **339,** Folded leaflets, *Fraxinus americana*: *Dasineura apicata*. **340,** Needle swellings, *Abies balsamea*: *Paradiplosis tumifex*. **341,** Cone scale swellings, *Picea glauca*: *Kaltenbachiola canadensis*. **342,** Twig swellings, *Picea glauca*: "*Mayetiola*" *piceae*. **343,** Swollen needles, with cross section in 2× to show larvae, *Pinus virginiana*: *Thecodiplosis brachynteroides*. **344,** Swollen base of needle, with sheath removed in 2× to show larvae, *Pinus resinosa*: *Thecodiplosis piniresinosae*. **345,** Brown-spotted needles, with larva on needle in 2×, *Pinus elliottii*: *Contarinia acuta*.

a
a
b
338
339
340
341
342
343
344
345
DLR

Phytolaccaceae

Rivina

Enlarged, unopened bud .... *Schizomyia rivinae* Felt

Infested buds are reportedly globular, downy, and mostly green, with whitish or rose tips. Adults were reared in June from larvae that left the galls the previous October. Host: *R. humilis*. Distr.: Florida.

Pinaceae

Abies (fir)

1a Larva in swelling of needle (340) .... *Dasineura balsamicola* (Lintner) and *Paradiplosis tumifex* Gagné

D. balsamicola. This species is an inquiline. Females lay eggs on needles in spring shortly after *P. tumifex* has laid its eggs but before galls have formed. One or more larvae of *D. balsamicola* can be found in each gall, where they grow faster than the gall maker, which eventually dies. The larvae of the inquiline and their needles drop to the duff a little later in the fall than those of *P. tumifex*. Pupation takes place the following spring. Host: *A. balsamea*. Distr.: Canadian Maritime Provinces to Wisconsin. Refs.: Akar and Osgood (1987), Bergdahl and Mazzola (1985), Osgood and Gagné (1978), Shorthouse and West (1986).
P. tumifex. This species is responsible for initiating the gall. Females lay their eggs in spring when needles are elongating. The newly hatched larva crawls to a spot on the needle, settles, and begins to feed. The needle then grows around the larva. Only one larva of this species is found in a gall. Larvae vacate the galls in late summer and drop to the duff, where they pupate the following spring. Affected needles die and dehisce soon after the larvae leave. Hosts: *A. balsamea*, *A. fraseri*, *A. concolor*. (Adults have not been reared from larvae from *A. concolor*. The species could be *P. tumifex* or a close relative). Distr.: Canadian Maritime Provinces to Wisconsin, south along Applachians to North Carolina, and California (the *A. concolor* record). Refs.: Akar and Osgood (1987), Bergdahl and Mazzola (1985), Osgood and Gagné (1978), West and Shorthouse (1982).

1b Larva in cone... 2
2a Larva between cone scales.. 3
2b Larva inside seed or cone scale.................................. 4
3a Larva orange, warty (84), with spatula .... *Resseliella conicola* (Foote) and *Resseliella coryloides* (Foote)

The "scale midge" of Hedlin (1967) on *Abies grandis* in British Columbia is one of these two species, which cannot be differentiated from larvae only. Larvae feed along the surface of the cone scales and cause discolored necrotic areas. Larvae appear singly or gregariously and are present throughout the summer. Hosts: *A. concolor*, *A. grandis*, *A. magnifica*, and possibly other conifers. Distr.: British Columbia to California. Ref.: Hedlin (1967).

3b Larva white, smooth, without spatula (27). *Asynapta* spp.

A. hopkinsi Felt. Larvae are often gregarious and probably feed on fungi growing on already damaged cones. Several generations may occur per year. Hosts: *A. concolor*, *A. grandis*, *A. magnifica*, and also other conifers. Distr.: western and southern United States and western Canada.
A. strobilophila Foote. Remarks for *A. hopkinsi* apply for this species also. Hosts: *A. concolor*, *A. magnifica*. Distr.: Oregon and California.

4a Larva in scale swelling. . . . . *Dasineura abiesemia* Foote

This is presumably the species reported in Hedlin (1967) as the "cone scale midge." Galls are formed in the cone scales, usually adjacent to the seeds. Usually one larva occurs per swelling, but occasionally two occur, separated by a thin wall. Larvae overwinter in their cells and pupate there in spring. Cone scales sometimes fuse to the seeds, resulting in seed loss. An unidentified *Contarinia* has been reared from scales of *A. lasiocarpa* in British Columbia, but its particular niche is unknown. Hosts: *A. concolor*, *A. grandis*, *A. magnifica*, *A. shastensis*. Distr.: Oregon and California. Ref.: Hedlin (1967).

4b Larva in seed . . . . *Dasineura* sp.

This is an undescribed species, referred to in Hedlin (1967) as the "seed midge." One larva occurs per infested seed. Pupation occurs in the seed the following spring. Nearly all infected seeds are aborted. Hosts: *A. grandis*, *A. lasiocarpa*. Distr.: British Columbia. Ref.: Hedlin (1967).

Larix (larch)

Larva in seeds and between scales . . . . *Resseliella* sp.

Host: *L. laricina*. Distr.: Maine.

Picea (spruce)

1a Larva in cone. 2
1b Larva in bud, swelling at base of needles, or resin droplet on branches 7

2a Larva in seed 3
2b Larva in other part of cone 4
3a Seed deformed, discolored . . . . "*Mayetiola*" *carpophaga* (Tripp)

Eggs are laid in spring on the developing cones. Larvae settle in the ovules, which swell to larger than normal size and grow irregularly. The pink larvae are full-grown by summer and overwinter in a cocoon in the seed. Pupation takes place in April in Ontario, and adults appear in early May. Host: *P. glauca*. Distr.: Ontario. Ref.: Tripp (1955).

3b Seed apparently normal. . . . . *Plemeliella* sp.

Seeds must be cut open to find the orange larvae. This species is known only from larvae and is tentatively placed in *Plemeliella*. In biology and general appearance the larvae resemble *Plemeliella abietina* Seitner of spruce seeds in Europe. Host: *P. glauca*. Distr.: Ontario. Ref.: Tripp (1955, as "Species A").

4a Larva in cone scale or rachis 5
4b Larva free between scales and seed wings 6
5a Larva in tunnel in rachis . . . . *Kaltenbachiola rachiphaga* (Tripp)

The yellowish-orange larvae live gregariously in pockets and sometimes fill the rachis. In midsummer the larvae line the tunnel with silk and form cocoons. Pupation occurs the following spring, and adults emerge in mid-May. Host: *P. glauca*. Distr.: Quebec and Ontario. Ref.: Tripp (1955).

5b Larva in swelling of cone scale (341) . . . . *Kaltenbachiola canadensis* (Felt)

Larvae occur singly in swellings that are occasionally numerous enough to give the cone a warty appearance. The orange larvae are full-grown by mid-July when they form a cocoon in place. Pupation occurs the following spring and adults emerge around mid-May. Host: *P. glauca*. Distr.: Ontario. Ref.: Tripp (1955).

6a Larva orange-red, warty, with clove-shaped spatula (84) . . . . *Resseliella* sp.

These larvae may be secondary feeders in cones but can be fairly numerous. They drop to the ground when full-grown and several generations may occur during the summer. Similar species occur in many other conifers. Host: *P. glauca*. Distr.: Ontario. Ref.: Tripp (1955, as "Species B").

6b Larva white, smooth, without spatula (27) . . . . *Asynapta hopkinsi* Felt

These larvae may be only secondary feeders between the cone scales but may occur in great numbers. Hosts: *P. englemannii*; also, many other conifers. Distr.: widespread in North America.

7a Larva in droplet or flow of resin .... *Cecidomyia* sp.

Larvae of this undescribed species live completely submerged in resin except for the hind spiracles. Hosts: *P. glauca*, *P. pungens*. Distr.: Wyoming, Colorado, and British Columbia. Ref.: Gagné (1978b).

7b Larva in swelling at tip of buds or on twig............................ 8
8a Malformed, usually terminal bud................. *Dasineura swainei* (Felt)

Eggs are laid singly or in clusters between basal scales or needles of elongating shoots in spring. Larvae crawl to the tips of the shoots and become embedded in individual chambers. The pink full-grown larvae overwinter in the galls and pupate the following spring. Bud growth is arrested and lateral shoots elongate to cause the development of multiple leaders and subsequent bushy growth. Hosts: *P. abies*, *P. glauca*, *P. mariana*, *P. pungens*, *P. rubens*. Distr.: Canadian Maritime Provinces to Alberta. Refs.: Cerezke (1972), Clark (1952).

8b Swollen twig (342) .... "*Mayetiola*" *piceae* (Felt)

Eggs are laid at the base of new needles. The larvae are always in individual chambers. Full-grown larvae are reddish orange. Adults emerge from the galls in late May. Severe infestations cause twigs to bend and twist. Many needles drop off, and the twigs sometimes are killed. Felt (1940) indicated that *Mayetiola tsugae* Felt caused this damage, but that species is known from a specimen caught in flight and is not the same as "*M.*" *piceae*. Hosts: *P. glauca*, *P. rubens*. Distr.: Canadian Maritime Provices south to Connecticut and west to Manitoba. Ref.: Smith (1952).

Pinus (pine)

1a Larva in cone.. 2
1b Larva in or on needles and branches 3
2a Larva immersed in resin between cone scales, the cone becoming deformed and sealed closed (346, 347)...................... *Cecidomyia bisetosa* Gagné

Eggs are laid on developing cones as early as January in Florida. Larvae enter and develop in new cones, where their feeding causes resin to flow. Affected cone scales grow much larger than normal size. Cones are deformed and sometimes killed. This species has one generation per year, and larvae pupate in the cones in December and January. All other species of *Cecidomyia* live in resin on branches, twigs, and needles of pines and a few other conifers. Host: *P. elliottii*. Distr.: Florida. Refs.: Gagné (1978c), Williams and Fatzinger (1977).

2b Larva free between cone scales, often associated with secondary necrosis .... *Asynapta* spp., *Resseliella* spp., and other, mycophagous species

Asynapta spp. Included here are at least two species, *Asynapta hopkinsi* Felt and the less common *Asynapta strobilophila* (Foote). The white larvae are often found in great numbers in cones, where they may be feeding on diseased tissue. Full-grown larvae drop to the ground to pupate. Hosts: *Pinus* spp.; possibly other conifers also. Distr.: widespread across North America. Ref.: Franklin and Coulson (1968).

Resseliella spp. As many as 100 orange larvae may live in a cone and reportedly stimulate resin flow that sticks the cone scales together. Larvae pupate in the cone. Evidently they cause no deformation and are often found in association with *Asynapta* spp. Hosts: *Pinus* spp.; possibly other conifers also. Distr.: widespread across North America. Ref.: Franklin and Coulson (1968).

Mycophagous species. *Lobodiplosis triangularis* Felt, *Mycodiplosis thoracica* (Fitch), and *Hyperdiplosis recurvata* (Felt) were reared from pine cones in southeastern United States. *Lestodiplosis* spp., predators of mites and other arthropods, have often been reared from pine cones.

3a Larva inside fascicle, usually causing some deformation or bending of the needles 4

3b Larva immersed in droplet of resin at base of fascicle or in resin-filled pit or swelling on branches (348, 351)........ 16

4a Needles usually of normal length, sometimes bent or discolored, occasionally slightly swollen at base, larva usually free on needle or in open pocket...... 5

4b Needles much shortened, greatly enlarged at base, each larva usually enclosed in cell in needle 8

5a Larva embedded in slightly swollen needle bases (344) .... *Thecodiplosis piniresinosae* Kearby

Eggs are laid singly or in clusters on elongating shoots in spring. One to 11 larvae develop in individual pockets at the base of a fascicle, between or around the needles. In Wisconsin, larvae mature by fall, drop to the soil, and pupate in spring. Shortly after the larvae exit, the affected needles turn brown and drop. In heavy infestations, severe needle drop can cause branch mortality. Host: *P. resinosa*. Distr.: New York and Ontario to Minnesota. Ref.: Kearby and Benjamin (1964).

5b Larva not embedded in needle tissue 6

6a On white pine, *Pinus strobus*........ *Resseliella pinifoliae* (Felt)

Larvae reportedly live free at the base of the needle bundles, where they cause needle browning and drop in autumn. Larvae drop to the soil in autumn. This species has not been reported since its description in 1939, but there are some larvae presumably of this species in the USNM that were collected in Ontario in 1971. Similar damage to Scotch pine, *Pinus sylvestris*, was also attributed to this species, but specimens were not preserved. Host: *P. strobus*. Distr.: Connecticut and Ontario.

6b On red pines (345) 7
7a On red pine, *Pinus resinosae*, and Scotch pine, *P. sylvestris* . . . . *Contarinia baeri* (Prell)

Eggs are laid in spring at needle bases of new shoots. The yellow-orange larvae settle at the base of the needles and cause an elliptical yellowish area 2–6 mm long, which becomes brown after the larva drops to the ground toward the end of summer. The opposite surface of the needle continues its growth so that the needle becomes bent. The larva diapauses in the soil and pupates the following spring. This species occurs also in Europe on *Pinus sylvestris*. It may be an immigrant. The earliest North American record is from New Brunswick in 1953. Hosts: *P. resinosa*, *P. sylvestris*. Distr.: New Brunswick to Ontario south to Pennsylvania and Michigan. Ref.: Skuhravý (1973), Wilson et al. (1987).

7b On slash pine, *Pinus elliottii*, and loblolly pine, *Pinus taeda* (345) . . . . *Contarinia* spp.

C. acuta Gagné. Larvae feed at points along the surface of developing needles that are still within the sheath. Characteristic, 4 mm long, brown lesions are formed at the feeding sites. After the full-grown larvae leave, the needles frequently break off at the lesion. Pupation occurs in the surface litter on the ground. This species may have as many as six generations per year. Hosts: *P. elliottii*, *P. taeda*. Distr.: Florida and Mississippi. Ref.: Gagné and Beavers (1984).

Contarinia spp. Three unidentified species of this genus were noticed on new shoots of slash pine in Florida, but their feeding sites were not determined. Host: *P. elliottii*. Distr.: Florida. Ref.: Gagné and Beavers (1984).

8a In eastern North America 9
8b In western North America. 10
9a On Virginia pine, *Pinus virginiana* (343, color plate 3A, B) . . . . *Thecodiplosis brachynteroides* (Osten Sacken)

Infested needles are much shortened and usually coalesced basally. The gall is purplish, and the two needles growing out of it usually diverge. One to several reddish-orange larvae become embedded in the swollen bases of the leaves. Full-grown larvae crawl out through the top of the gall and drop to the soil in early to late October. There is one generation a year, with pupation and adult flight occurring in early spring. This species may possibly be the same as *T. pinirigidae* on pitch pine because the two species cannot be separated on anatomical characters. Host: *P. virginiana*. Distr.: Maryland to Virginia.

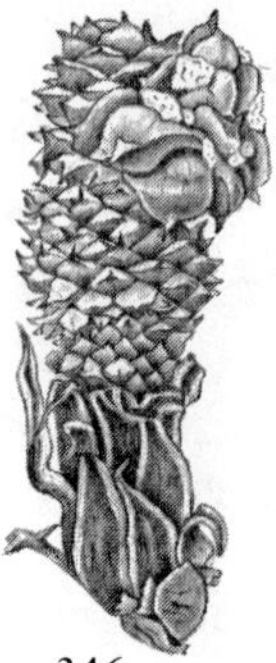

346

347

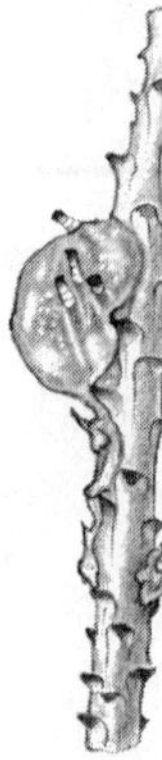

348

349

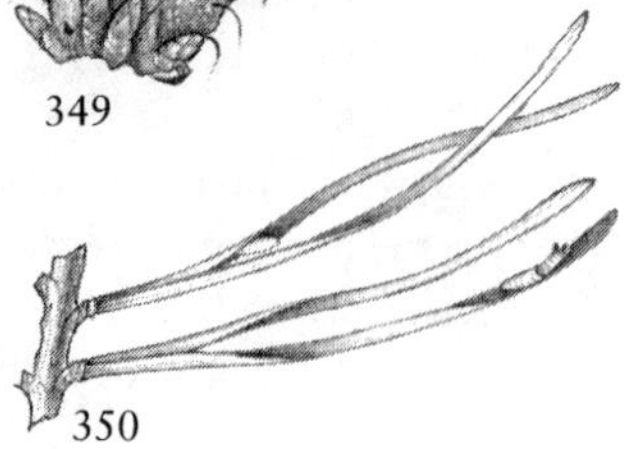

350

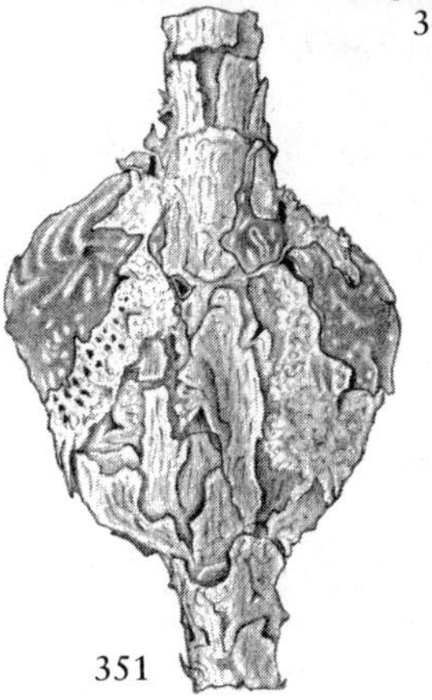

351

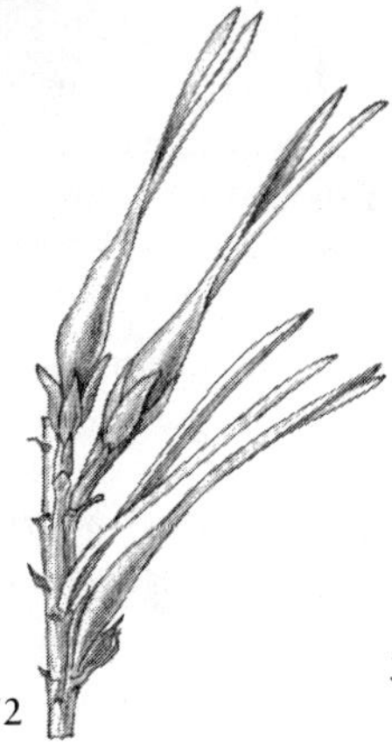

352

353

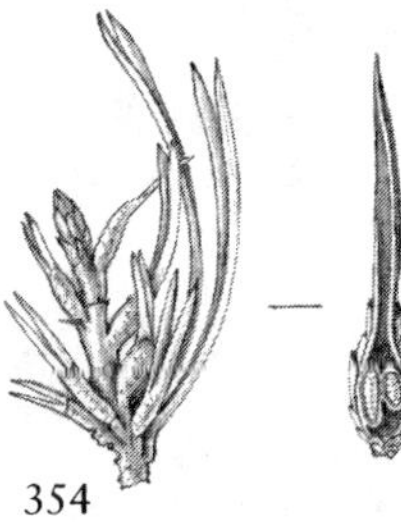

354

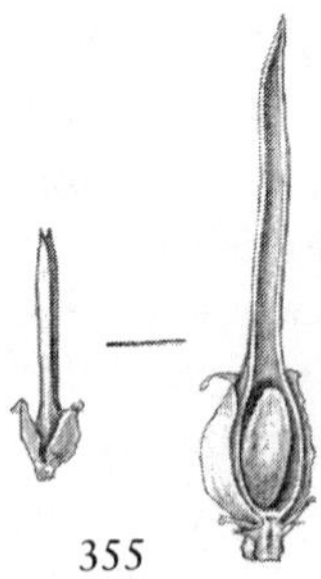

355

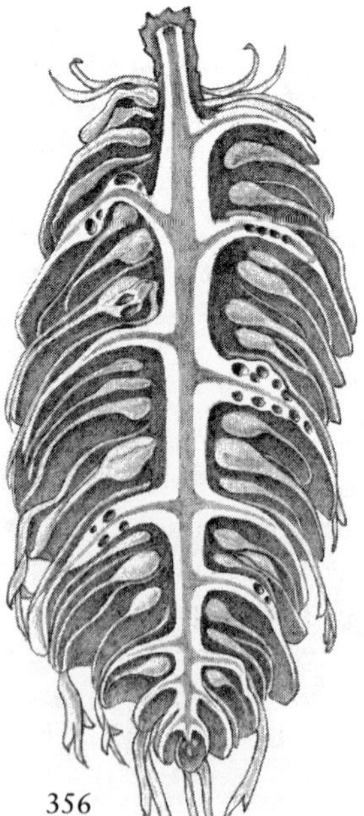

356

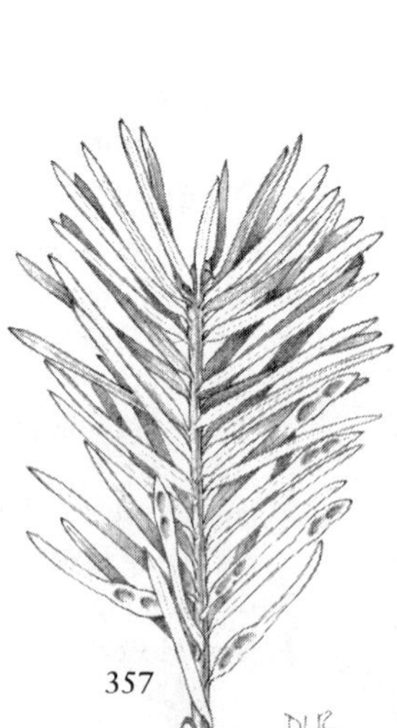

357

DLR

9b On pitch pine, *Pinus rigida*............. *Thecodiplosis pinirigidae* (Packard)

One or two larvae live in individual pockets between two of the three needles, which causes an uneven enlargement at the base of the needles. The much shortened needles are sometimes straight, sometimes recurved. The larvae drop to the ground in the fall, and there is one generation per year. Larvae were originally reported to crawl out on the leaves and form cocoons in which they pupated, but this is the habit of *Cecidomyia piniinopis* (q.v.) with which *T. pinirigidae* was confused. Host: *P. rigida*. Distr.: Maine.

10a On pinyon pine, *Pinus edulis* 11
10b On other species of pine 14
11a Needle fascicle swollen and misshapen, needles not developed beyond swelling (353) .... *Contarinia cockerelli* (Felt) and *Janetiella* sp.

C. cockerelli. This species is known from a single female, reportedly reared from galls such as those illustrated in figure 353. Such galls normally contain a species of *Janetiella*. The female has a stongly sclerotized, needlelike ovipositor, evidently for inserting eggs into plant tissue. Host: *P. edulis*. Distr.: Colorado. *Janetiella* sp. The biology of this species is probably similar to that of *J. coloradensis* of entry 13a of this key. Swellings harbor two larvae or pupae, each in a separate depression. Host: *P. edulis*. Distr.: Colorado.

11b Needle fascicle swollen at base, shorter than normal 12
12a Larvae gregarious between needles (352) .... *Pinyonia edulicola* Gagné

Eggs are laid on developing needles in mid-June to mid-July. Upon hatching, larvae crawl to the base of the needles. The needles expand into a spindleform enlargement around a chamber housing 5–40 larvae. Larvae grow during the summer and also the following spring. Pupation occurs in the gall, and adults emerge in early summer. Infested needles drop off later the same year, 6–8 years prematurely. Host: *P. edulis*. Distr.: Colorado and New Mexico. Refs.: Baskin and Brewer (1983), Houseweart and Brewer (1972).

12b Each larva in separate chamber, no more than one at base of each needle .. 13
13a Larva encased in removable capsule (355) .... *Janetiella coloradensis* Felt

Figures 346–357. Galls of Pinaceae. **346, 347,** First- and second-year cones, respectively, *Pinus elliottii*: *Cecidomyia bisetosa*. **348,** Resin masses on twigs, with pupal exuviae, *Pinus virginiana*: *Cecidomyia resinicola*. **349,** Larval cocoons on buds, *Pinus taeda*: *Cecidomyia piniinopis*. **350,** Cocoons on needles, one with pupal exuviae, *Pinus virginiana*: *C. piniinopis*. **351,** Swollen twig, *Pinus taeda*: *Cecidomyia reburrata*. **352–355,** Needle galls, *Pinus edulis*. **352,** *Pinyonia edulicola*. **353,** *Janetiella* sp. **354,** With section in 2×: *Janetiella* sp. **355,** With section in 2×: *Janetiella coloradensis*. **356, 357,** Galls of *Pseudotsuga menziesii*. **356,** Cone (longitudinal section) with scale swellings: *Contarinia oregonensis*. **357,** Needle swellings: *Contarinia pseudotsugae*.

Adults lay eggs on developing shoots in late May and early June. Ten or more larvae may enter the fascicle bundle, but usually only two, occasionally one or three, survive at the base of each needle bundle. Affected needles grow only to about one-third normal length, the base of the needle is enlarged, almost spherical, and the sheath enlarges to encase the gall and does not dehisce. A detachable capsule made up of needle tissue simulating two dwarfed needles surrounds each larva. Larvae grow partially in summer and complete growth the following spring. They then form a cocoon and pupate in the galls and emerge in May to early June. One generation occurs per year. Galled needles fall after the adults leave. Pinyon pine needles normally die after 6–8 years. This species has been reported from *P. virginiana* (Felt 1940 and elsewhere), but the gall on that pine is formed by *Thecodiplosis brachynteroides*. Host: *P. edulis*. Distr.: Colorado and Arizona. Ref.: Gillet and Brewer (1983).

13b Larva encased in simple depression at base of needles (354) .... *Janetiella* sp. near *coloradensis*

The biology of this species is similar to that of the related *J. coloradensis* above, except that the larvae are not encased in detachable capsules. The needle base is not as strongly swollen as for *C. coloradensis*. Host: *P. edulis*. Distr.: Colorado. Ref.: Brewer (1971).

14a On ponderosa pine, *Pinus ponderosa* .... *Contarinia coloradensis* Felt

Eggs are laid on new needles. Larvae settle at the bases of the needles, where they form chambers. The fascicle remains short and becomes greatly swollen at the base. The orange larvae drop to the soil in late summer or early fall. Adults emerge in spring from pupae in the duff. Needles die and drop off after their first year instead of after the normal 6–8 years. Host: *P. ponderosa*. Distr.: Colorado. Ref.: Brewer and Johnson (1977).

14b On other species of pine 15

15a On Monterey pine, *Pinus radiata* .... *Thecodiplosis piniradiatae* (Snow and Mills)

Eggs are laid on developing buds in January. Larvae crawl among the needles and cause them to become shortened and swollen at their bases. This species reportedly pupates on the plant in cocoons, but cocoons attributed to this species may be formed by a *Cecidomyia* mistaken for this species. Perhaps the larvae of *T. piniradiatae* dropped to the soil unnoticed. Browning of the needles is noticeable in the fall. Host: *P. radiata*; also reportedly *P. coulteri*, *P. sabiniana*, *P. sylvestris*, *P. tuberculata*. Distr.: California. Ref.: Snow and Mills (1900).

15b On *Pinus monophylla* .... *Janetiella* spp.

Two kinds of fascicle galls, each caused by a new species of *Janetiella*, are known from this host. The first causes a second needle to grow, while the other merely shortens the single needle. Larvae of both live in individual cells at the base of the fascicles. These flies pupate in the fascicles in spring and cause premature needle drop. Host: *P. monophylla*. Distr.: Arizona.

16a Larva without tubercles on dorsum of abdomen .... *Cecidomyia* spp.

Larvae are usually gregarious on red pines and pupate in the mass of resin in which they live (348, 351). Identification of these depends on microscopic characters of larvae, pupae, and males. See Gagné (1978b) for a review and technical key to these species:

C. fortunactus Gagné. Full-grown larvae of this species may be separated from the other species by the presence of a long, black, two-lobed spatula. Hosts: *P. elliottii*, *P. virginiana*. Distr.: Maryland and Florida.
C. reburrata Gagné. Hosts: *P. taeda*, *P. caribaea*. Distr.: Maryland and Florida, also Cuba.
C. resinicola (Osten Sacken). This species is the most widespread and probably the commonest North American species of this genus. It is known to injure the cambial layer and kill shoots. Hosts: known from almost all species of hard pine in North America. Distr.: coast to coast, Canada to Mexico and Florida, probably wherever pines naturally grow. Ref.: Reeks (1960).
C. resinicoloides Williams. Hosts: *P. banksiana*, *P. contorta*, *P. elliottii*, *P. ponderosa*, *P. radiata*, *P. virginiana*. Distr.: Ontario to Alberta south to Florida and California.
C. tortilis Gagné. Hosts: *P. clausa*, *P. virginiana*. Distr.: Maryland to Florida.

16b Larva with tubercles on dorsum of abdomen (87)...................... 17

Larvae may be single to gregarious. They pupate in a white cocoon some distance away from the resin mass in which they feed. Identification of most of these depends on microscopic characters of larvae, pupae, and males. See Gagné (1978b) for a review and technical key to these species.

17a Larva in running resin; on white pine, *Pinus strobus* .... *Cecidomyia candidipes* Foote

Several larvae are usually found in the same pitch mass. Host: *P. strobus*. Distr.: Quebec and Ontario, south to Illinois and Maryland.

17b Larva in inconspicuous pits at base of needle fascicles, usually on hard, or two- and three-needle pines (349, 350) .... *Cecidomyia* spp.

C. brevispatula Gagné. Hosts: *Pinus clausa*, *P. echinata*, *P. elliottii*, *P. radiata*, *P. virginiana*. Distr.: Maryland to Florida, and California.
C. piniinopis Osten Sacken. This species has been found on white pine as well as on many hard pines. It and *C. resinicola* have the greatest host distribution of the genus. Hosts: *P. banksiana*, *P. contorta*, *P. elliottii*, *P. ponderosa*, *P. rigida*, *P. strobus*, *P. taeda*, *P. virginiana*. Distr.: coast to coast, Canada to Mexico and Florida, probably wherever pines naturally grow. Ref.: Reeks (1960).

Pseudotsuga (Douglas-fir)

1a Larva in cone. 2
1b Larva in needle or twig 5
2a Larva in tunnel or chamber in cone scale 3
2b Larva between cone scales. 4
3a Larva in spherical chamber in swollen cone scale (356) . . . . *Contarinia oregonensis* Foote

Adults emerge from the ground in April and May. Eggs are laid in clusters near the base of cone scales on newly opened flowers. Larvae migrate along the inner surface of scales and become embedded singly in tissue, which begins to swell. In late summer or autumn, the full-grown orange larvae drop to the duff. Pupation occurs the following spring. Host: *P. menziesii*. Distr.: British Columbia and Oregon. Ref.: Hedlin (1961).

3b Larva in tunnel in apparently normal cone scale . . . *Camptomyia pseudotsugae* Hedlin and Johnson and *Contarinia washingtonensis* Johnson

Camptomyia pseudotsugae. This species is an inquiline in tunnels of *C. washingtonensis*. It occurs only in cones infested by *C. washingtonensis* and lives in close association with it. Eggs are laid in May and June, and the full-grown orange larvae drop to the duff in fall. Host: *P. menziesii*. Distr.: British Columbia to California. Ref.: Hedlin and Johnson (1968).
Contarinia washingtonensis. Eggs are laid on developing cones in May and June. Larvae live in tunnels among the cone scale fibers and can kill the scales and damage seeds. Full-grown larvae drop to the ground in late summer and fall. Pupation occurs in spring. Host: *P. menziesii*. Distr.: British Columbia to Oregon. Ref.: Hedlin and Johnson (1963).

4a Larva orange, warty, with spatula (as in 84) . . . . *Resseliella coryloides* (Foote)

This species lives between cone scales and on developing seed surfaces. It is usually found in great numbers and may have more than one generation per year.

Pupation occurs in the cone. Host: *P. menziesii* and other conifers. Distr.: western North America. Ref.: Hedlin et al. (1980).

4b Larva white, smooth, without spatula (27) . . . . *Asynapta hopkinsi* Felt

Larvae are usually gregarious and may feed only on damaged cones. They sometimes occur in great numbers in a cone. Full-grown larvae drop to the ground to pupate. Host: *P. menziesii* and other conifers. Distr.: widely distributed in North America. Ref.: Hedlin et al. (1980).

5a Larva at base of needles in drop of resin . . . . *Cecidomyia* sp.

Larvae live immersed in resin. When full-grown, larvae crawl up a needle and form a cocoon in which they pupate. Host: *P. menziesii*. Distr.: Oregon.

5b Larva in needle (357) 6

6a Only upper surface of needle discolored; needle often distorted or bent . . . . *Contarinia cuniculator* Condrashoff

This and the two species in couplet 7 are closely related and differ only in details of their biology and anatomy. Larvae are white to yellow. In September they escape from the needles and drop to the ground, where they diapause. Pupation occurs in early spring. Adults emerge in May, fly upward along a tree trunk and mate near the periphery of branches. Eggs are laid on the upper surface of the needles. Infested needles usually contain one larva, but up to four have been found. All three species can cause heavy needle loss. Host: *P. menziesii*. Distr.: British Columbia to California. Refs.: Condrashoff (1961a, b, 1962).

6b Both surfaces of needle discolored; needle usually not distorted or bent 7

7a Needle swollen surrounding feeding site (357) . . . . *Contarinia pseudotsugae* Condrashoff

This species appears to be the most serious pest of the three contarinias in Douglas-fir needles. Its biology is generally similar to that of *C. cuniculator* (see entry 6a of this key) except that eggs are laid on the lower epidermis of the needles and larvae escape from the lower side of the needle. Up to 12 larvae can be found in one needle. Host, distribution, and references are as for *C. cuniculator* (entry 6a of this key).

7b Needle flattened surrounding feeding site . . *Contarinia constricta* Condrashoff

The biology of this species is generally similar to *C. cuniculator* (see 6a) except that eggs are laid on the lower surface of the needles and larvae escape from the lower side. Usually one, occasionally two to three larvae are found in a needle. Host, distribution, and references are as for *C. cuniculator* (entry 6a of this key).

Platanaceae

Platanus (sycamore)

Rolled leaf margin or distorted leaf .... *Prodiplosis platani* Gagné

Larvae feed gregariously on new leaf growth. When full-grown, they spin cocoons and pupate on the leaves or in the ground. Several generations occur per year. Host: *P. occidentalis*. Distr.: New York to Maryland. Ref.: Neal and Gott (1988).

Poaceae (Gramineae)

Grasses harbor a distinctive cecidomyiid fauna of several genera and species groups. *Calamomyia* spp. and *Hybolasioptera* sp. live in silken tubes in stems. The only exterior sign of their presence is the black fungus associated with them that sometimes shows through the stems (fig. 362). *Astictoneura* spp., *Edestochilus*, and *Chilophaga* spp. cause a foreshortening of the stem, the resulting imbricated leaves producing a broom- or onion-shaped gall (363–365). *Edestosperma* sp. lives in seeds. The remaining cecidomyiids on grasses comprise *Contarinia* spp. of the *sorghicola* species group (at least two of them introduced), two or three species of *Dasineura* (probably all introduced), two species of *Mayetiola*, (one, the Hessian fly, introduced), *Sitodiplosis mosellana* (introduced), a few *Clinodiplosis* spp., and a few unidentified seed-infesting species that cannot be placed to genus. *Calamomyia* and *Contarinia* are probably very rich in species, but their presence is inconspicuous. Most North American grasses may eventually be found to harbor a *Calamomyia* and a *Contarinia* of the *sorghicola* group. To avoid repetition about the biology of species in those two genera, general information on them is summarized under their respective genera in chapter 4 instead of in the key.

Agropyron

1a Larva in spikelet.. *Contarinia* sp.

Hosts: *A. repens*, *A. smithii*. Distr.: Ontario and Nebraska.

1b Larva in imbricated stem gall (as in 365) ..... *Chilophaga* or *Astictoneura* sp.

The stem is foreshortened and the leaves are tightly imbricated. Host: *A. repens.* Distr.: Pennsylvania.

Agrostis

Larva on inflorescence .... *Dasineura graminis* Felt

This species is based on specimens caught in early June laying eggs on flower heads. The only other *Dasineura* on grasses in North America, *D. alopecuri* from *Alopecurus pratensis,* is an immigrant from Europe. It is possible that *D. graminis* is also an introduced species. A *Dasineura* identified as *D. graminis* was reared from seeds of this host in England. Larvae of that record fed on the seed as it was being formed and had at least two generations per year. Host: *A. tenuis* (as *vulgaris*). Distr.: New York; ?Europe. Ref.: Barnes (1946b).

Alopecurus

1a Larva stationary in floret .... *Contarinia geniculati* (Reuter) and *Dasineura alopecuri* (Reuter)

Both species are immigrants from Europe, possibly introduced during Colonial times but noticed here only shortly before 1935. *C. geniculati* prevents seed formation; *D. alopecuri* feeds on the developing seeds. Larvae of both species overwinter in the host. Larvae of *Contarinia* have no spatula; those of *Dasineura* have a spatula in the last instar. *Contarinia* is multivoltine, *Dasineura* usually univoltine. Hosts: *A. geniculatus, A. pratensis.* Distr.: New Brunswick and Ontario; Europe and New Zealand. Ref.: Barnes (1946b).

1b Larva active on inflorescence .... *Clinodiplosis setariae* (Felt)

Larvae crawl freely among the flowers. They may be saprophagous or mycophagous but have been found repeatedly on this and other grasses. Larvae drop to the soil to pupate. Several generations may occur per year. Hosts: *Alopecurus* sp.; also *Cenchrus* sp. Distr.: New York and Florida.

Ammophila (beach grass)

Enlarged stem base .... *Mayetiola ammophilae* Gagné

The damage is similar to that made on wheat by the Hessian fly. Infestation promotes subsequent rot and lodging. Larvae pupate in the stems. Two generations may occur per year. Host: *A. breviligulata.* Distr.: New Jersey.

Andropogon

Larva in floret .... *Contarinia wattsi* Gagné and unidentified cecidomyiid

C. wattsi. This species is a serious pest. Hosts: *A. gerardii*, *A. hallii*, *A. scoparius*, and *A. stolonifera*. Distr.: Florida, South Dakota, Nebraska, and New Mexico. Ref.: Watts and Bellotti (1967).
Unidentified cecidomyiid. This species is known only from a series of larvae found in seeds in fall. Larvae (89, 90) bear some resemblance to those taken from florets of *Bouteloua*, except that those on *Andropogon* have a spatula. Host: *A. gerardii*. Distr.: New York.

Aristida

1a Larva in seed............................ *Edestosperma aristidae* Gagné

Adults were reared in spring from stored, apparently normal seed. Host: *A. longiseta*. Distr.: Idaho.

1b Larva in imbricated stem gall (364) .... *Chilophaga gyrantis* Gagné

Galls are usually found close to the base of the stem. Each contains several larvae, usually one at the base of each leaf sheath. Full-grown larvae form a cocoon and pupate in the gall. Several generations occur per year. Host: *A. gyrans*. Distr.: Florida. Ref.: Gagné and Stegmaier (1971), Stegmaier (1973).

Bouteloua

Larva with seed .... cecidomyiid

Larvae (98) of an unidentified cecidomyiid, collected with seeds in early October, apparently belong to the supertribe Cecidomyiidi but cannot be identified further until adults are reared. Host: *B. curtipendula*. Distr.: Oklahoma.

Bromus (bromegrass)

Larva in floret .... *Contarinia bromicola* (Marikovskij and Agafonova)

This is evidently an immigrant species, discovered in the Great Plains as early as 1948 but described only in 1962 from Russia. It is at least bivoltine, and infestation rates can run over 50%. Hosts: *B. inermis*, *B. riparius*. Distr.: Great

Plains from Saskatchewan and Wisconsin, south to Oklahoma. Ref.: Neiman and Manglitz (1972).

Cenchrus

Larva on inflorescence .... *Clinodiplosis setariae* (Felt)

Larvae crawl freely among the flowers. When full-grown they drop to the soil to pupate. Probably several generations occur per year. Most species of *Clinodiplosis* appear to feed on fungi, but *C. setariae*, at least, seems regularly associated with grasses. Hosts: *C. echinatus*; also *Alopecurus* sp. Distr.: New York, Florida.

Cinna

Larva in apparently normal, sometimes discolored stem .... *Calamomyia* sp.

Host: *C. arundinacea.* Distr.: Maryland.

Danthonia

Larva in apparently normal, sometimes discolored stem .... *Calamomyia danthoniae* (Felt)

Host: *Danthonia* sp. Distr.: New York.

Dichanthelium

Larva in apparently normal, sometimes discolored stem .... *Calamomyia panici* (Felt)

This species is known from a single female caught in June while apparently laying eggs on its host. Host: *D. latifolium* (as *Panicum macrocarpa*). Distr.: New York.

Echinochloa

Larva in apparently normal, sometimes discolored stem .... *Calamomyia echinochloa* (Felt)

Host: *E. crusgalli.* Distr.: South Dakota.

Elymus

1a Larva in floret . . . . cecidomyiid

Larvae were taken from florets in July. This species cannot be identified to genus without the adult stage. Host: *Elymus* sp. Distr.: Arkansas.

1b Larva in stem 2
2a Larva in apparently normal, sometimes discolored stem . . . . *Hybolasioptera elymi* Gagné

The occasional discoloration of the stem is evidently due to the black fungus associated with this gall midge. Larvae overwinter in the stems and pupate there in spring. Host: *E.* "*americanus*." Distr.: Kansas.

2b Larva in swollen stem base (as in 358) . . . . *Mayetiola destructor* (Say)

This species normally occurs on wheat, but rye may be an accidental host. See under *Triticum* for further information. Host: *E.* "*americanus*" or *E. glaucus*. Distr.: North America; Europe, Australia, and New Zealand.

Eragrostis

Larva in apparently normal, sometimes discolored stem . . . . *Calamomyia panici* (Felt)

Host: *E. minor*. Distr.: unspecified.

Muhlenbergia

Imbricated stem gall (365) . . . . *Astictoneura muhlenbergiae* (Marten) and *Chilophaga colorati* (Felt)

Galls may contain one to many larvae, each larva enveloped by the base of a leaf sheath. Pupation occurs in the gall in spring. Larvae of the two species are distinct. That of *A. muhlenbergiae* has a smoothly rounded terminal segment (as in 63); that of *C. colorati*, a terminal segment with two projections (59). Hosts: *M. mexicana*, *M.* sp. Distr.: Connecticut to South Dakota and Iowa.

Panicum

Larva in apparently normal, sometimes discolored stem . . . . *Calamomyia inustorum* (Felt)

Host: *P. virgatum*. Distr.: South Dakota and Oklahoma.

Paspalum

Swollen stem base (363) .... *Edestochilus allioides* (Pritchard)

Adults were reared in August from the swellings. Host: *P. distichum*. Distr.: California and Utah.

Phragmites

Larva in apparently normal, sometimes discolored stem .... *Calamomyia phragmites* (Felt)

Host: *P. communis*. Distr.: Michigan. Ref.: Frohne (1939).

Setaria

Larva in floret .... *Contarinia* sp.

Host: *S. glauca*. Distr.: unspecified.

Sorghastrum

Larva in apparently normal, sometimes discolored stem .... *Calamomyia* sp.

Host: *S. nutans*. Distr.: Maryland.

Sorghum (sorghum, johnson grass)

Larva in floret .... *Contarinia sorghicola* (Coquillett)

This is an African species introduced into North America in the nineteenth century with seed of Johnson grass and sorghum. Perhaps the greatest pest of sorghum, it has several generations per year (9–12 in Texas) and prevents seed development. A *Clinodiplosis* sp., discovered in sorghum florets in Michigan (Webster and Kohls 1972), is probably a secondary invader. Hosts: *S. bicolor* and *S. halepense*. Distr.: United States and Mexico, wherever the hosts grow; also worldwide wherever sorghum is grown. Refs.: Harris (1970), Teetes (1985).

Spartina

1a Larva in floret .... *Contarinia* sp.

This undescribed species belongs to the *sorghicola* species group. Host: *S. alterniflora*. Distr.: North Carolina.

1b Larva in apparently normal, sometimes discolored stem (362) .... *Calamomyia* spp.

Calamomyia alterniflorae Gagné. Host: *S. alterniflora*. Distr.: Maine and North Carolina. Ref.: Gagné (1981b).
Calamomyia sp. This species is anatomically distinct from *C. alterniflorae*. Host: *S. cynosuroides*. Distr.: North Carolina. Ref.: Gagné (1981b).

Tridens

Larva in floret .... *Contarinia albescentis* Gagné

Hosts: *T. albescens*, *T. flavus*. Distr.: South Carolina to northern Mexico.

Tripsacum

Imbricated stem gall .... *Chilophaga tripsaci* (Felt)

Larvae live among the leaf sheaths. Adults were reared in August. Host: *T. dactyloides*. Distr.: Texas.

Triticum (wheat)

1a Larva in stem swelling (358–361) .... *Mayetiola destructor* (Say)

Figures 358–372. Galls of Orchidaceae, Poaceae, Polygonaceae, and Rhamnaceae. **358–361,** Stems of wheat, *Triticum aestivum*: *Mayetiola destructor*. **358,** Outer, basal leaves removed to show massed puparia. **359,** Puparium above node. **360,** Outer leaves removed, puparium in stem depression. **361,** Pupal exuviae in exit hole. **362,** Stems, *Spartina alterniflora*: *Calamomyia spartinae*; one in partial section to show larvae in silken tunnels among black fungal excrescence, the other with pupal exuviae in exit hole, a slight discoloration due to the fungus appearing around hole. **363–365,** Foreshortened, enlarged culms of various grasses: **363,** *Paspalum distichum*: *Edestochilus allioides*. **364,** *Aristida gyrans*: *Chilophaga gyrantis*. **365,** *Muhlenbergia mexicana*: *Astictoneura muhlenbergiae*. **366,** Leaf blister galls, *Coccoloba diversifolia*: *Ctenodactylomyia watsoni*. **367,** Swollen root nodules of orchid: *Neolasioptera* sp. **368–372,** Galls of *Ceanothus*. **368,** Swollen bud: *Asphondylia* sp. **369,** Leafy node gall, one in cross section: *Asphondylia ceanothi*. **370,** Globular leaf gall: *Asphondylia* sp. **371,** Normal and pedicellate, galled flowers: *Asphondylia* sp. **372,** Vein folds: *Contarinia* sp.

358

359

360

361

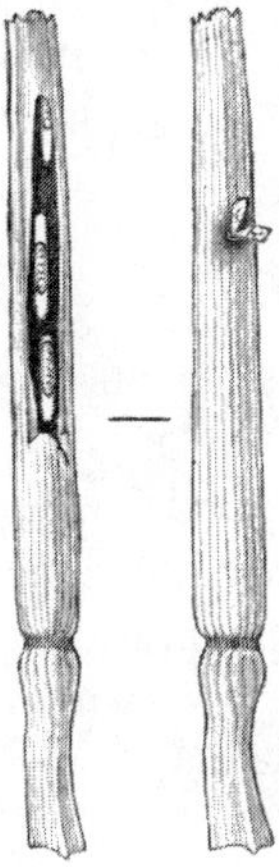
362

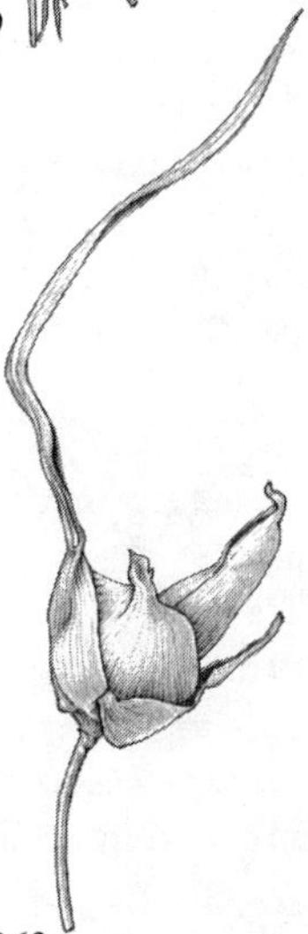
363

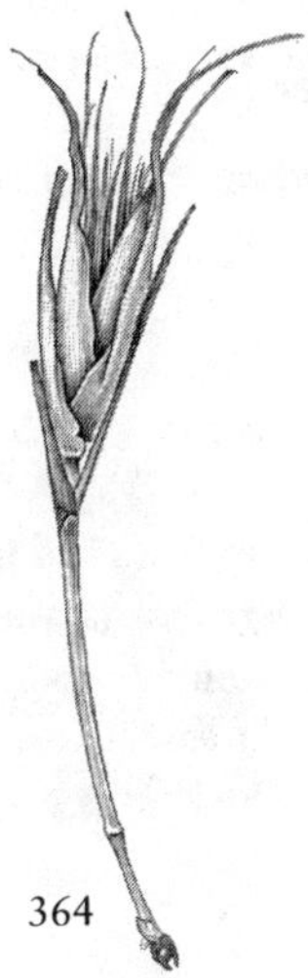
364

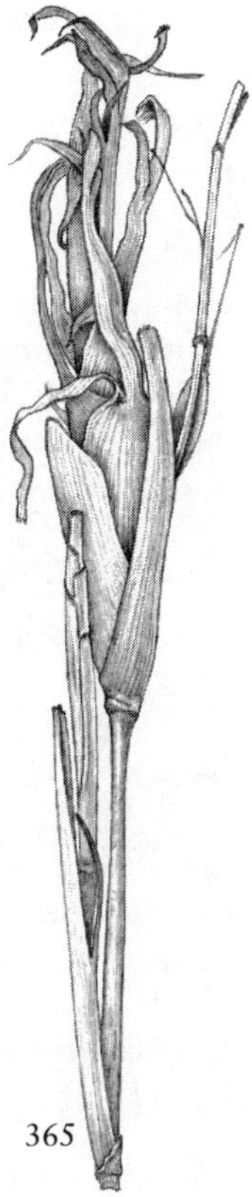
365

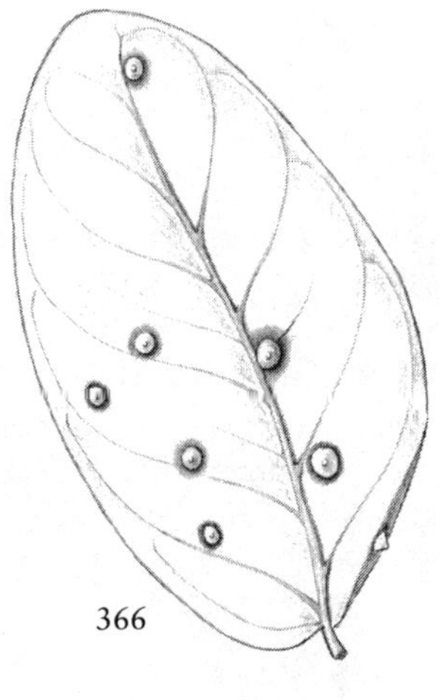
366

368

369

370

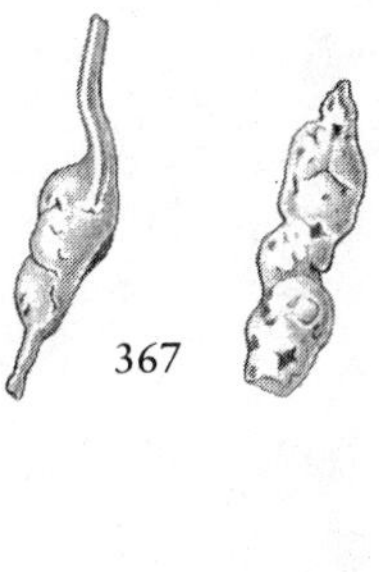
367

371

372

DLR

Eggs are laid on the upper surface of the leaf blade. Newly hatched larvae migrate to the base of a leaf sheath in 12–25 hours, settle, and begin to feed. Larvae diapause and pupate inside the next to last larval skin, which becomes hard. Several generations occur per year. Ears of infested plants become shrivelled, and lodging occurs just above the nodes. This major pest of wheat is an immigrant from Europe. The Hessian fly was so named because it supposedly came to the New World in straw used for bedding by the Hessian troops during the Revolutionary War. Hosts: *T. aestivum*; also *Elymus* in North America, and possibly other grasses as accidental hosts (Hayhurst 1909, Noble 1931); in Europe also on rye and barley (*Hordeum vulgare* and *Secale cereale*) (Ertel 1975). Distr.: coast to coast in North America, wherever wheat is grown; Europe, Middle East, North Africa, New Zealand. Refs.: Barnes (1956), Gagné and Hatchett (in press).

1b Larva on seed .... *Sitodiplosis mosellana* (Géhin)

The bright orange to brick red larvae live exposed on the surface of developing kernels. When full-grown, larvae drop to the soil. This species is multivoltine and is an immigrant from Europe. It appears to be associated with wheat glume blotch, a fungus disease. Host: *T. aestivum*. Distr.: coast to coast in North America, wherever wheat is grown; Europe. Refs.: Barnes (1956), Wellso and Freed (1982).

Polemoniaceae

Phlox (phlox)

Larva in terminal buds (234) .... *Clinodiplosis phlox* (Greene)

Larvae are gregarious and live among the developing leaves in terminal and lateral buds. Buds are killed, and infested plants do not flower well. This species begins feeding in early June and continues with several overlapping generations into October. The last, overwintering generation overwinters as full-grown larvae. *Lestodiplosis maculipennis* Greene was regularly found preying on *H. phlox*. Host: *Phlox* sp. and cultivars. Distr.: Ohio. Ref.: Neiswander (1962).

Polygonaceae

Coccoloba

Circular leaf blister (366) .... *Ctenodactylomyia watsoni* Felt

The monothalamous blisters are slightly nippled. Pupation takes place in the galls, and several generations may occur per year. Host: *C. diversifolia*. Distr.: Florida. Ref.: Mead (1970).

Eriogonum

Swollen flower bud .... *Prodiplosis falcata* Gagné

Adult midges were reared in spring from buds. This species may have an extensive host range, as do many other *Prodiplosis* spp. Hosts: *E. latifolium*; also *Ericameria* sp. (Asteraceae). Distr.: California.

Polygonum

1a Stem swelling .... *Lasioptera* sp.

The cecidomyiid on which this identification was based is not available but would more likely belong to *Neolasioptera*. Host: *P. scandens*. Distr.: unspecified.

1b Rolled leaf margin .... cecidomyiid

Specimens on which this record is based are not available. Host: *P. lapathifolium*. Distr.: unspecified.

Rumex (dock, sorrel)

Deformed, reddish seed capsules ....
.... *Contarinia acetosellae* (Rübsaamen) and *C. rumicis* (Loew)

Both species are probably introductions from Europe into North America during colonial times and occur on immigrant species of *Rumex*.

C. acetosellae. Host: *R. acetosella*. Distr.: Georgia and Illinois; Europe.
C. rumicis. Host: *R. crispus*. Distr.: New York; Europe.

Polypodiaceae

Cystopteris (bladder-fern)

Root gall .... cecidomyiid

Specimens on which this record is based are not available. Host: *C. fragilis*. Distr.: unspecified.

Portulacaceae

Portulaca

1a Succulent, tapered stem swelling . . . . *Neolasioptera portulacae* (Cook)

Swellings are 12–36 mm long and 8–12 mm in greatest diameter. Adults emerge in July and August in Florida. Host: *P. oleracea*. Distr.: Florida, Mexico, West Indies, and South America.

1b Swollen, unopened flower. . . . . *Asphondylia portulacae* Möhn

Adults emerged in November and December in Florida. Host: *P. oleracea*. Distr.: Florida; El Salvador.

Primulaceae

Lysimachia (loosestrife)

Swollen terminal bud . . . . *Dasineura lysimachiae* (Beutenmüller)

The swelling is composed of the young, inwardly rolled leaves. Larvae are gregarious. Full-grown larvae of at least the first generation form cocoons in the buds, and adults emerge soon after. Infested buds are usually killed. The damage appears similar to that caused by *Dasineura oxycoccana* on cranberry, and an early report (Smith 1890) noted that the two plants grow together. Hosts: *L. quadrifolia*, *L. terrestris*. Distr.: New York, New Jersey, and Indiana.

Ranunculaceae

Anemone (windflower)

Slightly enlarged bud . . . . *Dasineura anemone* Felt

Damaged buds contain numerous, gregarious, orange larvae. Host: *A. canadensis*. Distr.: New York.

Clematis

1a Tapered stem or tendril swelling . . . . *Neolasioptera clematidis* (Felt)

The single specimen known of this species was found apparently laying eggs in a slit of a stem. It was presumed to be the species responsible for the swellings found on clematis. Host: C. *virginiana*. Distr.: Massachusetts and New York.

1b Swollen bud. 2
2a From southwestern North America . . . . *Asphondylia clematidis* Felt

Swellings are subglobular, woolly, a little over 1 cm in diameter, and may be mono- or polythalamous. Galls appear in July, and adult flies emerge from them in August. Host: *C. drummondii*. Distr.: Texas. Ref.: Painter (1935).

2b From eastern North America 3
3a Swollen leaf bud. . . . . *Contarinia clematidis* Felt and *Dasineura clematidis* Felt

Swellings are about 1 cm in diameter, greenish, and usually in a leaf axil. Adults of both species were reared in August from galls collected in July. One may be an inquiline of the other. Host: C. *virginiana*. Distr.: New York.

3b Enlarged flower bud . . . . *Prodiplosis floricola* (Felt)

Infested flower buds fail to open. Adults were reared in August. Hosts: C. *virginiana*; also on *Spiraea* (Rosaceae) (q.v.) and *Caryocar* (Clusiaceae) in Brazil. Distr.: New York; Brazil.

Thalictrum (meadow rue)

Irregularly enlarged seed pod . . . . *Asphondylia thalictri* Felt

Infested seed pods are about 5 mm in diameter. Adults emerge from pods in August. A *Trotteria* sp. was reared from these galls as well. Host: *Thalictrum* sp. Distr.: New Hampshire and Massachusetts.

Rhamnaceae

Ceanothus

1a Bud or flower gall. 2
1b Leaf gall. 4
2a Pedicellate flower gall (371). . . . . *Asphondylia* sp.

Galls with pupae were collected in early August. Host: C. *velutinus*. Distr.: Montana.

2b Enlarged bud or leafy node gall 3
3a Enlarged bud (368).. *Asphondylia* sp.

Adults were reared in June and July. Hosts: *C. cuneatus*, *C. jepsonii*, *C. sanguineus*, *C. thyrsiflorus*. Distr.: Idaho, Oregon, and California.

3b Leafy node gall (369) .... *Asphondylia ceanothi* Felt

Leaves making up the gall are linear, slightly hairy basally, green at first, but turning brown and decaying after the adults emerge. Adults were reared in June. Host: *C. velutinus*. Distr.: Oregon and California.

4a Simple vein fold (372).. *Contarinia* sp.

Swellings are visible from beneath the leaf and are usually covered with fuzz; only a slit is evident above. Full-grown larvae escape through the slit in July. This species is probably univoltine. Hosts: *C. americanus*, *C. integgerimus*, *C. velutinus*. Distr.: Connecticut and Oregon.

4b Complex globular gall (370)............................... *Asphondylia* sp.

Galls with pupae were collected in mid-July. Host: *C. velutinus*. Distr.: Montana.

Rosaceae

Adenostoma

Larva in seed .... *Asphondylia adenostoma* Felt

Adults emerged from apparently unmodified seeds in early June. Host: *A. fasciculatum*. Distr.: southern California.

Agrimonia (agrimony)

Swollen flower .... *Contarinia agrimoniae* Felt

This species was reared in early September from larvae found in August. It may be multivoltine, as are many other contarinias living in flower heads. Host: *A. striata*. Distr.: New York.

Amelanchier (shad-bush, service-berry)

This plant genus is host to many leaf galls, some of bizarre shape, all probably made by species of *Blaesodiplosis*. A sample of kinds from the USNM collection is shown here (373), and two others have been illustrated elsewhere (Wells 1921b). Galls sometimes occur in such numbers as to cover the surface of a leaf. Each gall contains a single larva that leaves from an opening on one side of the leaf when full-grown. The species is probably univoltine. Adults have been reared of only one species, *Blaesodiplosis canadensis*.

1a Enlarged fruit without stone. . . . . *Contarinia* sp.

Infested fruit grow faster and larger, do not have stones, and drop sooner than uninfested fruit. This species is univoltine, the larva overwintering in the soil. Adults have not been reared. Similar damage is done to *Prunus* spp. by *C. virginianae* (q.v.). Host: *A. canadensis*. Distr.: New York.

1b Pouch-shaped leaf galls 2
2a Tubular, lipped galls (373d) . . . . *Blaesodiplosis canadensis* (Felt)

The gall illustrated here for this species is slightly different from that described by Felt in that this one protrudes about 5 mm from both sides of the leaf instead of only from the bottom. Host: *A. canadensis*. Distr.: Massachusetts and New York.

2b Gall conical, lipped on upper surface of leaf, convex and white-haired below (373b) . . . . *Blaesodiplosis* sp. and *Clinodiplosis canadensis* (Felt)

The *Clinodiplosis* was probably a secondary feeder in the galls. Host: *A. canadensis*. Distr.: Massachusetts.

Cercocarpus (mountain mahogany)

Enlarged bud (374) . . . . *Dasineura cercocarpi* Felt

Adults were reared in April in Colorado. Galls in southern California contained second-instar larvae in mid-August. Hosts: *C. betuloides*, *C.* spp. Distr.: Colorado, Oregon, and California.

Crataegus (hawthorn)

1a Circular leaf spot (375a) . . . . *Lasioptera excavata* Felt

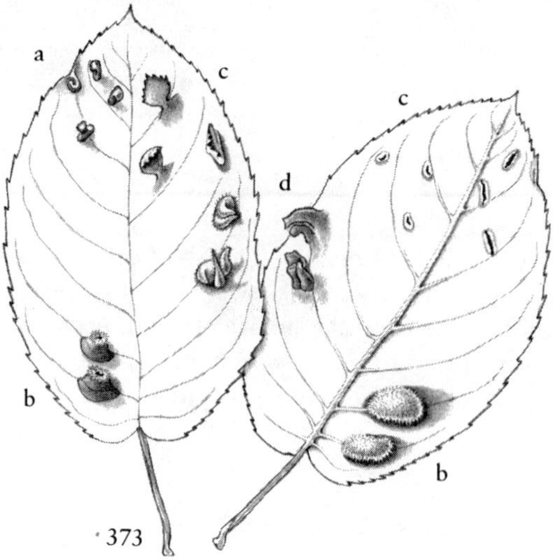

373

374

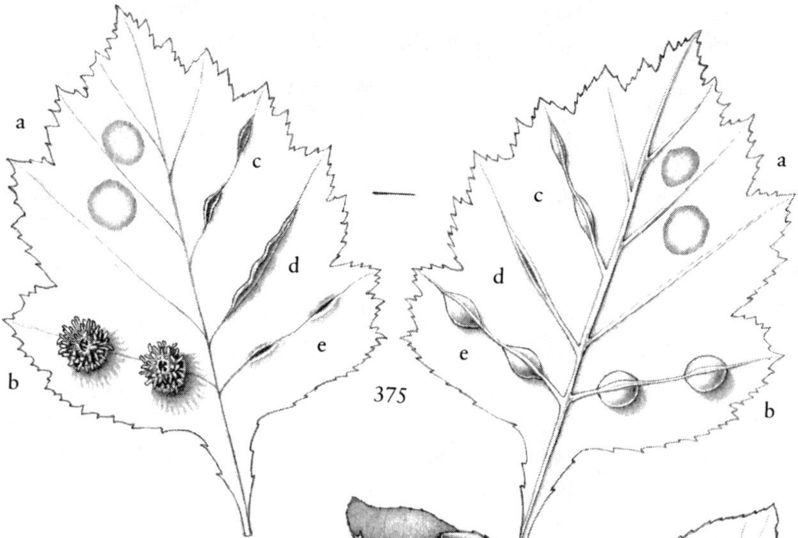

375

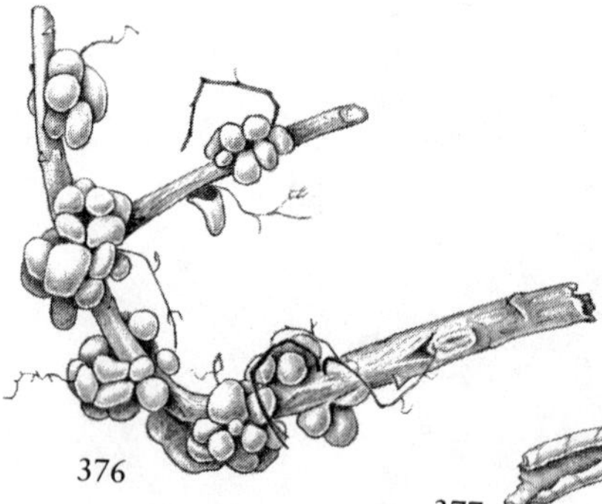

376

377

378

DLR.

Spots are pale green to purplish, often surrounded with red, and were found in summer. An adult was reared in August. Host: *Crataegus* sp. Distr.: New York.

1b Vein swelling or complex gall on leaf 2

2a Swelling with tubular papillae on at least one surface..................... 3

2b Swelling smooth, at most with short hair 4

3a Gall globular, lip of opening surrounded with fringe of numerous, fine, papillae (375b) .... *Dasineura crataegibedeguar* (Osten Sacken)

Galls develop in early spring, are monothalamous and 4–5 mm in diameter. They usually appear on the leaf lamina, occasionally also on the petiole or at the tip of young twigs, and most often open on the upper leaf surface. Larvae develop quickly and in spring drop to the ground, where they enter diapause until the following year. The pea-shaped gall with the pointed apex reported and figured in Felt (1940) belongs to this species; the tubular growths do not always appear until after the gall reaches full size. Hosts: *C. macracantha*, *C. pruinosa*, *C. tomentosa*, *C.* sp. Distr.: Massachusetts to Virginia, west to Illinois, including Ontario. Ref.: Wellhouse (1922).

3b Gall cylindrical, covered basally with nippled bumps and spiny processes .... *Dasineura* sp.

Unlike the other complex galls of hawthorn leaves, these are detectable only on one side of the leaf and resemble a pineapple. They are green or red, about 3 mm in diameter, 5 mm high, and usually on the upper surface of the midvein. This species is known only from a series of larvae from Illinois. Hosts: *C. punctata* and *C.* sp. Distr.: New York and Illinois. Ref.: Wellhouse (1922).

4a Gall resembling a cockscomb, bilaterally flattened, wavy, often red (375d) .... *Blaesodiplosis crataegifolia* (Felt)

This gall begins development in spring, but larvae grow slowly. In summer, the full-grown larvae drop to the soil, where they enter diapause until the following spring. A *Lestodiplosis* sp. was taken from these galls. Hosts: *C. coccinea*, *C. macrosperma*, *C. mollis*, *C. pruinosa*, *C.* sp. Distr.: Massachusetts to Ontario, south to Virginia. Ref.: Wellhouse (1922).

Figures 373–378. Galls of Rosaceae. **373,** Leaf galls, upper surface on left, lower on right, *Amelanchier canadensis*: **a–c:** *Blaesodiplosis* spp.; **d,** *Blaesodiplosis canadensis*. **374,** Enlarged bud, *Cercocarpus* sp.: *Dasineura cercocarpi*. **375,** Leaf galls, upper surface on left, lower on right, *Crataegus* sp.: **a,** leaf spots: *Lasioptera excavata*; **b,** fringed galls: *Dasineura crataegibedeguar*; **c,** *Blaesodiplosis* sp.; **d,** cockscomb gall: *Blaesodiplosis crataegifolia*; **e,** *Blaesodiplosis venae*. **376,** Subterranean galls, *Licania michauxii*: cecidomyiid. **377, 378,** Galls of pear, *Pyrus communis*. **377,** Curled leaf: *Dasineura pyri*. **378,** Enlarged, aborted fruitlets: *Contarinia pyrivora*.

4b Gall a simple, convex swelling, either globular or cylindrical and tapering, occasionally tinged with red 5

5a Gall globular (375e) .... *Blaesodiplosis venae* (Stebbins)

Galls and larvae develop quickly in early spring. Galls are usually monothalamous, 5–8 mm in diameter, usually slightly longer than wide, and most often open on the upper leaf surface. The same species may also be responsible for the elongate, tapering swelling in entry 5b. Full-grown larvae drop to the soil to diapause until the next spring. Hosts: *C. punctata*, *C.* sp. Distr.: Massachusetts, New York, Virginia, and Oregon.

5b Gall cylindrical and tapering (375c) .... *Blaesodiplosis* sp.

These galls may also be made by *B. venae*. Both this and the globular gall occur on the same leaf, but the former tend to be slightly thicker walled. Their variable length probably depends on the number of larvae in the gall. Hosts: *C. punctata*, *C.* sp. Distr.: New York, Virginia, and Oregon. Refs.: Larew and Capizzi (1983), Wellhouse (1922).

Fragaria (strawberry)

Kidney-shaped petiole gall, 1–2 cm long ...... *"Cecidomyia" reniformis* Stebbins

Galls contain several individual larval cells. Adults have not been reared; larvae belong to the Cecidomyiidi but cannot be placed to genus. Host: *F. virginiana*. Distr.: Massachusetts.

Heteromeles (toyon, Christmas berry)

Aborted berries .... *Asphondylia photiniae* Pritchard

Berries are slightly swollen, sterile, and remain green through the winter instead of turning the normal red. Damage is analagous to that of *Asphondylia ilicicola* on holly berries. Adults emerge from the berries in early spring. Host: *H. arbutifolia*. Distr.: California.

Holodiscus (ocean-spray)

Swollen leaf vein fold (391) .... *Dasineura* sp.

Folds are elongate, 5–10 mm long. Only a slit shows on the upper side of the leaf; on the bottom side the galls are convex and yellowish. Full-grown larvae escape through the slit on the upper side. Host: *H. discolor*. Distr.: Oregon.

Licania (ground-nut)

Globular subterranean gall (376) .... cecidomyiid

Galls are found up to 5 cm below the soil surface and may be clustered. Each contains one larva. Pupation occurs in the galls. This species is known from all life stages, but does not fit in any known genus. Host: *L. michauxii*. Distr.: Florida.

Physocarpus (ninebark)

1a Deformed seed.. *Dasineura* sp.

Host: *Physocarpus* sp. Distr.: Oregon.

1b Leaf gall... 2
2a Pea-sized, globular, succulent, green gall, the slitlike opening lined with white hairs... *Contarinia* sp.

Galls are visible chiefly on the upper surface of the leaves. They develop in May and early June. Adults of this gall midge have not been reared. Host: *P. opulifolius*. Distr.: Pennsylvania. Ref.: Wheeler and Hoebeke (1985).

2b Elongate, hard, vein swelling, the slitlike opening without hairs... cecidomyiid

Galls develop in spring and are visible chiefly on the upper surface of the leaves. Full-grown larvae and adults of this species are unknown. Host.: *P. opulifolius*. Distr.: Pennsylvania. Ref.: Wheeler and Hoebeke (1985).

Potentilla (cinquefoil)

Swollen stem .... *Neolasioptera potentillaecaulis* (Stebbins)

This species is known only from a description of the gall. It is tentatively placed in *Neolasioptera*. Host: *P. canadensis*. Distr.: Massachusetts.

Prunus (cherry, plum)

1a Swollen fruit (380) .... *Contarinia virginianae* (Felt) and *Dasineura pergandei* Felt

C. virginianae. Infested fruit are variously shaped and grow to twice normal size but do not develop a stone. Larvae live gregariously in a central cell and drop to the ground when full-grown. Pupation occurs the following spring. Larvae closely resemble *C. racemi* and *C. cerasiserotinae*, the other two contarinias on *Prunus*. Several other gall midges have been reared from infested fruit, or at least from containers in which the fruit were held. These include *Clinodiplosis canadensis* (Felt), presumably a fungivore; *Arthrocnodax cincta* (Felt), a mite predator; and *Rhizomyia absobrina* (Felt), its feeding habits unknown. Hosts: *P. virginiana*, *P. serotina*, *P.* sp.; Felt's unpublished notebook lists also the introduced *P. cerasus* as a host. Distr.: Maine and Ontario to Manitoba and Idaho, south to Virginia and Utah.

D. pergandei. This species was reared once from enlarged fruit and may be an inquiline. Host: *P. melanocarpa*. Distr.: Colorado.

1b Swollen rachis, bud, or leaf. 2

2a Swollen rachis (379) . . . . *Contarinia racemi* (Stebbins)

Swellings are green, soft, 1–3 cm long, 0.5–1 cm in diameter, and tapered. Larvae live among tunnels in the swollen rachis. Full-grown larvae drop to the ground, where they presumably diapause until the following spring. Hosts: *P. virginiana*, *P. serotina*. Distr.: Massachusetts to Maryland.

2b Leaf or bud gall. 3

3a Leaf fold or swelling. 4

3b Swollen or deformed bud 5

4a Simple leaf fold . . . . *Dasineura* sp.

The gall is a simple leaf fold. Although Felt reported adults of a rust-feeding cecidomyiid, *Mycodiplosis thoracica* (Fitch), from these galls, the white larvae he retained from leaf rolls belong to *Dasineura*, the probable leaf roller. Host: *P. virginiana*. Distr.: New Jersey.

4b Globular, hard, folded leaf (382) . . . . *Putoniella* sp.

The two halves of a leaf form a capsule, which eventually develops a longitudinal slit on the upper side through which full-grown larvae escape. Swellings appear in May and larvae leave in June. Larvae are gregarious. Adults have not been reared. This species is univoltine. Host: *P. angustifolia*. Distr.: Georgia.

Figures 379–391. Galls of Rosaceae. **379–382,** Galls of *Prunus*. **379,** Rachis swelling, *P. serotina*: *Contarinia racemi*. **380,** Normal and swollen (enlarged) fruit, one cut open to show larval cell, *P. serotina*: *Contarinia virginianae*. **381,** Swollen bud, *P. serotina*: *Contarinia cerasiserotinae*. **382,** Hard, leaf fold gall, one opened to show larvae, *P. angustifolia*: *Putoniella* sp. **383–385,** Galls of *Purshia tridentata*. **383,** Swollen flower receptacle with aborted seed in 2×: cecidomyiid. **384,** Normal receptacle and seed in 2×. **385,** Filamentous node gall, with section in 2×: *Contarinia* sp. **386,** Leaf vein swelling, *Sorbus americana*: cecidomyiid. **387–390,** Galls of *Rosa* sp. **387,** Leafy bud gall: *Dasineura* sp. **388,** Aborted flower bud: *Dasineura rhodophaga*. **389,** Curled leaves: *Dasineura* sp. **390,** Fuzzy, convex, leaf galls: *Blaesodiplosis* sp. **391,** Leaf vein swellings, *Holodiscus discolor*: *Dasineura* sp.

379

380

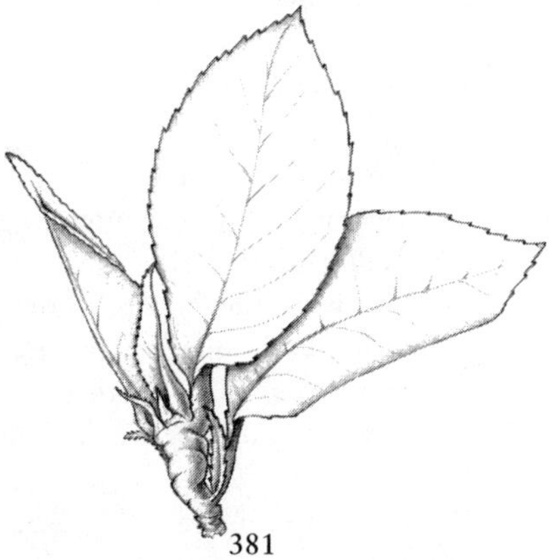
381

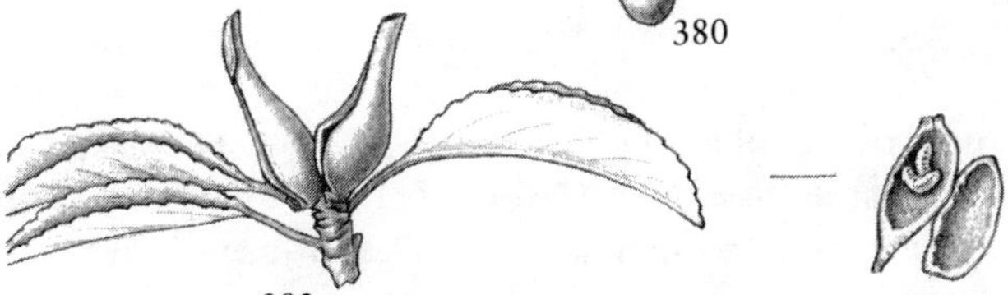
382

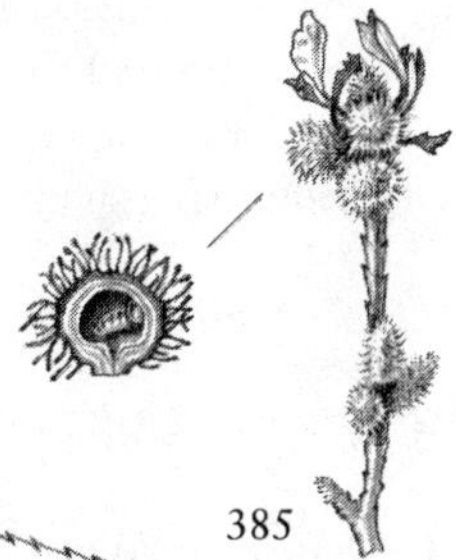
385

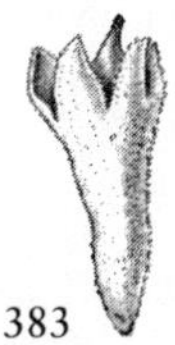
383

384

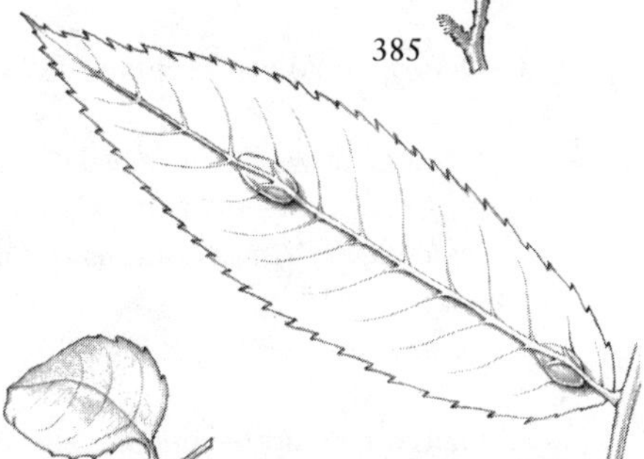
386

387

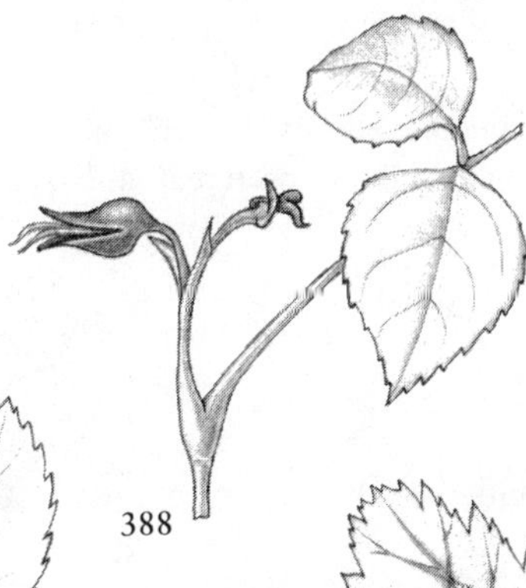
388

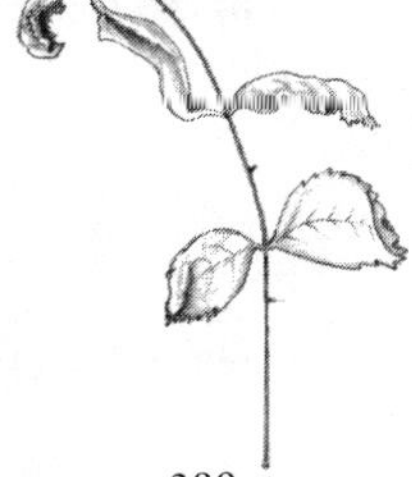
389

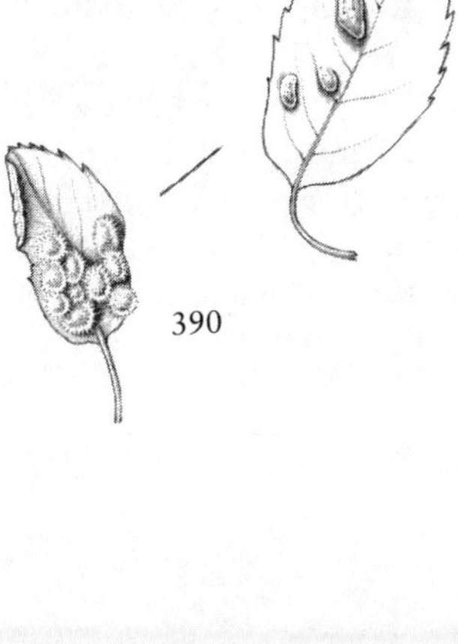
390

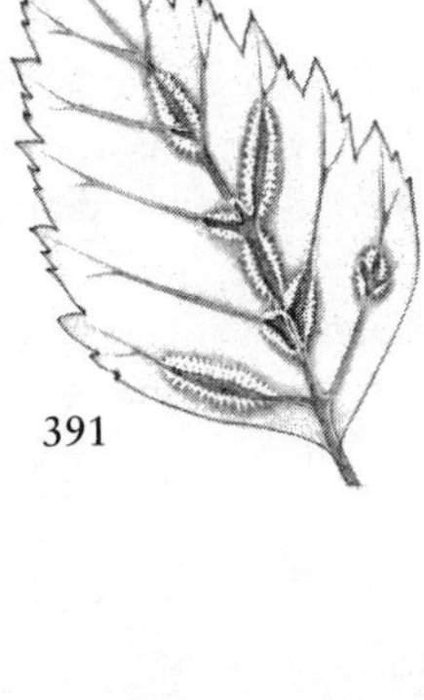
391

DLR

5a Swollen terminal bud (381)........ *Contarinia cerasiserotinae* (Osten Sacken)

Swollen buds are ovoid, 1–2 cm in length, and up to 1 cm in diameter. The swelling is soft and often bright red when it appears at the time the buds are leafing out in early spring. Leaves beyond the swelling and the bud usually die. Multiple branches may later appear below the base of the gall. Larvae live gregariously in a central cell and drop to the ground when full-grown. They presumably diapause there until the following spring. Host: *P. serotina*. Distr.: New York to Virginia, and Nebraska.

5b Leafy, apical bud gall...................................... *Contarinia* sp.

This is an undescribed species responsible for a foreshortened, leafy, cabbagelike stem growth. Larvae live free at the base of the leaves and probably drop to the ground to pupate. The one series of galls known was collected June 6. Host: *Prunus* sp. ("plum"). Distr.: Georgia.

Purshia (antelope bush)

1a Deformed flower receptacle (383, 384)......................... cecidomyiid

Infested flowers are aborted, the receptacles enlarged. Full-grown larvae in puparia were found in the plant in August. Adults have not been reared. Host: *P. tridentata*. Distr.: Montana to Oregon.

1b Filamentous node gall (385)............................... *Contarinia* sp.

Galls are monothalamous, yellow to brown. Larvae belong to *Contarinia* only in the broad sense. Adults have not been reared. Host: *P. tridentata*. Distr.: Montana to Oregon.

Pyrus (apple, pear)

1a Enlarged, sterile fruitlet (378) .... *Contarinia pyrivora* (Riley)

Infested fruit grow more quickly than normal until about 1 cm long, at which time the yellowish, gregarious larvae inside are full-grown and drop to the ground. The larvae enter diapause until the following spring. This is an immigrant from Europe, first discovered in North America in 1877 on introduced stock in Connecticut. Host: *P. communis*: pear. Distr.: widespread in northeastern United States and eastern Canada. Ref.: Barnes (1948a).

1b Curled leaf...2
2a On pear (377)................................*Dasineura pyri* (Bouché)

Eggs are laid on opening leaf buds in spring. Feeding by the gregarious larvae causes the leaves to curl under toward the midrib along their length. Infested leaves drop prematurely. Full-grown larvae drop to the soil. Several generations occur per year. This species is an apparent immigrant from Europe found in North America for the first time in Connecticut in 1927. Host: *P. communis.* Distr.: Connecticut and New York. Ref.: Barnes (1948a).

2b On apple ....*Dasineura mali* (Kieffer)

This species is an immigrant from Europe and was first recorded in North America from Massachusetts in 1928. Its biology is similar to that of *D. pyri* on pear. Host: *P. malus.* Distr.: New Brunswick, New England, and New York. Ref.: MacPhee and Finnamore (1978).

Rosa (rose)

1a Bud gall 2
1b Leaf gall...3
2a Curled and stunted buds (388)..........*Dasineura rhodophaga* (Coquillett)

This gall midge is a pest of cultivated roses. It has several overlapping generations per year. Populations build up until early fall, and a generation time can be as short as two weeks. The last generation overwinters in the ground in cocoons and emerges the following May. Eggs are laid inside the sepals of flower buds or leafy tips. Full-grown larvae leave the damaged tips, after which the buds wither, blacken, and die. Hosts: cultivated roses. Distr.: widespread in North America; possibly Europe also. Ref.: Smith and Webb (1976).

2b Large, rosette gall (387)..................................*Dasineura* sp.

Galls are 2–3 cm long. This record is based only on the gall. The damage has been attributed to *Rabdophaga rosacea* Felt, a species reared from willow (Felt 1915a). Larvae taken from among the leaves of the gall do not fit *Rabdophaga*, although they do belong to the Oligotrophidi. Host: *Rosa* sp. Distr.: Illinois and South Dakota.

3a Folded terminal leaflet (389) ....*Contarinia* sp. and *Dasineura* sp.

The single adult specimen reared from this damage was first provisionally identified as *Dasineura rosarum* (Hardy) (Felt 1908), which is a European species that causes similar damage to roses in Europe. The identification has not been confirmed. Taken from the same leaflets were three larvae of a *Contarinia*. Both genera contain species that roll leaves on various plants. *Dasineura* larvae were found in such galls in western North America. In late August those larvae were full-grown and were dropping to the ground. Hosts: *R. humilis*, *Rosa* sp. Distr.: New York and British Columbia to Oregon.

3b Fuzzy vein swellings (390). . . . . *Blaesodiplosis* sp.

Adults of this species have not been reared; larvae are tentatively placed in *Blaesodiplosis*. Galls are formed by an outpocketing along the vein of a leaflet. Full-grown larvae leave the gall through a slit on the lower side of the leaflet. Host: *Rosa* sp. Distr.: Oregon.

Rubus (blackberry, bramble, raspberry)

1a Aborted flower buds . . . . *Contarinia* sp. and *Dasineura rubiflorae* Felt

Contarinia sp. Larvae were taken in spring. Adults have not been reared. Host: *Rubus* sp. Distr.: Virginia.
D. rubiflorae. Larvae were found in May, and adults were reared the following spring. Larvae completely destroyed the infested buds. This species has not been reported since 1886. Host: *Rubus* sp. Distr.: Virginia.

1b Leaf or stem gall 2
2a Swollen stem (393) . . . . *Neolasioptera nodulosa* (Beutenmüller)

Swellings are usually on the smaller, apical branches, are sometimes strongly ridged, and contain several yellow larvae. Larvae overwinter in the galls. Hosts: *R. allegheniensis*, *R.* "*villosus*," *Rubus* sp. Distr.: New York and Ontario, west to Illinois, south to Maryland.

2b Leaf gall 3
3a Simple, smooth, marginal or vein fold. 4
3b Warty swelling or filamentous outgrowth. 5
4a Marginal leaf roll . . . . *Camptoneuromyia rubifolia* Felt

Although attributed to this species, of which one female was reared, the gall maker is probably a *Dasineura* or *Contarinia*, which form most leaf rolls. *Camptoneuromyia* spp. are not known to be responsible for any galls. Host: *Rubus* sp. Distr.: Massachusetts.

4b Folded leaf base (392) . . . . *Dasineura* sp.

Infested leaves can be severely deformed. Larvae are gregarious. When full-grown they form cocoons on the leaf and pupate there a short time later. Several generations may occur per year. Adults were reared in September from larvae found in late August. Host: *Rubus* sp. Distr.: Oregon.

5a Warty, mealy swelling (394) .... *Neolasioptera farinosa* (Osten Sacken)

Swellings occur on the leaf lamina or petiole. They are soft and fleshy in summer but become woody by late fall. Larvae overwinter in the leaves and pupate in spring. Host: *R. allegheniensis*, *R.* "*villosus*," *Rubus* sp. Distr.: Massachusetts to Ontario, south to North Carolina.

5b Filamentous outgrowth (231) .... "*Cecidomyia*" *muscosa* Stebbins

Galls were found in August. They were 1 cm in diameter, green at first, then brown, and occurred at the juncture of leaflets or on midveins. This gall may be formed by a gall wasp. Host: *R. allegheniensis*. Distr.: Massachusetts.

Sorbus (mountain-ash)

Pea-shaped swelling on leaf midrib (386) .... cecidomyiid

Swellings are mainly on the underside of the leaflets. Full-grown larvae leave the galls through the longitudinal, slitlike opening on the upper side. Host: *S. americana*. Distr.: New York.

Spiraea (spirea)

Spirea is host to a large array of cecidomyiids. Among them are *Parallelodiplosis* spp. and *Clinodiplosis* spp., which are apparently primary feeders on these plants, although most species in these genera are usually secondary gall inhabitants or fungivores. In addition to the gall midges keyed below, four additional species were reared from unspecified bud galls on *Spiraea salicifolia* in Massachusetts and New York (Felt 1918a). These are *Contarinia spiraeina* Felt, *Clinodiplosis spiraeina* (Felt), *Contarinia clarkei* (Felt), and an undescribed species of *Asphondylia*. The last is notable as one of the few records of an *Asphondylia* on Rosaceae in North America.

1a Leaf or stem gall 2
1b Bud gall 6
2a Swollen stem .... *Neolasioptera* sp.

This species is known only from an item in Felt's unpublished notebook and a larval specimen in the Felt collection. Host: *Spiraea* sp. Distr.: New York.

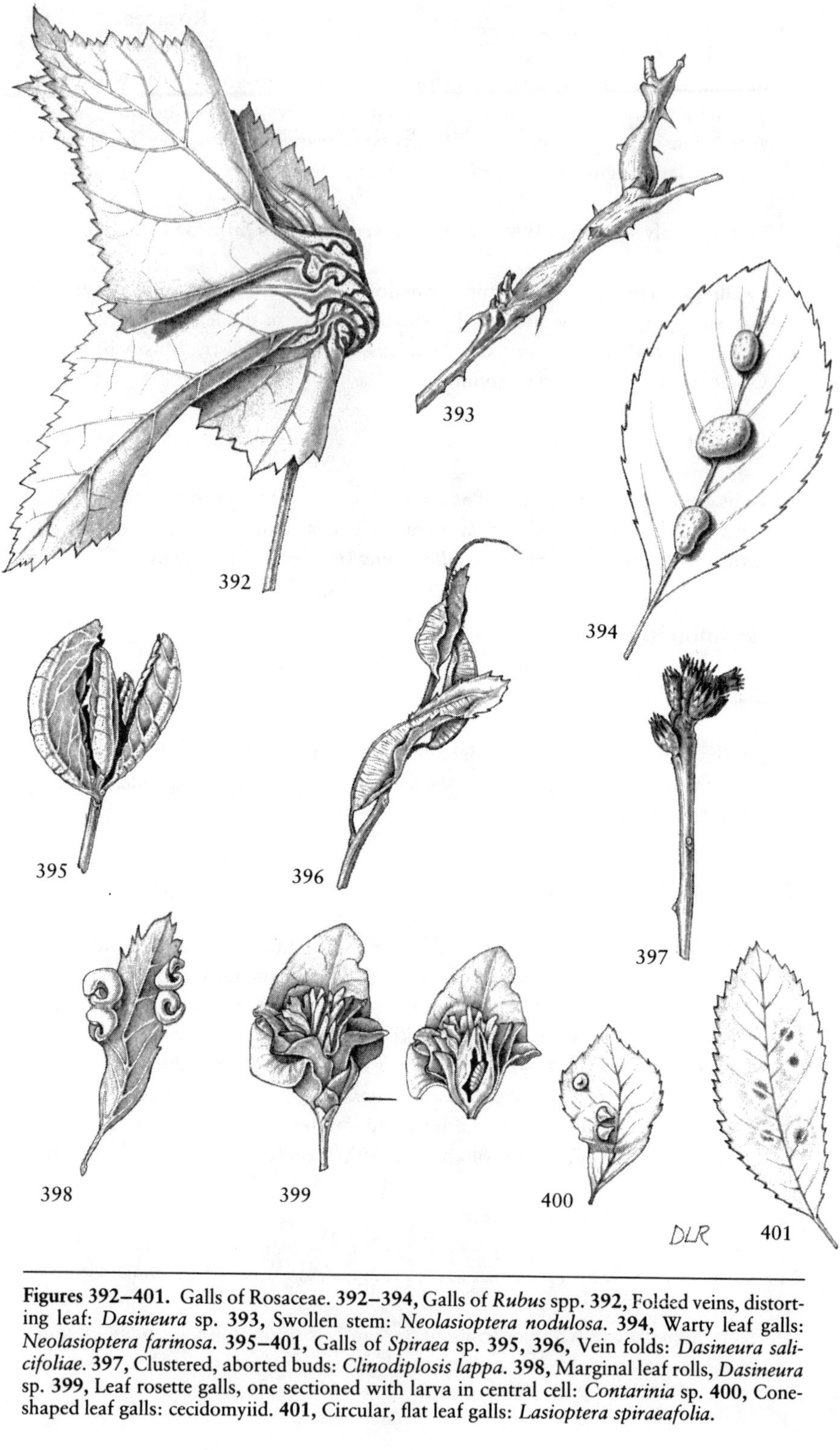

Figures 392–401. Galls of Rosaceae. **392–394,** Galls of *Rubus* spp. **392,** Folded veins, distorting leaf: *Dasineura* sp. **393,** Swollen stem: *Neolasioptera nodulosa.* **394,** Warty leaf galls: *Neolasioptera farinosa.* **395–401,** Galls of *Spiraea* sp. **395, 396,** Vein folds: *Dasineura salicifoliae.* **397,** Clustered, aborted buds: *Clinodiplosis lappa.* **398,** Marginal leaf rolls, *Dasineura* sp. **399,** Leaf rosette galls, one sectioned with larva in central cell: *Contarinia* sp. **400,** Cone-shaped leaf galls: cecidomyiid. **401,** Circular, flat leaf galls: *Lasioptera spiraeafolia.*

2b Leaf gall 3
3a Circular, flat spot (401) . . . . *Lasioptera spiraeafolia* Felt

Spots usually are surrounded by some discoloration. Adults emerged two weeks after galls were collected in June. Larvae evidently leave the galls to pupate. Hosts: *S. densiflora*, *S. douglasii*, *S. salicifolia*. Distr.: Maine to New York, Maryland, and Washington.

3b Swelling, roll, or cone-shaped gall 4
4a Cone-shaped gall on upper surface of leaf (400) . . . . cecidomyiid

Only larvae of this gall maker are known, and they cannot be placed into any known genus. Galls were collected in May. Host: *S. salicifolia*. Distr.: Massachusetts and New York.

4b Midrib swelling or marginal roll 5
5a Globular midrib swelling (395, 396) . . . . *Dasineura salicifoliae* (Osten Sacken) and *Parallelodiplosis* sp.

D. salicifoliae. Galls are especially visible from below. This is the species responsible for the damage. Larvae are gregarious. When full-grown, they leave the gall through a dorsal longitudinal slit. Hosts: *S. betulifolia*, *S. latifolia*, *S. salicifolia*, *S. tomentosa*. Distr.: New England to Ontario and British Columbia to Oregon.
Parallelodiplosis sp. This species may be only an inquiline in the leaf rolls. A larva in the Felt collection associated with this record and the adult *Parallelodiplosis* belongs to *Dasineura*. Host: *S. salicifolia*. Distr.: Massachusetts and New York.

5b Marginal roll (398) . . . . *Dasineura* sp. and *Parallelodiplosis spirae* Felt

Dasineura sp. Larvae from marginal galls are usually of this genus. Host: *S. salicifolia*. Distr.: Massachusetts.
P. spirae. The gall was attributed to this species (Felt 1921), but I have found only *Dasineura* larvae in galls. Possibly this species enters the tubelike galls after the *Dasineura* larvae depart. Host: *S. salicifolia*. Distr.: Massachusetts.

6a Slightly enlarged, reddened flowers . . . . *Clinodiplosis spiraeaflorae* (Felt) and *Prodiplosis floricola* (Felt)

C. spiraeaflorae. An adult was reared in August. Host: *S. salicifolia*. Distr.: New York.
P. floricola. Infested flower buds fail to open. Adults were reared in August. Hosts: *S. salicifolia*; also on *Clematis* (Ranunculaceae), and *Caryocar* (Clusiaceae) in Brazil. Distr.: New York; Brazil.

6b Bud gall 7
7a Globular, axil bud gall, bud scales appressed.................. cecidomyiid

The gall maker is unknown. Host: *Spiraea* sp. Distr.: unspecified.

7b Linear or rosette, usually apical bud gall................................ 8
8a Narrow, elongate, clustered buds (397)........ *Clinodiplosis lappa* (Stebbins)

Infested buds are made up of long leaf scales. The buds are aborted. Host: *S. salicifolia*. Distr.: Massachusetts.

8b Leafy, rosette galls (399).................................. *Contarinia* sp.

Larvae belong to *Contarinia* only in the broad sense. Adults have not been reared. Galls are monothalamous. Larvae drop to the soil in fall. Apparently only one generation occurs per year. Larvae of *Clinodiplosis* spp. are often found among the outer leaves of the gall. Host: *S. douglasii*. Distr.: Washington.

Rubiaceae

Borreria

Asphondylia spp. in the USNM have been reared from two kinds of galls on *Borreria terminalis* in Florida, one from an unspecified stem gall, the other from apical bud galls. Adults of both species were reared in October and November.

Cephalanthus (buttonbush)

Swollen stem .... *Rabdophaga cephalanthi* Felt

Adults were reared in spring from stems collected the previous fall. Host: *C. occidentalis*. Distr.: New York.

Galium (bedstraw)

Swollen, aborted flower bud .... *Contarinia* sp. and *Dasineura americana* Felt

Contarinia sp. This undescribed species was reared from larvae from the same kind of gall. Host: *G. asprellum*. Distr.: New York.
D. americana. This species was reared in early September from yellowish larvae. Galls were reportedly irregularly fusiform, up to 13 mm long, and 7–10 mm in diameter. Host: *G. asprellum*. Distr.: Massachusetts and North Carolina.

Psychotria (wild coffee)

Swollen stem and leaf . . . . cecidomyiid

The gall maker is known only from pupal exuviae protruding from the galls. Swellings of the stems were 10 mm long and 8 mm at the widest point. Host: *Psychotria* sp. Distr.: Florida.

Rutaceae

Citrus

Larva in flower . . . . *Prodiplosis longifila* Gagné

Larvae are gregarious; over 40 larvae have been counted in some flowers. Larvae damage the ovary and cause the fruit to abort. Hosts: *Citrus* sp. (Tahiti lime); also wild cotton in Florida and various crop plants in Peru. Distr.: Florida; Peru.

Salicaceae

Populus (aspen, cottonwood, poplar)

Gall midges of several genera occur on poplars, but one, *Harmandia*, is evidently a large and poorly known genus restricted to these plants. One gall midge species not treated in the key is "*Rabdophaga*" *populi* Felt, which is responsible for some unspecified damage to *P. tremuloides*. The only known series of *populi* was reared in spring in New York from cocoons found at the base of buds.

1a Enlarged winter bud (403). . . . . *Dasineura* sp.

Infested buds are conspicuous in summer. The red larvae are immersed in a sticky exudate between the bud scales. Larvae form cocoons in fall, diapause as larvae, and pupate in spring. Infested buds burst, but the leaves do not develop. Host: *P. trichocarpa*. Distr.: Washington and Oregon. Ref.: Larew and Capizzi (1983).

1b Leaf gall 2

2a Variously shaped, blister or globular gall on petiole or leaf lamina (404) . . . . *Harmandia* spp.

Several kinds of these leaf galls have been collected, but only two Nearctic cecidomyiids have been described from particular galls. *Harmandia* galls are monothalamous and yellow, green, or reddish. In late spring or summer full-grown larvae drop to the ground to overwinter (Whitney and Baranyay 1968). Various other cecidomyiids have also been reared from these galls. These include a *Clinodiplosis* sp., *Mycodiplosis thoracica*, and *Rhizomyia absobrina*, all presumably mycophages.

H. helena (Felt). This species was taken from a globular gall reportedly on the underside of the leaf. Host: *P. tremuloides*. Distr.: Massachusetts.

H. stebbinsae (Gagné). The gall (404a) is a blister, convex and reddish on the upper surface of the leaf, flat and light green below, occurring between or on veins. Host: *P. grandidentata*. Distr.: Massachusetts and Maryland.

2b Rolled leaf or curled leaf margin (402). . . . . *Prodiplosis morrisi* Gagné

Eggs are laid in newly developing leaf buds. Larvae are gregarious within the leaf rolls of marginal curls. Full-grown larvae drop to the ground to pupate. Up to five generations occur per year in Ontario. Infested leaves become deformed and later turn brown. Severe damage can stunt terminal leaders. Hosts: *P. deltoides*, *P. tremuloides*, and hybrid poplars. Distr.: eastern North America to Kansas and Texas. Ref.: Morris (1981).

Salix (willow)

Numerous species of gall midges are known from willows in North America, but comprehensive work has been done on only a few of them. Many species of *Rabdophaga*, the preponderant group of cecidomyiids on willows, have been described from only one or a few specimens taken from a variable twig enlargement. For many species in the key that follows, only the original collection state and host(s) are listed, although many of the galls may be common and widespread.

Many species were reared from willow but without definite reference to particular damage. These species are only listed here and not treated in the key below. Reference to the few details known of their biology can be found in Felt (1915a, b, 1916, 1940). These species are: *Rabdophaga americana* (Felt), *Rabdophaga caulicola* (Felt), *Rabdophaga caudicis* Gagné, *Rabdophaga essigi* Felt, *Rabdophaga globosa* Felt, *Rabdophaga latebrosa* Felt, *Rabdophaga latipennis* (Felt), *Rabdophaga normaniana* Felt, *Rabdophaga occidua* Gagné, *Rabdophaga perocculta* (Cockerell), *Rabdophaga persimilis* Felt, *Rabdophaga podagrae* Felt, *Rabdophaga ramuscula* Felt, *Rabdophaga racemi* Felt, *Rabdophaga sodalidatis* Felt, *Rabdophaga timberlakei* (Felt), *Rabdophaga tumidosae* (Felt), *Rabdophaga walshii* (Felt), and *Asphondylia*

Figures 402–404. Galls of Salicaceae: *Populus* sp. **402,** Rolled leaves and curled leaf margins, *P. tremuloides*: *Prodiplosis morrisi*. **403,** Enlarged buds, scales removed from one to show larvae, *P. trichocarpa*: *Dasineura* sp. **404,** Leaf galls, upper leaf surface on left, lower surface on right, *Populus* sp.: **a,** slightly convex blister: *Harmandia stebbinsae*; **b,** hemispherical galls: *Harmandia* sp., possibly *helena*.

402

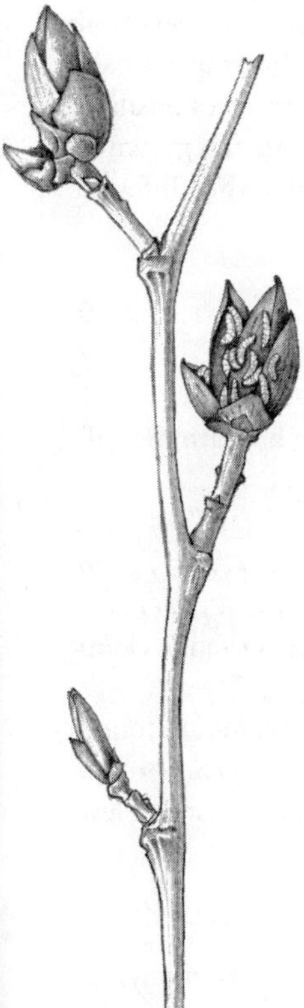

403

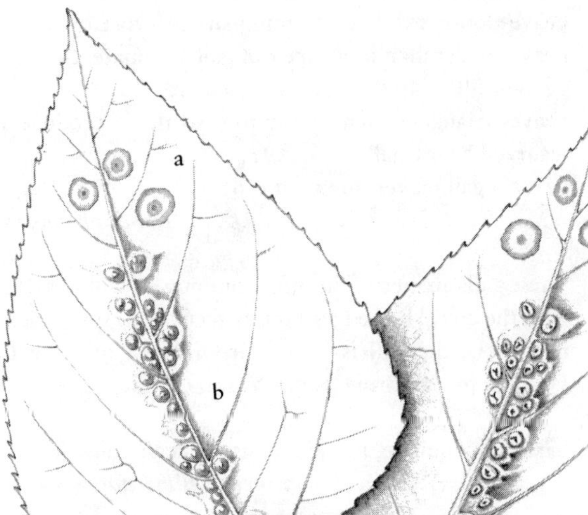

404

DLR

salictariae Felt. The last species may not have come from willow. The single known specimen is labelled from Indiana, supposedly from willow, but the original, unpublished collection data associated with Felt's specimen refers to probable rabdophagas reared from willow in Washington, D.C. Since no other *Asphondylia* has ever been reared from willow, *A. salictariae* may be from a different host.

1a Bud gall 2
1b Leaf gall or simple twig enlargement. 10
2a Bud gall comprising many modified leaves 3

Leaves of these galls are typically largest on the outside of the gall but narrow considerably toward the center of the gall. In this key only the outer leaf shape of the mature gall is described. The gall formers are always at the heart of the gall, surrounded by linear leaves. The larva feeds upside down and reverses position before pupation. The gall makers have one generation per year, are full grown in fall, overwinter in the gall in a cocoon, and pupate the following spring. Infested twigs are considerably stunted. Ref.: Wilson (1968c).

2b Bud gall not comprising imbricated, modified leaves. 9
3a Leaves generally appressed, imbricate, simulating a pine cone. 4
3b Leaves loose, exfoliate, forming an open rosette 7
4a Leaves wider than long, apex of gall attenuate, closed, the twig beneath the gall straight (405–407) 5
4b Leaves triangular, narrowing toward tip to produce an open "beak," or twig recurved below gall 6
5a Larva a gall maker, found at center of gall (405, 406) . . . . *Rabdophaga strobiloides* (Osten Sacken)

These galls are one of the most common on willow. Occasionally some are found with the apical leaf edges slightly recurved. Host: originally found on *S. cordata* in Illinois, the gall is known from many other willow species. Distr.: New England to California. Ref.: Wilson (1968c).

5b Larva an inquiline, found among the leaf scales . . . . *Dasineura albovittata* (Walsh) and other cecidomyiids

---

Figures 405–413. Galls of Salicaceae: *Salix* spp. All except dissected galls of **405** and **413** drawn from galls out of original type series. **405, 406,** "Cone" galls: *Rabdophaga strobiloides*. **405,** Terminal, branch galls, one with leaves removed to show central larval cell. **406,** Clustered, lateral branch galls possibly attributable to *R. strobiloides*. **407,** "Cone" gall with recurved stem: *Rabdophaga salicisgnaphaloides*. **408,** "Cabbage" gall: *Rabdophaga salicisbrassicoides*. **409,** Foreshortened, swollen stem: *Rabdophaga salicistriticoides*. **410,** "Rose" gall: *Rabdophaga salicisrhodoides*. **411,** "Horn" galls: *Rabdophaga saliciscornu*. **412,** "Hazel" gall: *Rabdophaga saliciscoryloides*. **413,** "Beaked" galls, one in section: *Rabdophaga rigidae*.

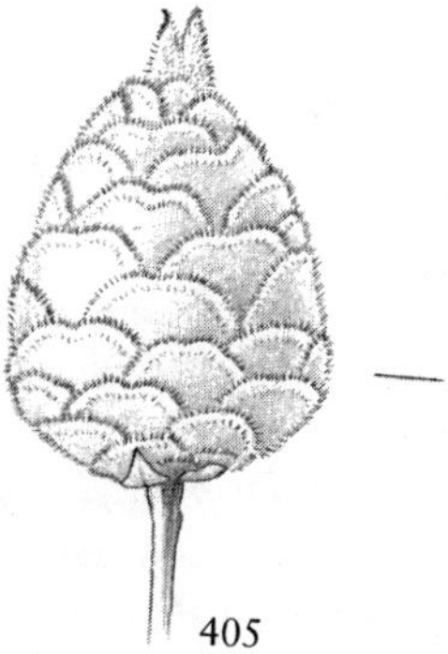
405

406

407

408

409

410

412

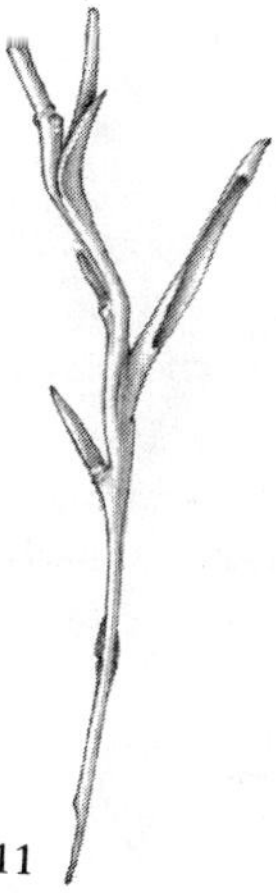
411

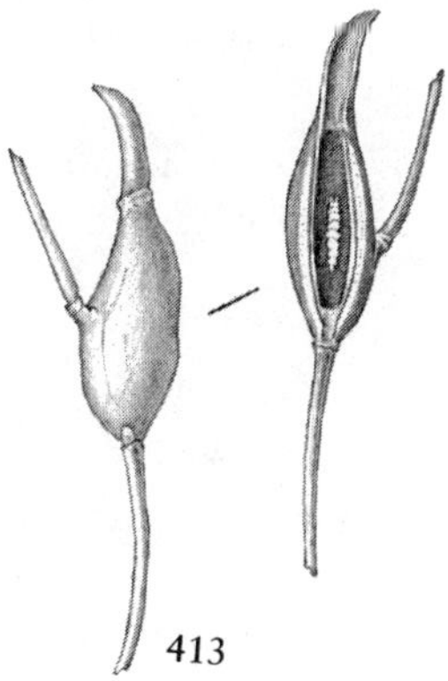
413

DLR

D. albovittata. Red larvae of this species are commonly found in large numbers. The adults are much smaller than the gall former, and their legs have alternating light and dark bands. At least two generations occur each year, one in spring when the gall forms and another that overwinters in cocoons between leaves. *Dasineura gemmae* Felt from *Salix* sp. in California has the same habits. Hosts: *Salix* spp. Distr.: Massachusetts to Maryland, west to California. Ref.: Walsh (1864).

Other cecidomyiids. Predaceous *Lestodiplosis* spp. have been reared from these galls as well as four species whose types are lost and cannot now be identified with certainty: "*Cecidomyia*" *atrocularis* Walsh, "*Cecidomyia*" *atricornis* Walsh, "*Cecidomyia*" *annulipes* Walsh, "*Cecidomyia*" *decemmaculata* Walsh.

6a Tip of gall beaklike; twig beneath gall not recurved .... *Rabdophaga salicisstrobiliscus* (Osten Sacken)

The original reporter of this gall (Walsh 1864) suggested that it might only be an aberrant form of that formed by *R. strobiloides*. Hosts: *S. discolor*, *S. rostrata*. Distr.: Illinois.

6b Tip of gall not beaklike; twig beneath gall recurved (407) .... *Rabdophaga salicisgnaphaloides* (Osten Sacken)

Galls are smaller than those of *strobiloides*, only about 5–10 mm long and 5 mm wide. Hosts: *S. candida*, *S. humilis*. Distr.: Illinois.

7a Leaves conspicuously longer than wide (408) .... *Rabdophaga salicisbrassicoides* (Packard)

Galls range from 5–25 mm in diameter, the size evidently dependent on the size of the affected twig. A predaceous cecidomyiid *Lestodiplosis septemmaculata* (Walsh) has been reared from these galls. Host: *S. longifolia*. Distr.: Illinois. Ref.: Walsh (1864).

7b Leaves almost as wide as long... 8

8a Gall leaves open, flared, the gall shorter than wide (412) .... *Rabdophaga saliciscoryloides* (Osten Sacken)

Galls are 5–7 cm long and 5–10 cm in diameter. The difference between these and galls of *R. salicisbrassicoides* and *R. salicisrhodoides* may be only superficial. Host: *S. ?discolor*. Distr.: Illinois. Ref.: Walsh (1864).

8b Gall leaves more sessile, the gall longer than wide (410) .... *Rabdophaga salicisrhodoides* (Osten Sacken)

The gall simulates an elongated rose blossom. It is 1–2 cm in diameter and 2–5 cm long. Host: *S. humilis*. Distr.: Illinois. Ref.: Walsh (1864).

9a Complex gall; lateral bud deformed into elongate, thin-walled, hornlike structure (411)........................ *Rabdophaga saliciscornu* (Osten Sacken)

Galls are monothalamous, 10–20 mm long, about 2 mm wide, pointed at the tip, and slightly curved. Host: *S. humilis*. Distr.: Illinois. Ref.: Walsh (1864).

9b Simple gall; terminal bud greatly swollen (413) .... *Rabdophaga rigidae* (Osten Sacken)

This is probably the most common of the willow galls. Galls are monothalamous, subglobular to tapered, 1–3 cm long, and 0.5–1.0 cm wide. The galls are first evident in May. A few leaves often begin to grow beyond the gall but soon die. After the adults emerge from the galls the following spring, the buds usually die, but new growth may develop from around the base of the gall. Hosts: many species of willow. Distr.: New England to the West Coast; eastern Asia, including Japan. Refs.: Nijveldt and Yukawa (1982), Smith et al. (1983), Wilson (1968b).

10a Swollen twigs 11
10b Leaf gall... 13
11a Foreshortened twig, the buds close together (409) .... *Rabdophaga salicistriticoides* (Osten Sacken)

Buds may be so close as to resemble kernels on a wheat head. The enlargement is 2–4 cm long. Larvae overwinter in the galls. In spring the pupae escape from the kernellike buds. Hosts: *S. cordata* and *S. humilis*. Distr.: New York and Illinois. Ref.: Walsh (1864).

11b Internodal twig swelling without foreshortening the stem 12
12a Twig swelling very large, soft, sometimes resembling a potato in shape (416) .... *Rabdophaga salicisbatatas* (Osten Sacken)

This gall is variable but usually polythalamous and bulging abruptly out of the twig. One generation occurs per year. Adults emerge from the galls in spring. An apparently inquilinous cecidomyiid, "*Cecidomyia*" *orbitalis* Walsh, was reportedly reared from these galls. The type specimens are lost, so the species cannot be identified to genus. Hosts: *S. humilis* and many other species of *Salix*. Distr.: New England to California. Ref.: Walsh (1864).

12b Twig swelling tapered, woody, sometimes barely noticeable (417) .... *Dasineura corticis* Felt, *Lygocecis* spp., *Neolasioptera* sp., *Rabdophaga* spp., as well as sawflies

D. corticis. This species was reared in May from small cavities beneath bark. It may be an inquiline or a secondary feeder. The name "*Cecidomyia*" *orbitalis* Walsh might be applicable to this *Dasineura*. Host: *Salix* sp. Distr.: Massachusetts.

L. packardi (Felt), *L. porterae* (Cockerell), and two other, undescribed species of *Lygocecis* in the USNM, were reared from enlarged willow stems. These species have sawlike ovipositors evidently used to insert their eggs into willow shoots. *L. packardi* comes from several species of willow from Massachusetts to Texas; *L. porterae* was reared from a willow in New Mexico.

Neolasioptera sp. A species was reared in May from *Pontania* (Hymenoptera: Tenthredididae) galls after the sawflies had emerged. The flies may have been inquilines. Host: *S. humilis*. Distr.: Ontario.

R. cornuta (Walsh). This is a borer in larger willow stems. Walsh (1864) found it associated with stems affected by *R. salicisbrassicoides*. Hosts: *Salix* spp. Distr.: Illinois. Ref.: Walsh (1864).

R. ?salicis (Schrank). The damage this species does is apparently similar to that of *R. cornuta*. *R. salicis* occurs on many willows in both Europe and Japan; the specimens identified as this species in North America were found on imported nursery stock. They could well be this or some other European species, but American specimens identified in the past as *R. salicis* do not compare exactly with European specimens identified as *salicis*. Ref.: Nijveldt and Yukawa (1982).

R. salicisnodulus (Osten Sacken). This was originally reared from monothalamous galls, but similar galls are often polythalamous, some harboring more than one species. Heavily infested shoots can be killed. The exit holes seen in spring are cut by the pupae. This and other *Rabdophaga* and *Lygocecis* spp. listed probably have only one generation per year. Host: *Salix* spp. Distr.: New England to Illinois. Ref.: Wilson (1968a).

13a Rolled leaf (415)................................ *Dasineura plicata* (Felt)

Leaves are tightly rolled into a tube. Full-grown larvae form cocoons in the leaves. Adults were reared in May and August. Host.: *Salix* sp. Distr. Massachusetts and New York.

13b Spherical or tubelike leaf gall (414)................................ 14

14a Spherical gall protruding from only one side of leaf (414b) .... *Iteomyia salicifolia* (Felt)

Galls are monothalamous and about 2 mm in diameter. Full-grown larvae leave the galls in late summer and spin cocoons in which they overwinter. Host: *Salix* sp. Distr.: New York.

14b Tube gall, beaked on one side of leaf and convex on the other (414a) .... *Iteomyia salicisverruca* (Osten Sacken)

Galls are monothalamous, 2–3 mm in diameter, and often coalesced. The biology is unknown, but larvae presumably leave the galls in the fall, as do larvae of *I. salicifolia*. Host: *S. humilis* and other willows. Distr.: New York, Illinois, Washington, and Oregon. Ref.: Walsh (1864).

Sapotaceae

Bumelia

Irregular twig swelling (420) .... *Asphondylia bumeliae* Felt

Swellings are 2–5 cm long and polythalamous. Each larva lives in a separate cell in the cambium layer. Galls are present throughout the year. Adults of the type series from Texas emerged from the galls in May. Host: *B. lanuginosa*. Distr.: South Carolina, Louisiana, and Texas.

Saxifragaceae

Fendlera

Enlarged, woody bud .... cecidomyiid

Infested buds are woody, with a central cell inhabited by several larvae. Only early-instar larvae are known, so they cannot be identified further. Host: *F. rupicola*. Distr.: Arizona.

Hydrangea

Enlarged bud (419) .... *Asphondylia hydrangeae* Felt

Infested buds are monothalamous and about 1 cm long by 1 cm in diameter. Adults emerge from the buds in spring. One generation occurs per year. Host: *H. arborescens*. Distr.: Maryland, Virginia, and Missouri.

Ribes (gooseberry)

The damage caused by "*Mayetiola*" *californica* Felt was not specified, so that species is not treated in the key. It was reared in February from larvae collected near tips of the previous year's growth on *R. menziesii* in California.

1a Enlarged and deformed fruit.............. "*Cecidomyia*" *grossulariae* Fitch

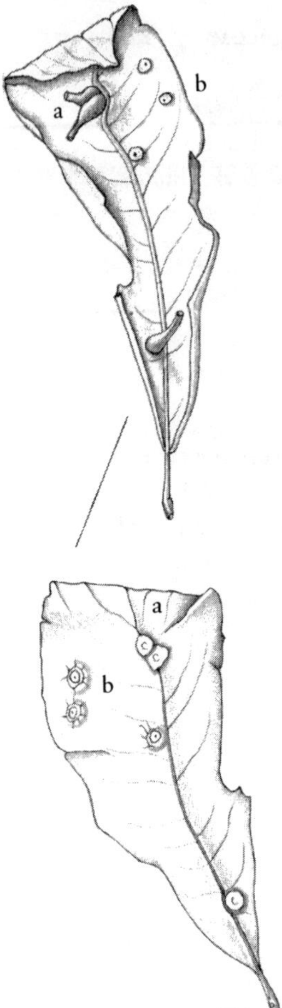

414

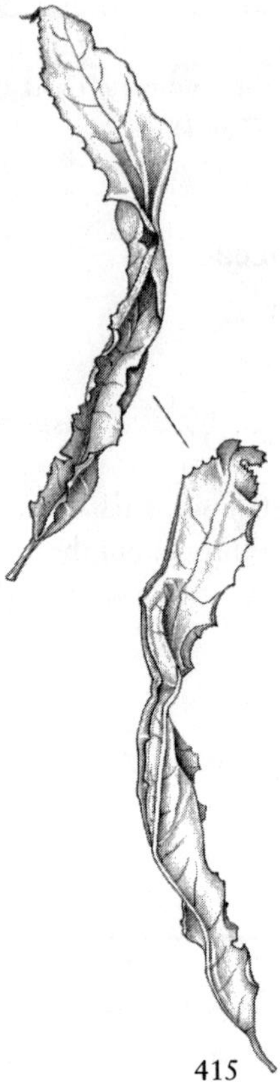
415

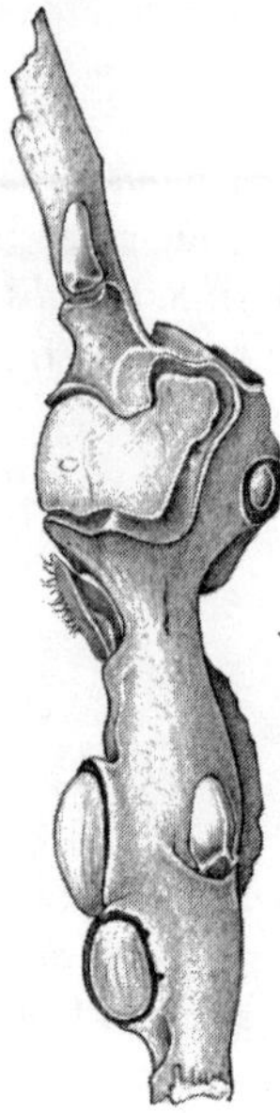
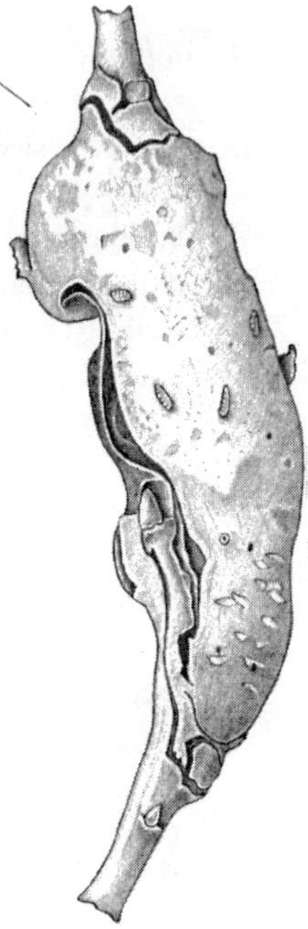
416

417

418

419

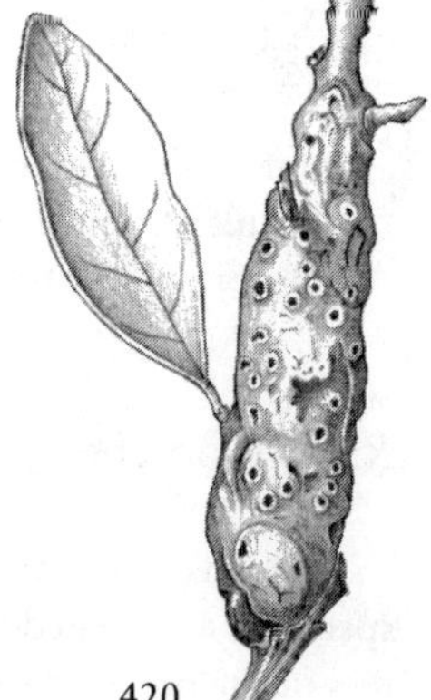
420

Adults that form part of the reared type series belong to a species of *Hyperdiplosis* and may have been secondary fungus-feeders in the fruit. Larvae from typical galls are a species of Oligotrophidi and are most likely the gall maker. Infested fruit develop and drop earlier than normal fruit. Host: *R. grossularia*. Distr.: New York and Wyoming.

1b Bud, leaf, or stem gall 2
2a Enlarged, deformed bud (418) .... *Rhopalomyia grossulariae* Felt

This species forms an enlarged cluster of buds with greatly increased size and number of scales. Infested buds are killed. Witches' brooms may later develop when secondary buds at the base of a gall send forth new growth. Larvae overwinter in the buds and pupate there in spring. Adults emerge in May. Host: *R. grossularia*. Distr.: Ohio. Ref.: Houser (1912).

2b Leaf or stem gall 3
3a Leaf blister .... cecidomyiid

Larvae from the blisters belong to the Oligotrophini. They do not fit into any known genus. The leaf blisters, about 3 mm in diameter, are reportedly common and several may occur on a leaf. Hosts: *R. inerme* and *R. longifolium*. Distr.: Colorado.

3b Irregular stem gall .... *Neolasioptera* sp.

The gall is reportedly about 2 cm long and 1 cm in diameter. The gall maker was listed elsewhere as *Cecidomyia* sp. (Felt 1940). A series of adults in the USNM from "currant" in Washington belong to *Neolasioptera* and may be the species reported earlier. Host: *Ribes* sp. Distr.: Washington.

Scrophulariaceae

Diplacus

1a Cabbagelike, woolly, apical bud gall .... *Asphondylia diplaci* Felt

---

Figures 414–420. Galls of Salicaceae, Sapotaceae, and Saxifragaceae. **414,** Leaf galls, lower leaf surface at top, upper surface at bottom, *Salix* sp.: **a,** tube galls: *Iteomyia salicisverruca*; **b,** spherical galls: *Iteomyia salicifolia*. **415,** Leaf fold, upper leaf surface at top, lower at bottom, *Salix* sp.: *Dasineura plicata*. **416,** Soft stem swellings, *Salix* sp.: *Rabdophaga salicisbatatas*. **417,** Hard stem swelling, half in section, *Salix* sp.: *Rabdophaga salicisnodulus*. **418,** Enlarged, deformed buds, *Ribes grossularia*: *Rhopalomyia grossulariae*. **419,** Swollen buds, one in section, *Hydrangea arborescens*: *Asphondylia hydrangeae*. **420,** Stem swelling with exit holes, *Bumelia lanuginosa*: *Asphondylia bumeliae*.

Host: *D. longiflorus*. Distr.: California.

1b Swollen stem. . . . . *Neolasioptera diplaci* (Felt)

Adults emerged in November. Host: *D. longiflorus*. Distr.: California.

Gerardia

Larva in seed pods . . . . *Contarinia* sp.

This species is the same that was elsewhere recorded under *Cecidomyia* sp. (Felt 1940). Larvae were collected in September. Host: *G. flava*. Distr.: New Jersey.

Keckiella

Swollen stem . . . . *Lasioptera tibialis* Felt

This species may be an inquiline. It has been reared also from *Atriplex* sp. in association with a *Neolasioptera* gall. Adults were reared in April. Host: *K. antirrhinoides*. Distr.: California.

Leucophyllum

Larva on leaf . . . . *Contarinia* sp.

Larvae of this unidentified species were intercepted on the host at the United States–Mexico border. Host: *L. frutescens*. Distr.: Mexico.

Mimulus

Elongate stem swelling . . . . *Neolasioptera mimuli* Felt

Adults were reared in November. Host: *M. "glutinosus."* Distr.: California.

Scrophularia

Distorted flower bud . . . . *Contarinia* sp.

A male *Lestodiplosis scrophulariae* (Felt) was reared from this damage, but the larva associated with it belongs to a *Contarinia* sp., which probably causes the damage. The larva was collected in July. Host: *S. marilandica*. Distr.: New York.

Simmondsaceae

Simmondsia (jojoba)

Deformed, aborted fruit .... *Asphondylia websteri* Felt

Hosts: *S. chinensis*; also *Medicago sativa* and *Parkinsonia* spp. (Fabaceae). Distr.: on jojoba in California, Arizona, and Baja California. Refs.: Gagné and Woods (1988), Pinto and Frommer (1980).

Solanaceae

Capsicum (pepper)

1a Swollen stem.................... *Neolasioptera* sp., probably *capsici* Möhn

This record is based only on galls and larvae intercepted at the United States–Mexico border. *Neolasioptera capsici* is known from *Capsicum baccatum* in El Salvador. Its larvae appear similar to those of this record, so the two records could represent the same species. Host: *Capsicum* sp. Distr.: Mexico.

1b Bud, flower, or fruit gall................................ *Asphondylia* sp.

This species is known from adults and pupae intercepted at the United States–Mexico border. The particular gall it forms is unknown, but is likely to be in the flowers or fruit. Host: *C. annuum*. Distr.: Mexico.

Lycium (matrimony vine)

Swollen stem .... ?*Neolasioptera* sp.

This record is based only on the gall, which may likely be formed by a *Neolasioptera*. Host: *L. torreyi*. Distr.: New Mexico.

Lycopersicon (tomato)

Larva in fruit .... *Lasioptera kallstroemia* Felt and *Neolasioptera* sp.

These two species have been found in association with one another. The *Lasioptera* also has been found on *Kallstroemia*, so is probably an inquiline. Hosts: *L. lycopersicum*; also *Kallstroemia* (Zygophyllaceae). Distr.: Sinaloa and elsewhere in Mexico.

Solanum

Swollen stem .... *Lasioptera solani* (Felt) and *Neolasioptera* sp.

Both species have been reared from stem galls, but the latter is the more common. Larvae overwinter in the gall, and adults emerge in spring. Host: *S. carolinense.* Distr.: Maryland and North and South Carolina.

Taxodiaceae

Taxodium (bald cypress)

1a Swollen, aborted seed.......................... *Sequoiomyia taxodii* (Felt)

Larvae overwinter in the seeds, and adults emerge in spring. Host: *T. distichum.* Distr.: Illinois to Arkansas, and from a Pleistocene fossil from Maryland.

1b Branchlet or needle galls (421a–d) .... *Taxodiomyia* spp. 2

These species form complex galls on bald cypress. Many kinds of galls are known, but cecidomyiids have been described from only three of them, which are keyed below. Galls develop quickly in spring while the larvae are still in an early stage; they drop to the ground with the branchlets in autumn. Larvae overwinter in the galls.

2a Greatly swollen, polythalamous, elongate-ovoid branchlet gall (421a) .... *Taxodiomyia cupressiananassa* (Osten Sacken)

The gall is spongy and has a whitish bloom when fresh. Many rudimentary needles emerge from it to give it the look of a pineapple. This species has two generations per year in Illinois. Host: *T. distichum.* Distr.: Maryland to Florida, Illinois to Louisiana. Refs.: Chen and Appleby (1984a, b).

2b Monothalamous needle gall 3

3a Bluish-white, obconic gall (421d, color plate 4F) .... *Taxodiomyia cupressi* (Schweinitz)

Hosts: T. *ascendens* and T. *distichum.* Distr.: Maryland to Florida, Illinois to Louisiana.

3b Brown, swollen gall resembling a sausage (421c) .. *Taxodiomyia taxodii* (Felt)

Hosts: T. *ascendens* and T. *distichum.* Distr.: Maryland to Florida, Illinois to Louisiana.

Tiliaceae

Tilia (basswood, linden)

1a Warty, circular gall on various plant parts (423) .... *Contarinia verrucicola* (Osten Sacken)

Galls may occur on leaves, including the petioles, fruit, and twigs. Each spot usually contains one larva, occasionally two. Galls develop in spring. Full-grown larvae exit from the underside of the leaf to overwinter in the soil. The species is probably univoltine. Adults have not been reared. This species was found on both American and European lindens growing within 150 feet of one another in Illinois (note 2686, Felt unpublished notebook). No gall similar to this is known from Europe. Hosts: *T. americana*, *T. europaea*. Distr.: Massachusetts to Minnesota, south to Maryland and Missouri.

1b Irregularly swollen young bud or twig (422) .... *Contarinia citrina* (Osten Sacken)

Galls are succulent when young, occasionally grow to 2.5 cm in diameter, and are polythalamous. Full-grown larvae drop to the soil to overwinter. Adults have not been reared. Generally similar galls are made in Europe by *Contarinia tiliarum* Kieffer on *Tilia* spp. Host: *T. americana*. Distr.: Massachusetts to Maryland, and Illinois.

Turneraceae

Piriqueta

Larva in flower heads .... *Contarinia* sp.

Adults were reared in fall from flower heads. Host: *P. caroliniana*. Distr.: Florida.

Ulmaceae

Celtis (hackberry)

1a Swollen twig .... ?*Lasioptera* sp.

This gall is known only from a listing by Felt (1940). The swelling should be woody and solid except for the larval cells and unlike the hollow twig swelling

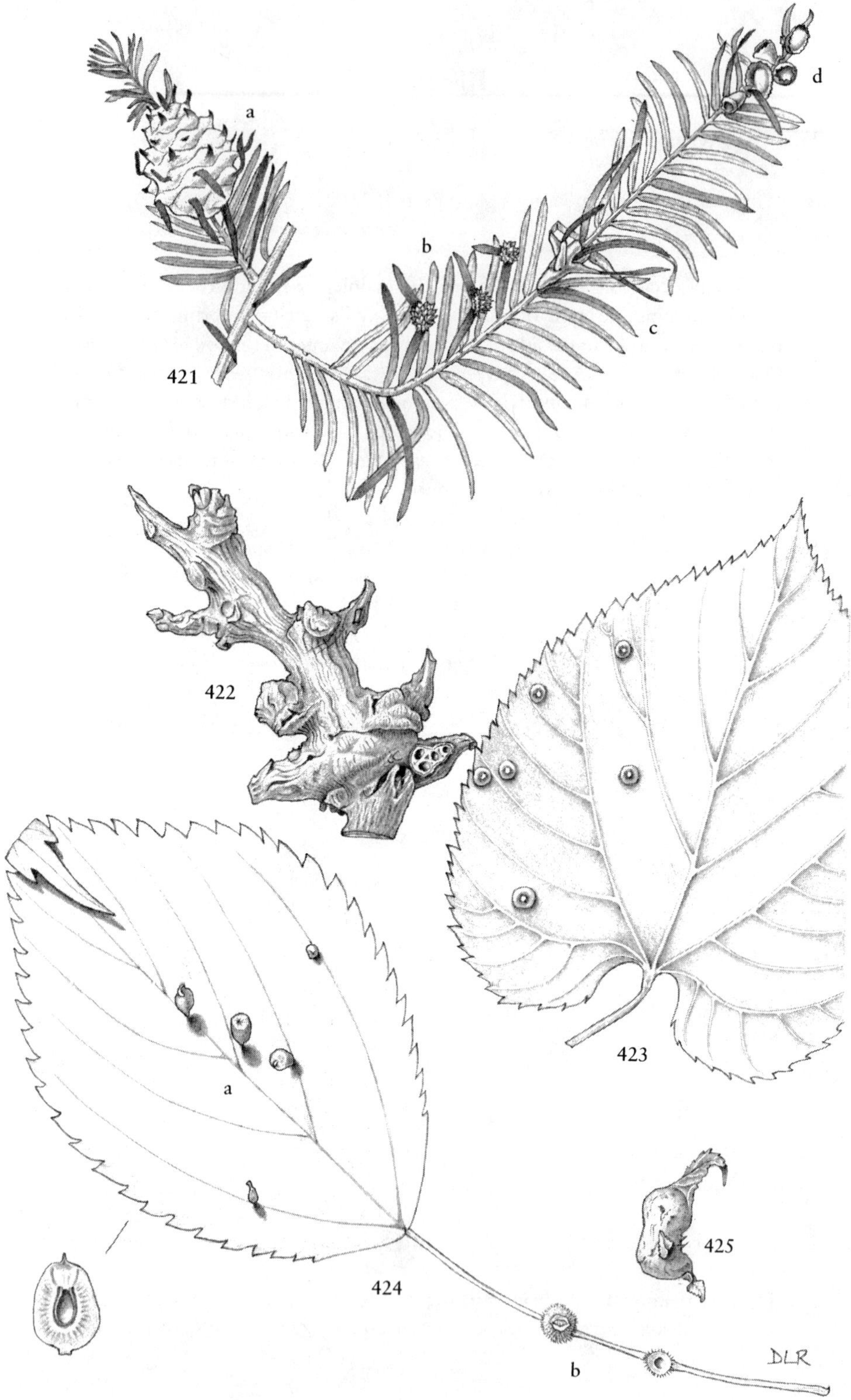
421
a
b
c
d
422
423
424
a
b
425
DLR

caused in early spring by the agromyzid fly *Cecidomyiaceltis deserta* Patton (Gagné 1983a).

1b Complex galls on twigs, petioles, leaves, or fruit 2

2a Nondehiscent galls made by *Pachypsylla* (Homoptera: Psyllidae) (430j); inquilinous cecidomyiid .... *Parallelodiplosis acernea* (Felt)

Eggs are laid through early summer near the *Pachypsylla* galls. The gall shown in 430j is made by *Pachypsylla celtidismamma* (Riley). Upon hatching, the inquilines crawl into the orifice of the gall. Three to 10 early-instar larvae may occur in a gall, but only one will live to maturity. The gall midge larvae cause the death of the psyllid nymphs, presumably through successful competition for food. About 1% of psyllid galls are affected. Host: *C. occidentalis*. Distr.: New York to Minnesota and North Carolina. Ref.: Moser (1965).

2b Dehiscent gall (430a–i) .... *Celticecis* spp.

Ten species of *Celticecis* have been described from North America east of the Rocky Mountains. Five are known from their galls only and, of the remaining five, adults from only three are known. Because so little is known of any except *C. spiniformis*, all are treated as a unit here. Galls of nine of them are illustrated in 430. *Celticecis capsularis* (Patton) (f), *C. oviformis* (Patton) (d), *C. pubescens* (Patton) (g), *C. semenrumicis* (Patton) (i), and *C. spiniformis* (Patton) (h) were described from unspecified hackberries from Texas; *C. unguicula* (Beutenmüller) (e) and *C. wellsi* (Felt) (a) from *Celtis occidentalis* in Ohio; *C. celtiphyllia* (Felt) (b) from *Celtis occidentalis* from Iowa; *C. texana* (Felt) (c) from *Celtis reticulata* in Texas. *C. painteri* (Felt) was described from *Celtis reticulata* and *Celtis mississippienis* from Texas. The gall of *Celticecis painteri* (Felt) was not well characterized and is not illustrated here, but its original description indicates that it may be similar to *C. capsularis*. Additional undescribed kinds of *Celticecis* galls are known from museum specimens but are not illustrated here. Galls are monothalamous and usually appear on the under surface of the leaves, but can occur also on the upper surface, the petioles, or on twigs. Most gall growth occurs in early spring when larvae are still in their earlier instars. Full-grown larvae form a cocoon in the gall. Galls dehisce at various times during the summer. Larvae pupate the following spring. Many kinds of galls occur on more

---

Figures 421–425. Galls of Taxodiaceae, Tiliaceae, and Urticaceae. **421,** Branchlet and needle galls, *Taxodium distichum*: *Taxodiomyia* spp.: **a,** branchlet gall: *T. cupressiananassa*; **b,** starburst needle galls: *Taxodiomyia* sp.; **c,** sausagelike needle galls: *T. taxodii*; **d,** obconic needle galls: *T. cupressi*. **422,** Swollen twig, one part sectioned to show larval cells, *Tilia americana*: *Contarinia citrina*. **423,** Warty leaf gall, *Tilia americana*: *Contarinia verrucicola*. **424,** Leaf galls, *Laportea canadensis*: **a,** conical, nippled galls of various ages, one sectioned in 2×: "*Dasineura*" sp.; **b,** fuzzy, spherical gall, one showing socket only: "*Dasineura*" sp. **425,** Amorphous leaf gall, *Laportea canadensis*: "*Dasineura*" sp.

than one species of hackberry, but the greatest diversity appears to be in the southcentral United States. Considerable variation occurs within larvae of *C. spiniformis* across its range (Moser 1965), so any revision will have to take into consideration the whole geographic range of *Celticecis* flies. Hosts: *Celtis* spp. Distr.: North America east of the Rocky Mountains in the range of *Celtis* spp. Refs.: Moser (1954, 1965), Wells (1916).

Ulmus (elm)

1a Leaf pouch, occasionally greatly distorting leaf (428) .... *Janetiella ulmii* (Beutenmüller)

Larvae live singly or gregariously in concave, often closed depressions or pouches open to the upper surface of the leaf and visible as convex swellings on the under surface of the leaf. These galls first appear in early spring. Full-grown larvae drop to the soil, and adults of the next generation emerge a few weeks later. Several generations may occur per year. Host: *U. americana*. Distr.: Ontario to Maryland.

1b Bud or samara gall 2

2a Malformed samara (426b, 427, color plate 1A–C) .. *Dasineura porrecta* (Felt)

Adults of this species emerge in early March and lay their red eggs on opening flower buds. Hatching larvae settle on developing samaras, where they form closed, hair-lined pockets. One or more larvae live in a pocket. In early April the full-grown larvae drop to the soil, where they remain until the following spring. Host: *U. americana*. Distr.: Maryland, Wisconsin, and Iowa.

2b Enlarged leaf or stem bud 3

3a Enlarged leaf bud (426a)........................... *Dasineura ulmea* Felt

A damaged bud is completely filled by a single larva or pupa in spring. It is apparent that the buds are dead only after the adults have emerged from them in early spring at about the time normal buds are bursting. This species is univoltine. Host: *U. americana*. Distr.: Massachusetts to Maryland, west to Illinois. Ref.: Houser (1916).

Figures 426–430. Galls of Ulmaceae. **426,** Buds, *Ulmus americana*: **a,** swollen leaf buds, one in 2× with larva exposed: *Dasineura ulmea*; **b,** eggs on flower bud: *Dasineura porrecta*. **427,** Malformed seeds, two in 2×, *Ulmus americana*: *Dasineura porrecta* **428,** Vein swellings, *Ulmus americana*: *Janetiella ulmii*. **429,** Swollen branch bud, *Ulmus rubra*: *Dasineura* sp. **430,** Twig and leaf galls, *Celtis* sp.: *Celticecis* spp.: **a,** *Celticecis wellsi*; **b,** *Celticecis celtiphyllia*; **c,** *Celticecis texana*; **d,** *Celticecis oviformis*; **e,** *Celticecis unguicula*; **f,** *Celticecis capsularis*; **g,** *Celticecis pubescens*; **h,** *Celticecis spiniformis*; **i,** *Celticecis semenrumicis*; **j,** galls shown from both leaf surfaces, *Pachypsylla celtidismamma* (Psocoptera: Psyllidae).

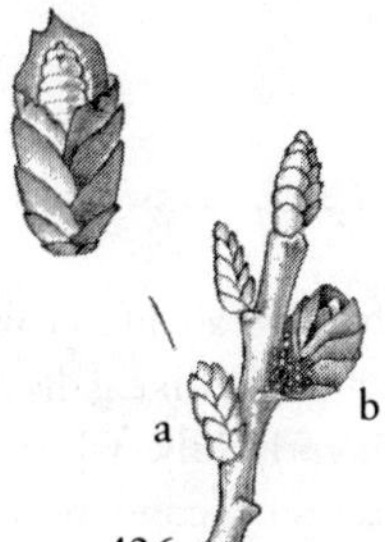

426

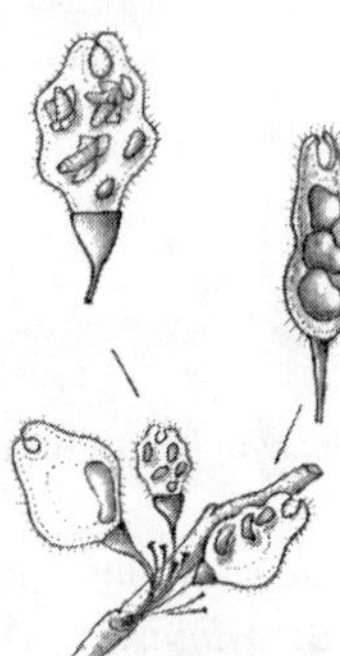
427

428

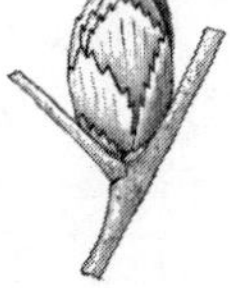
429

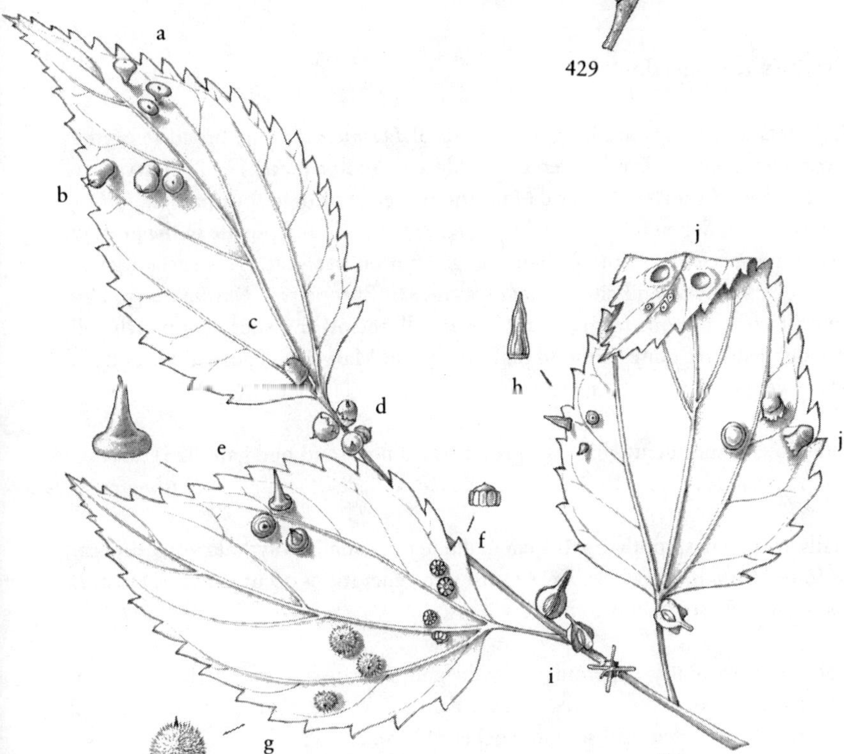

430

DLR

3b Enlarged stem bud (429) .... *Dasineura* sp.

Buds fail to elongate normally in early spring, and develop into a leafy growth about 1 cm in length and diameter. The active larvae live in bulbletlike galls and grow very slowly. They remain in the first instar through early July, when the outer leaves of the buds turn brown and rough. Larvae enter the second instar in about mid-August and the third toward the end of September. In spring they form a cocoon in which they pupate. Host: *U. rubra*. Distr.: Maryland.

Urticaceae

Boehmeria (false nettle)

Tapered stem swelling .... *Neolasioptera boehmeriae* (Beutenmüller)

Swellings are 1–2 cm long and turn woody in fall. Larvae overwinter in the stems, and adults emerge the following spring. Host: *B. cylindrica*. Distr.: Ontario to Maryland.

Laportea (wood-nettle)

Laportea and *Urtica* are host to a group of *Dasineura* in the broad sense that cause complex galls on leaves, stems, and flowers (color plate 4A, B). Larvae are characterized by a shortened spatula and the presence of only four papillae on the terminal segment. When full-grown, larvae leave the galls and pupate in the ground. At least some species have more than one generation per year. This species group appears to be Nearctic and closely host specific. In Europe two *Dasineura* spp. are known on *Urtica*, but one forms a simple leaf roll, the other a succulent pouch gall. Even though galls are quite common on *Laportea* in Maryland, I have not found any on *Urtica*; yet two kinds are recorded from *Urtica* in New York.

1a Indehiscent, succulent, globular, broad-based node and bud gall (425) .... *Dasineura* sp.

Galls are green, smooth, up to 1 cm in diameter, usually polythalamous, and can be found from June to October. At least two generations occur per year. Host: *L. canadensis*. Distr.: Ontario to New York.

1b Dehiscent, regularly spheroid or ovoid gall, narrowly connected to plant or situated in socket 2

2a Spherical, fuzzy stem gall set into socket (424b)............... *Dasineura* sp.

Galls are white to purplish and monothalamous. I found these only in October in Maryland, so this species may be univoltine. Host: *L. canadensis*. Distr.: Maryland.

2b Ovoid, smooth, translucent gall with short terminal nipple, connected to plant by narrow pedicel (424a, color plate 4A–B) .... *Dasineura* sp.

Galls are green and monothalamous. Commonly found on leaves in June, the galls can be seen also on petioles, stems, and flower parts later in summer. At least two generations occur per year. Host: *L. canadensis*. Distr.: Ontario to Maryland.

Pilea

Tapered stem swelling .... *Neolasioptera* sp.

Swellings are 1–2 cm long and turn woody in fall. Larvae overwinter in the stems and adults emerge the following spring. Host: *P. pumila*. Distr.: Maryland.

Urtica (nettle)

See general remarks under *Laportea*.

1a Smooth, green leaf gall, elongate-pyriform, the broad end apical .... *Dasineura urnicola* (Osten Sacken)

Galls are about 3 mm high and monothalamous. They dehisce when the larvae inside are full-grown. Host: *U. dioica*. Distr.: New York.

1b Hairy, grayish, ovoid leaf gall .... *Dasineura* sp.

Galls are about 3 mm high, monothalamous, and dehiscent when the larvae are mature. Host: *U. dioica*. Distr.: New York.

Verbenaceae

Avicennia

Globular leaf gall projecting from both surfaces of the leaf .... cecidomyiid

Galls are 3–5 mm in diameter. Full-grown larvae evidently escape through an opening on the lower side of the leaf. Host: *A. germinans*. Distr.: Florida.

Duranta

Enlarged, aborted flowers .... *Asphondylia* sp.

This species is known from a series of adults in the USNM. Host: *D. repens*. Distr.: Florida.

Lantana

Enlarged, aborted flowers .... *Asphondylia* sp., possibly *camarae* Möhn

Host: *L. camara*. Distr.: Mexico and, if this gall midge is *A. camarae*, El Salvador.

Verbena

1a Enlarged, aborted flower .... *Asphondylia verbenae* Felt

Infested flowers are 5 mm long and 2 mm in diameter. The gall consists of the aborted ovary and the lower corolla tube. Galls remain on the plant until normal buds mature into fruit. Adults emerged from the galls in October. Host: *V. halei*. Distr.: Texas. Ref.: Painter (1935).

1b Stem swelling or rolled leaf 2
2a Stem swelling .... *Neolasioptera* spp.

N. verbenae (Felt). Swellings occur in the flower stalk and are 1 cm long. Host: *V. prostrata*. Distr.: California.
Neolasioptera sp. This species is known only from a larva distinct from that of *N. verbenae*. Swellings are 3–5 mm long and usually at the base of leaves. Host: *V. hastata*. Distr.: Ontario.

2b Marginal leaf roll (229) .... *Clinodiplosis verbenae* (Beutenmüller)

Infested young terminal leaves become deformed and twisted. Full-grown larvae drop to the ground to pupate. Several generations occur per year. This species is one of the few Clinodiplosini that appear to be primary plant-feeders in North America. Host: *V. hastata*. Distr.: New York and New Jersey.

Violaceae

Viola (violet)

1a Rolled, deformed leaf. *Prodiplosis violicola* (Coquillett)

Eggs are laid before leaves unfold. Larvae curl the upper surface of the leaves and when full-grown drop to the ground to pupate. Several generations occur per year. This species was once a major pest of cultivated violets in greenhouses. It occurs also on wild violets. Hosts: *Viola* spp. Distr.: Connecticut to Virginia, and Wisconsin and Minnesota. Refs.: Coquillett (1900), Garman (1922).

1b Enlarged, deformed fruit. . . . . *Dasineura semenivora* (Beutenmüller)

Infested fruit are onion-shaped, green, and 6–15 mm in diameter. Larvae live in individual cells within the fruit and overwinter within a cocoon. Evidently only one generation occurs per year. Hosts: *Viola* spp. Distr.: Ontario and Massachusetts to New Jersey.

Vitaceae

Parthenocissus (Virginia creeper, woodbine)

1a Swollen midrib fold . . . . *Dasineura parthenocissi* (Stebbins)

Swellings are thick-walled, turgid, usually about 15 mm long, and can be found throughout the summer. Larvae are gregarious within the folds and reportedly drop to the soil when full-grown. Host: *P. quinquefolia*. Distr.: Massachusetts to Ontario, south to Maryland.

1b Irregular, corky root swelling . . . . *Lasioptera psederae* Felt

Swellings are polythalamous, up to 25 mm long. Larvae overwinter in the roots. Host: *P. quinquefolia*. Distr.: Ohio.

Vitis (grape)

1a Deformed flower or swollen fruit. 2
1b Swelling or gall on other plant parts 3
2a Deformed flower. . . . . *Contarinia johnsoni* Felt

Eggs are laid in opening flower buds. Larvae live gregariously in developing flowers and prevent fruit set. Early this century, *C. johnsoni* was a serious pest of certain varieties of cultivated grape in New York. Full-grown larvae drop to the soil where they remain until the following spring. *Contarinia viticola* Rübsaamen damages grape similarly in Europe and may be the same species. Hosts: *Vitis* spp., cultivated American varieties. Distr.: New York. Ref.: Hartzell (1910).

2b Swollen fruit, the seed not developed...................... *Asphondylia* sp.

Damaged grapes with larvae of this apparently undescribed species have been intercepted at the United States-Mexico border. Adults were reared from the grapes in June. Hosts: *Vitis* spp. Distr.: Mexico.

3a Woody, elongate-tapered swelling of tendrils, petioles, or midveins (432f, g) .... *Neolasioptera vitinea* Felt

Swellings are 5–15 mm long, and several may occur in tandem. Larvae live along the pith. One generation occurs per year, the adults emerging from the galls in spring. Hosts: *Vitis* spp. Distr.: New England to Virginia, west to Missouri.

3b Complex gall on node or leaf, or leaf spot or blister 4

4a Complex, woody, node gall or dehiscent spine gall on leaf (432c, e, h, 433) .... *Schizomyia* spp.

Four *Schizomyia* spp. have been described from spinelike, nutlike, single or clustered, dehiscent galls on grape. The spinelike galls (432c, e) usually occur on the leaves and are monothalamous. The others are usually polythalamous, each larva in its own cell. Polythalamous galls always occur at a node. *S. viticola* (Osten Sacken) comes from leaf galls (432c); *S. vitiscoryloides* (Parkard) from clustered, "hazelnut" galls (432h); and *S. vitispomum* (Osten Sacken) from applelike galls (433). In addition, *S. petiolicola* Felt was reportedly reared from a tapered stem swelling attributable to *Neolasioptera vitinea*, but I have never seen *Schizomyia* larvae in these galls. All the galls vary in shape and hairiness. Hosts: *Vitis* spp. Distr.: eastern half of North America.

Figures 431–433. Galls of Vitaceae: *Vitis* spp. **431,** Amorphous, succulent swellings on inflorescence and leaves: *Janetiella brevicauda.* **432,** Leaves and tendril: **a,** black spot galls: *Janetiella* sp.; **b,** hairy, blister galls: *Janetiella* sp.; **c,** woody, spinelike galls: *Schizomyia viticola*; **d,** blister galls with hairy dimple, one in 2×: *Janetiella* sp.; **e,** woody, spinelike gall: *Schizomyia* sp.; **f, g,** woody petiole and tendril swelling, respectively: *Neolasioptera vitinea*; **h,** woody "hazelnut" node gall: *Schizomyia vitiscoryloides.* **433,** Woody "apple" node gall: *Schizomyia vitispomum.*

431

432

433

4b Irregular, succulent swellings on various plant parts or in indehiscent leaf spots or blisters (431, 432a, b, d) .... *Dasineura* spp., *Janetiella* spp., and miscellaneous cecidomyiids 5

Although placed in *Dasineura* or *Janetiella*, depending on whether the tarsal claws are toothed, the species responsible for the entire suite of these galls on grape form a monophyletic group. A separate genus will eventually be erected for them. It is now very difficult to associate any of the described species with particular galls. More than one species has been reared from one kind of gall. Ref: McGiffen and Neunzig (1985).

5a Globular, irregular, usually red swelling on leaf, tendril, or flower (431) .... *Dasineura* spp., *Janetiella* spp., and miscellaneous cecidomyiids

Brachineura vitis Felt. This species, presumably an inquiline, was also reared from a jar containing such galls. Distr.: New York.
D. vitis Felt. Adults were reared in July. Distr.: New York.
J. brevicauda Felt. This is the most commonly reared species from these galls. It is probably multivoltine. Full-grown larvae cut a hole in the gall and drop to the soil to pupate. Galls sometimes cover so much surface that leaves and flower parts are deformed. Distr.: New Hampshire to North Carolina, west to Ontario and Missouri. Ref.: Johnson (1952).
Lasioptera vitis Osten Sacken. This species has been reared several times with *Janetiella* sp. from these galls, but its numbers are always small. This *Lasioptera* is evidently an inquiline. Adults were reared in summer. Hosts: *Vitis* spp. Distr.: Massachusetts to Maryland. Ref.: Johnson (1952).
Rhizomyia vitis (Felt). Reared from a jar containing amorphous galls, this species could be an inquiline in the galls. *Rhizomyia* spp. are not known to be primary plant-feeders. Distr.: New York.

5b Regular, monothalamous, blister or spot gall on leaf (432a, b, d) .... *Janetiella* spp. and *Ledomyia arizonensis* (Felt)

Janetiella spp. Larvae of several distinct, undescribed species, are found in various kinds of leaf galls. Distr.: Massachusetts to Virginia.
L. arizonensis. This species was reared from small, warty leaf bumps, 2 mm in diameter. It was reared in spring from a container with leaves taken the previous fall. *Ledomyia* spp. may all feed on fungus. Distr.: Arizona.

Zygophyllaceae

Kallstroemia

Swollen stem or inflorescence .... *Lasioptera kallstroemia* Felt

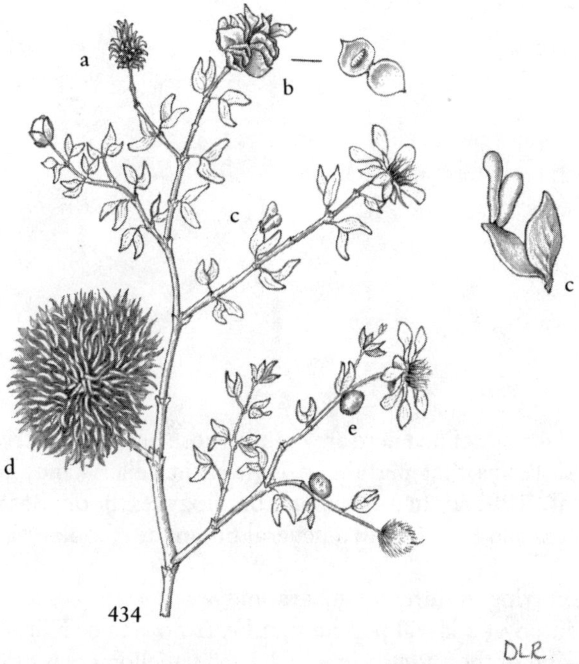

Figure 434. Galls of Zygophyllaceae: *Larrea tridentata* **a,** Conelike bud gall: *Asphondylia* sp.; **b,** widened pair of leaves stuck together with sticky substance, one in 2× opened to show larva: *Contarinia* sp.; **c,** leaf galls, examples in 2×: *Asphondylia* sp.; **d,** filamentous bud gall: *Asphondylia auripila*; **e,** black, resinous bud gall: *Asphondylia* sp.

Stem swellings are about 10 mm long and 5–9 mm in diameter and usually occur at the nodes from May to October. When they occur on the inflorescence in the fall, they cause the fruit to abort. Adults were reared in fall from both kinds of galls. This species has also been reared from *Solanum* in Mexico, so may be an inquiline. Host: *K. parviflora*; also *Solanum* (Solanaceae). Distr.: Texas. Ref.: Painter (1935).

Larrea (greasewood)

More than 15 kinds of galls have been discovered on *Larrea tridentata* in Arizona during a recent study (G. L. Waring, pers. comm.), but a report has not yet been published. All except one are formed by *Asphondylia* spp. Galls occur on the flowers, at the stem nodes, and as outgrowths of the leaves. Samples are shown here in 434. Only one of these species has been described: *Asphondylia auripila* Felt, which is responsible for the large, filamentous node gall (434d). An undescribed species of *Contarinia* is responsible for the one kind of gall not formed by an *Asphondylia*. The damage consists of a widened pair of leaves that are stuck together to form a flat, circular cell (434b).

Glossary

Anatomical terms defined here are mainly those peculiar to Cecidomyiidae or their galls. Terms that pertain to Diptera in general may be found in McAlpine et al. (1981) or in a general entomology textbook. Botanical terms not defined here can be found in a general botany textbook or a dictionary.

allopatric occurring in different geographic areas.
asetose condition of a larval papilla that bears no seta or hair.
bifilar condition of the antenna in which each flagellomere is girdled by two looped circumfila.
bivoltine having two generations per year.
blister as in "leaf blister"; a convex, ovoid to circular area caused by one or more larvae in a concavity on the underside of a leaf or inside leaf tissue.
circumfilum (pl.: *circumfila*) threadlike, continuous, sensory processes of antennae of adults, derived from forked nerve cells, each resulting tine connected at its apex with that of a neighboring cell; peculiar to Cecidomyiidae.
diapause a state of dormancy that interrupts development of an immature stage of the life cycle.
dimorphism condition of having two different kinds of structure.
empodium (pl.: *empodia*) the large pad situated between two tarsal claws.
fascicle the needle bundle of pines.
flagellum (pl.: *flagella*) the antenna exclusive of the basal two segments and formed of 8–25 or more *flagellomeres*, or articles.
fold an inpocketing of a leaf blade between veins (leaf fold) or along a vein (vein fold).
gall a predictable and consistent plant deformation that occurs in response to feeding or other stimulus by foreign organisms; *complex gall*: the result of tissue that is completely reorganized into a shape that is fundamentally different from any structure normally found on the plant; *simple gall*: a gall that does not differ from normal plant tissue except that the tissue is swollen, rolled, or otherwise marked or distorted.

imbricated overlapping, as with the shingled leaves forming some galls.
inquiline an organism that may regularly be found in particular galls but does not cause the gall.
instar the larva between molts; gall midges have three larval instars.
monophagous feeding on one species of host.
monophyletic containing species or other groups derived from a single ancestor.
monothalamous condition of a complex gall that has a single larval cell.
multivoltine having three or more generations per year.
mycophagous feeding on fungi.
oligophagous feeding on a few, closely related species of hosts.
paedogenetic reproducing during the larval stage.
papilla (pl.: *papillae*) dermal sense organ of a larva.
paraphyletic not containing all descendants of a single ancestor.
pheromone an intraspecific chemical signal.
phytophagous feeding on plants.
polyphagous feeding on many species of hosts.
polyphyletic containing species or other groups derived from unrelated ancestors.
polythalamous condition of a complex gall that has two or more discrete cells.
pouch gall a capacious vein or leaf fold.
puparium (pl.: *puparia*) in gall midges, the second-instar larval skin when it becomes the cocoon for the third-instar larva and pupa.
roll as in "leaf roll"; a leaf formed into a cylinder; *marginal leaf roll*: only the margin of the leaf is formed into a cylinder.
samara a winged fruit.
saprophagous feeding on dead or decaying organic matter.
sensoria sensory receptors.
setiform condition of a larval papilla that bears a seta or hair.
spatula (pl.: *spatulae*) a sclerotized, usually elongate, epidermal structure found on the venter of the prothorax of most full-grown cecidomyiid larvae.
spot as in "leaf spot"; a flat, ovoid to circular area; usually with one or more larvae in a concavity on the underside of a leaf or inside leaf tissue.
symbiont (adj.: *symbiotic*) an organism living in close association with another.
sympatric occurring in the same geographic area.
trichome a three-forked plant hair.
trifilar condition of the antenna when each flagellomere is girdled by three looped circumfila.
univoltine having one generation per year.

Literature Cited

Abbas, S. R. 1965. Biology of *Pipaldiplosis* Mani (Diptera: Cecidomyiidae). Indian J. Entomol. 27: 21–30.

Aderkas, P. von, and B. V. Peterson. 1987. *Chirosia betuleti* (Ringdahl) (Diptera: Anthomyiidae) a gall-former on the ostrich fern, *Matteuccia struthiopteris*, with notes on other insect-fern associates. Proc. Entomol. Soc. Wash. 89: 532–547.

Åhman, I. 1981. The potential of some *Brassica* species as host plants of the *Brassica* pod midge (*Dasineura brassicae* Winn.). Entomol. Tidskr. 102: 111–119.

——. 1985. Oviposition behaviour of *Dasineura brassicae* on a high-versus a low-quality *Brassica* host. Entomol. Exp. Appl. 39: 247–253.

——. 1986. Oviposition in *Dasineura brassicae* Winn. (Dipt.: Cecidomyiidae). Adaptive, mechanistic, and applied aspects. Sver. Lantbruksuniv. Vaxtskyddsrapp. Avh. 9: 1–25 and variously paged addenda I–VIII.

Akar, H., Jr., and E. A. Osgood. 1987. The method by which larvae of *Dasineura balsamicola* (Lintner) (Diptera: Cecidomyiidae) gain access to the interior of galls induced by *Paradiplosis tumifex* Gagné (Diptera: Cecidomyiidae). Can. Entomol. 119: 395–396.

Anonymous. 1982. National List of Scientific Plant Names. Vol. 1, v and 416 pp; vol. 2, i and 438 pp. U.S.D.A. Soil Conserv. Serv. SCS-TP-159. Washington, D.C.

Appleby, J. E., and R. B. Neiswander. 1965. *Oligotrophus apicis* sp. n., a midge injurious to junipers; with key to species of *Oligotrophus* found in the United States. Ohio. J. Sci. 65: 166–175.

Askew, R. R., and J. M. Ruse. 1974. The biology of some Cecidomyiidae (Diptera) galling the leaves of birch (*Betula*) with special reference to their chalcidoid (Hymenoptera) parasites. Trans. R. Entomol. Soc. Lond. 126: 129–167.

Barnes, H. F. 1930a. On some factors governing the emergence of gall midges (Cecidomyiidae: Diptera). Proc. Zool. Soc. Lond. 1930: 381–393.

——. 1930b. On the resistance of basket willows to button gall formation. Ann. Appl. Biol. 17: 638–640.

——. 1931a. Grass seed production and gall midges. Herbage Rev. 1: 7–9.

——. 1931b. The sex ratio at the time of emergence and the occurrence of unisexual families in the gall midges (Cecidomyiidae, Diptera). J. Gen. 24: 225–234.

——. 1939. Some gall midges and their host plant range. Ann. Appl. Biol. 26: 318–347.

——. 1946a. Gall Midges of Economic Importance. Vol. 1: Gall midges of Root and Vegetable Crops. Crosby Lockwood & Son, London. 104 pp. and 10 pls.

——. 1946b. Gall Midges of Economic Importance. Vol. 2: Gall Midges of Fodder Crops. Crosby Lockwood & Son, London. 160 pp., frontispiece, and 3 pls.

——. 1948a. Gall Midges of Economic Importance. Vol. 3: Gall Midges of Fruit. Crosby Lockwood & Son, London. 184 pp. and 9 pls.

——. 1948b. Gall Midges of Economic Importance. Vol. 4: Gall Midges of Ornamental Plants and Shrubs. Crosby Lockwood & Son, London. 165 pp. and 10 pls.

——. 1949. Gall Midges of Economic Importance. Vol. 6: Gall Midges of Miscellaneous Crops. Crosby Lockwood & Son, London. 229 pp. and 14 pls.

——. 1951. Gall Midges of Economic Importance. Vol. 5: Gall Midges of Trees. Crosby Lockwood & Son, London. 270 pp. and 8 pls.

——. 1952. The gall midges of St. John's wort (*Hypericum* spp.) with descriptions of two new species. Bull. Entomol. Res. 42: 697–705.

——. 1956. Gall Midges of Economic Importance. Vol. 7: Gall Midges of Cereal Crops. Crosby Lockwood & Son, London. 261 pp. and 16 pls.

Baskin, R., and J. W. Brewer. 1983. Studies on emergence, parasitization and control on the pinyon spindle gall midges, *Pinyonia edulicola* Gagné, in Colorado. Z. Angew. Entomol. 96: 321–327.

Batra, L. R. 1964. Insect-fungus blister galls of *Solidago* and *Aster*. J. Kans. Entomol. Soc. 37: 227–234.

Batra, L. R., and R. W. Lichtwardt. 1963. Association of fungi with some insect galls. J. Kans. Entomol. Soc. 36: 262–278.

Batra, S. W. T., and L. R. Batra. 1967. The fungus gardens of insects. Sci. Am. 1967: 112–120.

Baxendale, F. P., C. L. Lippincott, and G. L. Teetes. 1983. Biology and seasonal abundance of hymenopterous parasitoids of sorghum midge (Diptera: Cecidomyiidae). Envir. Entomol. 12: 871–877.

Baxendale, F. P., and G. L. Teetes. 1983. Factors influencing adult emergence from diapausing sorghum midge, *Contarinia sorghicola* (Diptera: Cecidomyiidae). Envir. Entomol. 12: 1064–1067.

Baylac, M. 1986. Observations sur la biologie et l'écologie de *Lestodiplosis* sp. (Dipt.: Cecidomyiidae), prédateur de la cochenille du hêtre *Cryptococcus fagi* (Hom.: Coccoidea). Ann. Soc. Entomol. Soc. Fr. (N.S.) 22: 375–386.

——. 1987. Morphologie comparée de quelques larves du genre *Lestodiplosis* Kieffer (Diptera, Cecidomyiidae). Nouv. Rev. Entomol. (N.S.) 4: 195–209.

Bergdahl, D. R., and M. Mazzola. 1985. Some relationships between *Uredinopsis mirabilis*, *Paradiplosis tumifex*, and *Dasineura balsamicola* in the needles of *Abies balsamea*. Can. J. For. Res. 15: 551–554.

Beutenmüller, W. 1907. New species of gall-producing Cecidomyiidae. Bull. Am. Mus. Nat. Hist. 23: 385–400, pls. XIII–XVII.

Bishop, G. W. 1954. Life history and habits of a new seed midge, *Dasyneura gentneri* Pritchard. J. Econ. Entomol. 47: 141–147.

Borkent, A., and J. Bissett. 1985. Gall midges (Diptera: Cecidomyiidae) are vectors for their fungal symbionts. Symbiosis 1: 185–194.

Brandhorst, C. T. 1977. Plant galls of the vicinity of Seward, Nebraska. Trans. Nebr. Acad. Sci. 4: 13–20.

Breland, O. P., and L. H. Schmitt. 1948. The biology of two sunflower gall makers (Diptera: Cecidomyiidae; Lepidoptera: Lyonetiidae). Entomol. News 59: 225–234.

Brewer, J. W. 1971. Biology of the pinyon stunt needle midge. Ann. Entomol. Soc. Am. 64: 1099–1102.

Brewer, J. W., and P. R. Johnson. 1977. Biology and parasitoids of *Contarinia coloradensis* Felt. A gall midge on Ponderosa pine. Marcellia 39: 391–398.

Brewer, J. W., V. Skuhravý, and M. Skuhravá. 1984. Biology, distribution and control of *Monarthropalpus buxi* (Laboulbène) (Diptera, Cecidomyiidae). Z. Angew. Entomol. 97: 167–175.

Brown, V. K., and G. C. McGavin. 1982. The biology of some mine and gall forming Diptera on bracken, *Pteridium aquilinum* (L.) Kühn. J. Nat. Hist. 16: 511–518.

Caresche, L. A., and A. J. Wapshere. 1975. The *Chondrilla* gall midge, *Cystiphora schmidti* (Rübsaamen) (Diptera, Cecidomyiidae). II. Biology and host specificity. Bull. Entomol. Res. 65: 55–64.

Cerezke, H. F. 1972. Observations on the distribution of the spruce bud midge (*Rhabdophaga swainei* Felt) in black and white spruce crowns and its effect on height growth. Can. J. For. Res. 2: 69–72.

Chen, Chiou-Nan, and J. E. Appleby. 1984a. Biology of the cypress twig gall midge, *Taxodiomyia cupressiananassa* (Diptera: Cecidomyiidae), in central Illinois. Ann. Entomol. Soc. Am. 77: 203–207.

——. 1984b. Chemical control of the cypress twig gall midge, *Taxodiomyia cupressiananassa* in central Illinois. J. Environ. Hort. 2: 38–40.

Chiang, H. C. 1968. Ecology of insect swarms. V. Movement of individual midges, *Anarete pritchardi*, within a swarm. Ann. Entomol. Soc. Am. 61: 584–587.

Clark, J. 1952. The spruce bud midge, *Rhabdophaga swainei* Felt (Cecidomyiidae: Diptera). Can. Entomol. 84: 87–89.

Cockerell, T. D. A. 1902. Some gall insects. Can. Entomol. 34: 183–184.

Collin, J. E. 1922. Description of a new genus and two new species of Cecidomyiidae. Trans. Entomol. Soc. Lond. 1921: 504–517, pls. XIV–XVII.

Condrashoff, S. F. 1961a. Three new species of *Contarinia* Rond. (Diptera: Cecidomyiidae) in Douglas-fir needles. Can. Entomol. 93: 123–130.

——. 1961b. Description and morphology of the immature stages of three closely related species of *Contarinia* Rond. (Diptera: Cecidomyiidae) from galls on Douglas-fir needles. Can. Entomol. 93: 833–851.

——. 1962. Bionomics of three closely related species of *Contarinia* Rond. (Diptera: Cecidomyiidae) from Douglas-fir needles. Can. Entomol. 94: 376–394.

Coquillett, D. W. 1900. A new violet pest (*Diplosis violicola* n. sp.). U.S.D.A. Div. Entomol. Bull. (NS) 22: 48–51.

Coutin, R. 1959. Quelques particularités du cycle évolutif des cécidomyies. La diapause prolongée des larves et l'apparition différée des imagos. Ann. Epiphytes 4: 491–500.

——. 1982. La cécidomyie des écorces de lavande. Phytoma. No. 336: 31–32.

Coutin, R., and A. Faivre-Amiot. 1981. Interrelations écologiques et symbiotiques complexes entre insectes et végétaux, *Lasioptera donacis* n. sp. (Diptera: Cecidomyiidae) et *Arundo donax* L. Congr. Nat. Soc. Sav. Perpignan. 1981. Sci. Fasc. 11: 195–201.

Coutin, R., and K. M. Harris. 1969. The taxonomy, distribution, biology and economic importance of the millet grain midge, *Geromyia penniseti* (Felt), gen. n., comb. n. (Dipt., Cecidomyiidae). Bull. Entomol. Res. (1968) 59: 259–273.

——. 1974. Biologie de *Contarinia sorghi* (Harris) comb. nov. sur le mil au Sénégal (Dipt. Cecidomyiidae). Ann. Soc. Entomol. Fr. 10: 457–465.

Coutin, R., and H. Katlabi. 1986. Cecidomyiidae. Pp. 95–113 and 1 pl. *In* Y. Arambourg (ed.), Traité d'entomologie oléicole. Conseil Oléicole International, Madrid.

Coutin, R., and J. Riom. 1967. Dimorphisme des galles provoquées par *Mikiola fagi* Hartig. (Dipt. Cécid.) sur *Fagus silvatica* L. Androcécidie ey gynocécidie. C. R. Acad. Sci. Paris 265: 975–978.

Curry, P. S., R. P. Knowles, and J. Waddington. 1983. Seasonal occurrence and chemical control of the bromegrass seed midge, *Contarinia bromicola* (Diptera: Cecidomyiidae) in Saskatchewan. Can. Entomol. 115: 75–79.

DeBoo, R. F., and J. P. LaPlante. 1975. *Contarinia baeri*, a pest of scots pine. Envir. Can. Bimonthly Res. Notes 31: 1.

DeClerck, R. A., and T. A. Steeves. 1988. Oviposition of the gall midge *Cystiphora sonchi* (Bremi) (Diptera: Cecidomyiidae) via the stomata of perennial sowthistle (*Sonchus arvensis* L.). Can. Entomol. 120: 189–193.

Diatloff, G., and W. A. Palmer. 1987. The host specificity of *Neolasiopteralathami* Gagné (Diptera: Cecidomyiidae) with notes on its biology and phenology. Proc. Entomol. Soc. Wash. 89: 122–125.

Dickerson, E. L., and H. B. Weiss. 1920. The insects of the evening primroses in New Jersey. J. N.Y. Entomol. Soc. 28: 32–74, pls. I–III.

Driggers, B. F. 1926. The blueberry tip worm (*Contarinia vaccinii* Felt.), a new species of midge attacking cultivated blueberries. J. N.Y. Entomol. Soc. 34: 82–85.

Eaton, C. B., and Y. S. Yuill. 1960. Gouty pitch midge. U.S. For. Serv. For. Pest Leafl. 46: 1–8.

Ehler, L. E. 1982. Ecology of *Rhopalomyia californica* Felt (Diptera: Cecidomyiidae) and its parasites in an urban environment. Hilgardia 50: 1–32.

——. 1987. Ecology of *Rhopalomyia californica* Felt at Jasper Ridge (Diptera: Cecidomyiidae). Pan-Pac. Entomol. 63: 237–241.

Ertel, M. 1975. Untersuchungen zur Larvalsystematik der Gattung *Mayetiola* (Cecidomyiidae, Diptera). Stuttg. Beitr. Naturkd. (A) 267: 1–64.

Felt, E. P. 1907a. New Species of Cecidomyiidae. New York State Education Department, Albany. 53 pp. (A preprint, with some changes, of Felt 1907b.)

——. 1907b. Appendix: New species of Cecidomyiidae. N.Y. State Mus. Bull. 110: 97–165.

——. 1907c. Cecidomyiidae: a statement. Can. Entomol. 39: 197–198.

——. 1907d. New Species of Cecidomyiidae II. New York State Education Department, Albany. 19 pp. (as 5–23). (A preprint, with some changes, of pp. 286–304 of Felt 1908.)

——. 1908. Appendix D. New species of Cecidomyiidae II. N.Y. State Mus. Bull. 124: 286–422, pls. 33–44.

——. 1911a. A generic synopsis of the Itonidae. J. N.Y. Entomol. Soc. 19: 31–62.

——. 1911b. Hosts and galls of American gall midges. J. Econ. Entomol. 4: 451–475

——. 1913. A study of gall midges. N.Y. State Mus. Bull. 180: 127–226, pls. 11–14.

——. 1915a. A study of gall midges II. N.Y. State Mus. Bull. 180: 79–213, pls. 1–15.

——. 1915b. A study of gall midges III. N.Y. State Mus. Bull. 180: 127–288, pls. 2, 4–19.

——. 1916. A study of gall midges IV. N.Y. State Mus. Bull. 180: 101–172, pls. 13–18.

——. 1918a. Key to American insect galls. N.Y. State Mus. Bull. 200: 5–310, pls. 1–16.

——. 1918b. A study of gall midges V. N.Y. State Mus. Bull. 202: 101–252, pls. 3–8.

——. 1918c. A study of gall midges, VI. N.Y. State Mus. Bull. 202: 76–205, pls. 4–12.

——. 1921. A study of gall midges, VII. N.Y. State Mus. Bull. 231–232: 81–240, pls. 8–20.

——. 1925. Key to gall midges (a resumé of studies I–VII, Itonididae). N.Y. State Mus. Bull. 257: 1–239, pls. 1–8.

——. 1940. Plant Galls and Gall Makers. Comstock Publishing Co., Ithaca, N.Y. vii and 364 pp.

Fitch, A. 1845. Insects injurious to vegetation. No. 2. Insects of the genus *Cecidomyia*, including the Hessian and wheat-fly. Am. Quart. J. Agric. Sci. 1: 255–269 and 1 pl.

——. 1846. The Hessian fly: its history, character, transformation, and habits. Trans. N.Y. State Agric. Soc. 6: 316–373, pl. II.

——. 1859. Fifth report on the noxious and other insects of the state of New York. N.Y. State Agric. Soc. Trans. (1858) 18: 781–854.

——. 1861. Sixth report on the noxious and other insects of the state of New York. N.Y. State Agric. Soc. Trans. (1860) 20: 745–868, pls. 1–2.

Foote, R. H. 1965. Family Cecidomyiidae. *In* A. Stone, C. W. Sabrosky, W. W. Wirth, R. H. Foote, and J. R. Coulson (eds.), A catalog of the Diptera of America north of Mexico. U.S.D.A. Handb. 276: i–iv, 1–1696.

Forbes, S. A. 1909. Noxious and beneficial insects of the State of Illinois. Rept. Ill. State Entomologist 24: i–xxii, 1–124, pls. I–III.

Frankie, G. W., J. W. Brewer, W. Cranshaw, and J. F. Barthell. 1987. Abundance and natural enemies of the spindle gall midge, *Pinyonia edulicola* Gagné in Colorado (Diptera: Cecidomyiidae). J. Kans. Entomol. Soc. 60: 133–144.

Franklin, R. T., and R. N. Coulson. 1968. Insects affecting seed production of shortleaf pine in the Georgia piedmont. Can. Entomol. 100: 807–812.

Frohne, W. C. 1939. Biology of certain subaquatic flies reared from emergent water plants. Pap. Mich. Acad. Sci. Arts Lett. 24 (II, 1938): 139–147.

Fronk, W. D., A. A. Beetle, and D. G. Fullerton. 1964. Dipterous galls on the *Artemisia tridentata* complex and insects associated with them. Ann. Entomol. Soc. Am. 57: 575–577.

Gagné, R. J. 1967. The genus *Oligotrophus* Latreille (Diptera: Cecidomyiidae) in North America and a new species injurious to *Betula papyrifera* March. Entomol. News 78: 129–134.

——. 1968a. A taxonomic revision of the genus *Asteromyia* (Diptera: Cecidomyiidae). Misc. Publ. Entomol. Soc. Am. 6: 1–40.

——. 1968b. A catalogue of the Diptera of the Americas south of the United States. No. 23. Family Cecidomyiidae. Pp. 23.1–23.62. Departamento de Zoologia, Secretaria da Agricultura, São Paulo, Brazil.

——. 1969a. A revision of the genus *Walshomyia* and the description of a new species reared from *Cupressus* galls in California (Diptera: Cecidomyiidae). Pan-Pac. Entomol. 45: 16–19.

——. 1969b. Revision of the gall midges of bald cypress. Entomol. News 79: 269–274.

——. 1969c. A tribal and generic revision of the Nearctic Lasiopteridi (Diptera: Cecidomyiidae). Entomol. Soc. Am. 62: 1348–1364.

——. 1970. A new genus and new species of Cecidomyiidae on pinyon pine (Diptera). Entomol. News 81: 153–156.

——. 1971a. Two new species of North American *Neolasioptera* from *Baccharis* (Diptera: Cecidomyiidae—Compositae). Proc. Entomol. Soc. Wash. 73: 153–157.

——. 1971b. Taxonomic notes on two racemicolous gall-makers on *Solidago* spp. (Diptera: Cecidomyiidae—Compositae). Proc. Entomol. Soc. Wash. 73: 169.

——. 1972. New synonymy and homonymy in Cecidomyiidae (Diptera). Proc. Entomol. Soc. Wash. 74: 321–326.

——. 1973a. Cecidomyiidae from Mexican Tertiary amber (Diptera). Proc. Entomol. Soc. Wash. 75: 169–171.

——. 1973b. A review of *Karschomyia* Felt with descriptions of 7 new Nearctic species (Diptera: Cecidomyiidae). Proc. Entomol. Soc. Wash. 75: 345–354.

——. 1973c. A generic synopsis of the Nearctic Cecidomyiidi (Diptera: Cecidomyiidae). Ann. Entomol. Soc. Am. 66: 857–889.

——. 1973d. Family Cecidomyiidae, pp. 480–517. *In* D. E. Hardy and M. Delfinado. A Catalog of the Diptera of the Oriental Region. Vol. 1. Nematocera. Univ. Hawaii Press, Honolulu.

——. 1975a. The gall midges of ragweed, *Ambrosia*, with descriptions of 2 new species (Diptera: Cecidomyiidae). Proc. Entomol. Soc. Wash. 77: 50–55.

——. 1975b. A revision of the nearctic Stomatosematidi (Diptera: Cecidomyiidae: Cecidomyiinae). Ann. Entomol. Soc. Am. 68: 86–90.

——. 1975c. A redefinition of *Diarthronomyia* Felt as a subgenus of *Rhopalomyia* Rübsaamen (Diptera: Cecidomyiidae: Oligotrophidi). Ann. Entomol. Soc. Am. 68: 482–484.

——. 1975d. A review of the Nearctic genera of Oligotrophidi with piercing ovipositors (Diptera: Cecidomyiidae). Entomol. News 86: 5–12.

——. 1975e. A new species of *Mayetiola* (Diptera: Cecidomyiidae) on American beachgrass in New Jersey. Entomol. News 86: 85–87.

——. 1977. Cecidomyiidae (Diptera) from Canadian Amber. Proc. Entomol. Soc. Wash. 79: 57–62.

——. 1978a. A new species of *Rhopalomyia* (Diptera: Cecidomyiidae) from *Artemisia ludoviciana* Nuttall (Compositae) in Utah. Great Basin Nat. 37: 426–428.

——. 1978b. A systematic analysis of the pine pitch midges, *Cecidomyia* spp. (Diptera: Cecidomyiidae). U.S.D.A. Tech. Bull. No. 1575: 1–18.

——. 1978c. A new species of *Cecidomyia* injurious to cones of slash pine in Florida. Fla. Entomol. 61: 193–196.

——. 1981a. Cecidomyiidae, pp. 257–292. *In* J. F. McAlpine, B. V. Peterson, G. E. Shewell, H. J. Teskey, J. R. Vockeroth, and D. M. Wood (eds.), Manual of Nearctic Diptera. Vol. 1. Research Branch, Agriculture Canada. Monograph No. 27, vi and 674 pp.

——. 1981b. A new species of North American Cecidomyiidae (Diptera) from *Spartina alterniflora* (Poaceae). Proc. Entomol. Soc. Wash. 83: 129–132.

——. 1983a. *Celticecis* (Diptera: Cecidomyiidae), a new genus for gall makers on hackberries, *Celtis* spp. (Ulmaceae). Proc. Entomol. Soc. Wash. 85: 435–438.

——. 1983b. *Acericecis* Gagné, a new genus for *Cecidomyia ocellaris* Osten Sacken (Diptera, Cecidomyiidae), the maple leaf ocellate gall maker in North America. Proc. Entomol. Soc. Wash. 85: 704–709.

——. 1984. The geography of gall insects, pp. 305–322. *In* T. N. Ananthakrishnan (ed.), Biology of Gall Insects. Oxford and IBH Publishing, New Delhi.

——. 1985. Descriptions of six new Nearctic Cecidomyiidae (Diptera) that live in xylem vessels of fresh-cut wood, and a review of *Ledomyia* (s. str.). Proc. Entomol. Soc. Wash. 87: 116–134.

——. 1986a. A new Nearctic species of *Contarinia* (Diptera: Cecidomyiidae) recently introduced into Hawaii on *Cupressus* (Cupressaceae). Proc. Entomol. Soc. Wash. 88: 127–130.

——. 1986b. Revision of *Prodiplosis* (Diptera: Cecidomyiidae) with descriptions of three new species. Ann. Entomol. Soc. Am. 79: 235–245.

——. 1986c. The transition from fungus-feeding to plant-feeding in Cecidomyiidae (Diptera). Proc. Entomol. Soc. Wash. 88: 381–384.

——. In press. Cecidomyiidae. *In* N. Evenhuis (ed.), Catalog of Australasian/Oceanian Diptera. B. P. Bishop Museum, Honolulu, Hawaii.

——. In preparation. Cecidomyiidae. *In* F. C. Thompson (ed.), Systematic Database of Flies of America North of Mexico. U.S.D.A., Washington, D.C.

Gagné, R. J., and G. M. Beavers. 1984. *Contarinia* spp. (Diptera: Cecidomyiidae) from shoots of slash pine (*Pinus elliottii* Engelm.) with the description of a new species injurious to needles. Fla. Entomol. 67: 221–228.

Gagné, R. J., and P. E. Boldt. In press. A new species of *Neolasioptera* (Diptera: Cecidomyiidae) from *Baccharis* (Compositae) in southeastern United States and the Dominican Republic. Proc. Entomol. Soc. Wash.

Gagné, R. J., and G. W. Byers. 1985. A remarkable new Neotropical species of *Contarinia* (Diptera: Cecidomyiidae). J. Kans. Entomol. Soc. 57: 736–738.

Gagné, R. J., and J. H. Hatchett. In press. Larval instars of Hessian fly (Diptera: Cecidomyiidae). Ann. Entomol. Soc. Am.

Gagné, R. J., and B. A. Hawkins. 1983. Biosystematics of the Lasiopterini (Diptera: Cecidomyiidae) associated with *Atriplex* spp. (Chenopodiaceae) in Southern California. Ann. Entomol. Soc. Am. 76: 379–383.

Gagné, R. J., and C. E. Stegmaier, Jr. 1971. A new species of *Chilophaga* on *Aristida* (Gramineae) in Florida (Diptera: Cecidomyiidae). Fla. Entomol. 54: 335–338.

Gagné, R. J., and K. Valley. 1984. Two new species of Cecidomyiidae (Diptera) from honeylocust, *Gleditsea triacanthos* L. (Fabaceae). Proc. Entomol. Soc. Wash. 86: 543–549.

Gagné, R. J., and W. M. Woods. 1988. Native American plant hosts of *Asphondylia websteri* (Diptera: Cecidomyiidae). Ann. Entomol. Soc. Am. 81: 447–448.

Gagné, R. J., and A. L. Wuensche. 1986. Identity of the *Asphondylia* (Diptera: Cecidomyiidae) on guar, *Cyamopsis tetragonoloba* (Fabaceae) in the southwestern United States. Ann. Entomol. Soc. Am. 79: 246–250.

Gallun, R. L., and J. H. Hatchett. 1969. Genetic evidence of elimination of chromosomes in the Hessian fly. Ann. Entomol. Soc. Am. 62: 1095–1101.

Ganz, M. N. 1968. Ovule abortion and fruit proliferation instigated by an insect in *Opuntia macrorhiza* Engelm. Cactus Succulent J. 40: 249–252.

Garman, P. 1922. The violet gall midge. *Phytophaga violicola* (Coquillett). Conn. Agric. Exp. Stn. Bull. New Haven 234: 152–156.

Gillet, C., and J. W. Brewer. 1983. Biology of *Janetiella coloradensis* Felt, a needle gall midge on *Pinus edulis* Engelm. Z. Angew. Entomol. 95: 326–335.

Given, B. B., and G. F. Woods. 1964. Gall midge to assist in controlling St. John's wort. N. Z. J. Agric. 108: 61–63.

Gleason, H. A., and A. Cronquist. 1963. Manual of Vascular Plants of Northeastern United States and Adjacent Canada. D. Van Nostrand C., Princeton, N.J. li and 810 pp.

Gordon, A. J., and S. Neser. 1986. The seasonal history of the gall midge, *Zeuxidiplosis giardi* Kieffer (Diptera: Cecidomyiidae), an introduced natural enemy of St. John's wort, *Hypericum perforatum* L. J. Entomol. Soc. South. Afr. 49: 115–120.

Gossard, H. A. 1908. The catalpa midge. *Cecidomyia catalpae* Comstock. Bull. Ohio Agric. Expt. Stn. 197: 1–13.

Hagen, H. A. 1881. Entomological notes. Can. Entomol. 13: 37.

Hallberg, E., and I. Åhman. 1987. Sensillar types of the ovipositor of *Dasineura brassicae*: structure and relation to oviposition behaviour. Physiol. Entomol. 12: 51–58.

Haridass, E. T. 1987. Midge-fungus interactions in a cucurbit stem gall. Phytophaga 1: 57–74.

Harris, K. M. 1966. Gall midge genera of economic importance (Diptera: Cecidomyiidae). Part 1: Introduction and subfamily Cecidomyiinae; supertribe Cecidomyiidi. Trans. R. Entomol. Soc. Lond. 118: 313–358.

——. 1968. A systematic revision and biological review of the cecidomyiid predators (Diptera: Cecidomyiidae) on world Coccoidea (Hemiptera-Homoptera). Trans. R. Entomol. Soc. Lond. 119: 401–494.

——. 1970. The sorghum midge. PANS 16: 36–42.

——. 1975. The *Chondrilla* gall midge, *Cystiphora schmidti* (Rübsaamen) comb. n. (Diptera: Cecidomyiidae). I. Taxonomy and description. Bull. Entomol. Res. 65: 51–54.

——. 1979. Descriptions and host ranges of the sorghum midge, *Contarinia sorghicola* (Coquillett) Diptera: Cecidomyiidae, and of eleven new species of *Contarinia* reared from Gramineae and Cyperaceae in Australia. Bull. Entomol. Res. 69: 161–182.

——. 1980. 18. Family Cecidomyiidae, pp. 238–251. *In* R. W. Crosskey (ed.), Catalogue of the Diptera of the Afrotropical Region. British Museum (Natural History), London.

Hartnett, D. C., and W. G. Abrahamson. 1979. The effects of stem gall insects on life history patterns in *Solidago canadensis*. Ecology 60: 910–917.

Hartzell, A. 1910. The grape-blossom midge, *Contarinia johnsoni* Sling., pp. 514–530. *In* A preliminary report on grape insects. Bull. N.Y. Expt. Stn. Geneva 331.

Haseman, L., and S. R. McLane. 1940. The history and biology of the juniper midge (*Contarinia juniperina* Felt). Ann. Entomol. Soc. Am. 33: 612–614, pl. I.

Hawkins, B. A., and R. D. Goeden. 1982. Biology of a gall-forming *Tetrastichus*

(Hymenoptera: Eulophidae) associated with gall midges on saltbush in southern California. Ann. Entomol. Soc. An. 75: 444–447.
——. 1984. Organization of a parasitoid community associated with a complex of galls on *Atriplex* spp. in southern California. Ecol. Entomol. 9: 271–292.
Hawkins, B. A., R. D. Goeden, and R. J. Gagné. 1986. Ecology and taxonomy of the *Asphondylia* spp. (Diptera: Cecidomyiidae) forming galls on *Atriplex* spp. (Chenopodiaceae) in Southern California. Entomography 4: 55–107.
Hayhurst, P. 1909. Quack grass (*Agropyron*), a host of the Hessian fly. J. Econ. Entomol. 2: 231–234.
Hedlin, A. F. 1961. The life history and habits of a midge, *Contarinia oregonensis* Foote (Diptera: Cecidomyiidae) in Douglas-fir cones. Can. Entomol. 93: 952–967.
——. 1964a. Results of a six-year plot study on Douglas-fir cone insect population fluctuations. For. Sci. 10: 124–128.
——. 1964b. Life history and habits of a midge, *Phytophaga thujae* Hedlin (Diptera: Cecidomyiidae) in western red cedar cones. Can. Entomol. 96: 950–957.
——. 1967. Cone insects of grand fir, *Abies grandis* (Douglas) Lindley, in British Columbia. J. Entomol. Soc. B. C. 64: 40–44.
Hedlin, A. F., and N. E. Johnson. 1963. Life history and habits of a midge, *Contarinia washingtonensis* Johnson (Diptera: Cecidomyiidae) in Douglas-fir cones. Can. Entomol. 95: 1168–1175.
——. 1968. A new species of *Camptomyia* (Diptera: Cecidomyiidae) from Douglas-fir cones. Can. Entomol. 100: 532–535.
Hedlin, A. F., H. O. Yates III, D. C. Tovar, B. H. Ebel, T. W. Koerber, and E. P. Merkel. 1980. Cone and Seed Insects of North American Conifers. Environment Canada, Canadian Forestry Service, Ottawa. iv and 122 pp.
Hennig, W. 1973. Ordnung Diptera (Zweiflügler). Handb. Zool. 4 (2) 2/31 (Lfg. 20): 1–337.
Highland, H. A. 1964. Life history of *Asphondylia ilicicola* (Diptera: Cecidomyiidae), a pest of American holly. J. Econ. Entomol. 57: 81–83.
Holz, B. 1970. Revision in Mitteleuropa vorkommender mycophager Gallmücken der *Mycodiplosis*-Gruppe (Diptera, Cecidomyiidae) unter Berücksichtigung ihrer Wirtsspezifität. Paul Jllg Photo-Offsetdruck, Stuttgart. 238 pp.
Houard, C. 1908. Les zoocécidies des plantes d'Europe et du bassin de la Méditerranée. Vol. 1. Hermann et Fils, Paris. 569 pp.
Houser, J. S. 1912. The gooseberry gall midge or bud deformer (*Rhopalomyia* grossulariae Felt). J. Econ. Entomol. 5: 178–184, pls. 5–6.
——. 1916. *Dasyneura ulmea* Felt—a new elm pest. J. Econ. Entomol. 9: 82–84.
Houseweart, M. W., and J. W. Brewer. 1972. Biology of a pinyon spindle gall midge (Diptera: Cecidomyiidae). Ann. Entomol. Soc. Am. 65: 331–336.
Imle, E. P., and A. Hartzell. 1939. A cecidomyiid larva infesting flowering stems of lilies. Contrib. Boyce Thompson Inst. 10: 277–279.
Jarvis, T. D. 1909. A catalogue of the gall insects of Ontario. 39th Ann. Rept. Entomol. Soc. Ont. 1908: 70–98 and pl. A–R.
Jauffret, F. 1970. Observations sur les galles de *Rhopalomyia ptarmicae* Val. sur l'*Achillea ptarmica*. Marcellia 36: 187–192.
——. 1973. Etude ultrastructurale de la transformation du méristème apical de l'*Euphorbia cyparissias* L. tissu nourricier sous l'action du *Bayeria capitigena* (Bre.). C. R. Acad. Sci. Paris, Ser. D 276 (7): 1177–1180.

Jensen, D. D. 1946. The identity and host plants of blossom midge in Hawaii (Diptera: Cecidomyiidae: *Contarinia*). Proc. Haw. Entomol. Soc. 12: 525–534.

Johnson, R. B. 1952. The biology, ecology, and control of *Janetiella brevicauda* Felt. Ph.D. dissertation, Cornell Univ., Ithaca, N.Y. 151 pp., 17 pls.

Johnson, W. T., and H. H. Lyon. 1989. Insects That Feed on Trees and Shrubs, 2d ed. Cornell University Press, Ithaca, N.Y. 464 pp.

Jones, D. P. 1940. Gall midges (Cecidomyiidae) affecting grass-seed production in mid-Wales and west Shropshire, together with descriptions of two new species. Ann. Appl. Biol. 27: 533–544.

Jones, R. G., R. J. Gagné, and W. F. Barr. 1983. A systematic and biological study of the gall midges (Cecidomyiidae) of *Artemisia tridentata* Nuttall (Compositae) in Idaho. Contrib. Am. Entomol. Inst. 81: ii, 1–79.

Katlabi, H. S., and R. Coutin. 1979. Biologie et cycle évolutif de *Dasyneura oleae* (Dipt. Cecidomyiidae). Ann. Soc. Entomol. Fr. 15: 371–376.

Kaufmann, T. 1973. Preliminary observations on cecidomyiid midge and its role as a cocoa pollinator in Ghana. Ghana J. Agric. Sci. 6: 193–198.

Kearby, W. H., and D. M. Benjamin. 1964. The biology and ecology of the red-pine needle midge and its role in fall browning of red pine foliage. Can. Entomol. 96: 1313–1322.

Keifer, H. H., E. W. Baker, T. Kono, M. Delfinado, and W. E. Styler. 1982. An illustrated guide to plant abnormalities caused by eriophyid mites in North America. U.S.D.A., Agric. Handb. No. 573: 1–178.

Keller, S., and N. Wilding. 1985. *Entomophthora brevinucleata* sp. nov. (Zygomycetes, Entomophthoraceae), a pathogen of gall midges (Dip.: Cecidomyiidae). Entomophaga 30: 55–63.

Kieffer, J.-J. 1913. Diptera. Fam. Cecidomyidae. Genera Insectorum. Fasc. 152. 346 pp. and 15 pls.

Kleesattel, W. 1979. Beiträge zu einer Revision der Lestremiinae (Diptera, Cecidomyiidae) unter besonderer Berücksichtigung ihrer Phylogenie. Fellbach. 203 pp.

Küster, E. 1911. Die Gallen der Pflanzen. Ein Lehrbuch der Botaniker und Entomologen. Leipzig. 437 pp.

Larew, H., and J. Capizzi. 1983. Common Insect and Mite Galls of the Pacific Northwest. Oregon State University Press, Corvallis. 80 pp.

Larew, H., R. J. Gagné, and A. Y. Rossman. 1987. Fungal gall caused by a new species of *Ledomyia* (Diptera: Cecidomyiidae) on *Xylaria enterogena* (Ascomycetes: Xylariaceae). Ann. Entomol. Soc. Am. 80: 502–507.

Lintner, J. A. 1897. Reports of the injurious and other insects of the State of New York. Rpt. 11 (for 1895): 85–326, 16 pls. Albany, N.Y. "1896."

McAlpine, J. F., B. V. Peterson, G. E. Shewell, H. J. Teskey, J. R. Vockeroth, and D. M. Wood (eds.). 1981. Manual of Nearctic Diptera. Vol. 1. Research Branch, Agriculture Canada. Monog. No. 27. vi and 674 pp.

McGiffen, K. C., and H. H. Neunzig. 1985. A guide to the identification and biology of insects feeding on muscadine and bunch grapes in North Carolina. Bull. N. C. Agric. Res. Serv. 470: iv, 1–93.

McKay, P. A., and J. H. Hatchett. 1984. Mating behavior and evidence of a female sex pheromone in the Hessian fly, *Mayetiola destructor* (Say) (Diptera: Cecidomyiidae). Ann. Entomol. Soc. Am. 77: 616–620.

MacPhee, A. W., and D. B. Finnamore. 1978. The apple leaf midge, *Dasineura mali* (Diptera: Cecidomyiidae), in New Brunswick apple orchards. Can. Entomol. 110: 1363–1364.

Mamaev, B. 1968. Evolution of Gall Forming Insects: Gall Midges. Akademia Nauk, Leningrad. 235 pp. In Russian. Translated 1975. British Library, Boston Spa, United Kingdom. iii and 316 pp.

Mamaev, B. M., and E. P. Krivosheina. 1965. Larvae of Gall Midges. Diptera, Cecidomyiidae. Akademia Nauk USSR, Moscow. 278 pp.

Mani, M. S. 1964. Ecology of Plant Galls. Dr. W. Junk, The Hague. xii and 434 pp.

Mann, J. 1969. Cactus-feeding insects and mites. U.S. Nat. Mus. Bull. 256: i–x, 1–158.

Mansour, M. H. 1975. The role of plants as a factor affecting oviposition by *Aphidoletes aphidimyza* (Diptera: Cecidomyiidae). Entomol Exp. Appl. 18: 173–179.

Marchal, P. 1897. Les cécidomyies des céréales et leurs parasites. Ann. Entomol. Soc. Fr. 66: 1–105, pls. 1–8.

Maze, M. G. 1949. Animals that produce galls in Allegheny County. Proc. Pa. Acad. Sci. 23: 133–150.

Mead, F. W. 1970. *Ctenodactylomyia watsoni* Felt, a gall midge pest of seagrape, *Coccoloba uvifera* L., in Florida (Diptera: Cecidomyiidae). Fla. Dept. Agric. Entomol. Circ. 97: 1–2.

Meijere, J. C. H. de. 1935. Ueber *Craneiobia lawsonianae* de Meij., eine Gallmücke aus den Früchten von *Chamaecyparis lawsonianae* (Dipt. Itonididae). Tijdsschr. Entomol. 78: 129–133.

Metcalfe, M. M. 1933. The morphology and anatomy of the larva of *Dasyneura leguminicola* Lint. (Diptera). Proc. Zool. Soc. Lond. 1933: 119–130, pls. I–VI.

Meyer, H. 1984. Experimentell-ökologische Untersuchungen an Gallmücken (Cecidomyiidae-Diptera) in Salzwiesenbereichen Nordwestdeutschlands. Faun. Oekol. Mitt. Suppl. 5: 1–124.

Meyer, M. J. 1952. Cécidogenèse de la galle de *Lasioptera rubi* Heeger et rôle nourricier d'un mycélium symbiotique. C. R. Acad. Sci. 234: 2556–2558.

Miller, G. E. 1984. Biological factors affecting *Contarinia oregonensis* (Diptera: Cecidomyiidae) infestations in Douglas-fir seed orchards on Vancouver Island, British Columbia. Environ. Entomol. 13: 873–877.

Miller, G. E., and J. H. Borden. 1981. Evidence for a sex pheromone in the Douglas-fir cone gall midge. Can. For. Serv. Res. Notes 1: 9–10.

Milne, D. L. 1961. The function of the sternal spatula in gall midges. Proc. R. Entomol. Soc. Lond. 36: 126–131.

Mitchell, R. G., and W. P. Nagel. 1969. Tree selection for controlling midges on Douglas-fir. Am. Christmas Tree. J. 13 (4): 11–13.

Möhn, E. 1955. Beiträge zur Systematik der Larven der Itonididae (=Cecidomyiidae, Diptera). 1. Teil: Porricondylinae und Itonidinae Mitteleuropas. Zoologica 105 (1 & 2): 1–247 and 30 pls.

——. 1959. Gallmücken (Diptera, Itonididae) aus El Salvador. 1. Teil. Senckenbergiana Biol. 40: 297–368.

——. 1960. Gallmücken (Diptera, Itonididae) aus El Salvador. 2. Teil. Senckenbergiana. Biol. 41: 197–240.

——. 1961. Gallmücken (Diptera, Itonididae) aus El Salvador. 4. Zur Phylogenie der

Asphondyliidi der neotropischen und holarktischen Region. Senckenbergiana Biol. 42: 131–330.

——. 1962. Studien über neotropische Gallmücken (Diptera, Itonididae). 1. Teil. Broteria. 31: 211–239.

——. 1963. Studien über neotropische Gallmücken (Diptera, Itonididae). 1. Teil. Broteria. 32: 3–23.

——. 1964a. Gallmücken (Diptera, Itonididae) aus El Salvador. 6. Teil Lasiopteridi. Dtsch. Entomol. Z. 11: 47–143.

——. 1964b. Gallmücken (Diptera, Itonididae) aus El Salvador. 7. Teil: Lasiopteridi. Beitr. Entomol. 14: 553–600.

——. 1966–1971. Cecidomyiidae = (Itonididae). Vol. 2(2), part 6. L: 1–248 and pls. I–IV. *In* E. Lindner (ed.), Die Fliegen der Palaearktischen Region. Schweizerbart'sche Verlagsbuchhandlung, Stuttgart.

——. 1975. Gallmücken (Diptera, Itonididae) aus El Salvador. 8. Teil: Lasiopteridi. Stuttgarter Beitr. Naturk. (A) Nr. 276: 1–101.

Morrill, W. L. 1982. Hessian fly: host selection and behavior during oviposition, winter biology, and parasitoids. J. Ga. Entomol. Soc. 17: 156–167.

Morris, R. C. 1981. *Prodiplosis morrisi* Gagné a new pest of poplars (Diptera: Cecidomyiidae). Ont. Min. Nat. Res. Pest Control Rep. No. 13: i, 1–7.

Moser, J. C. 1954. A preliminary study on the gall makers of hackberry with a description of a new parasite, *Torymus vesiculus* n. sp. (Hymenoptera: Torymidae). M.S. thesis, Ohio State Univ. ii and 60 pp., and 16 unnumbered pages with photographs.

——. 1965. The interrelationshops of three gall makers and their natural enemies, on hackberry (*Celtis Occidentalis* L.). N.Y. State Mus. Bull. 402: iv and 1–95.

Neal, J. W., Jr., and K. M. Gott. 1988. Evidence for multivoltinism in *Prodiplosis platani* Gagné (Diptera: Cecidomyiidae), a leaf curl midge of American sycamore.

Needham, J. G. 1925. Observations on a flower gall of the chaparral. J. Entomol. Zool. 17: 17–20.

——. 1946. An insect community lives in flower heads. Nat. Geogr. 1946: 340–356.

Neger, F. W. 1908. Ambrosiapilze. Ber. Dtsch. Bot. Ges. 16: 735–754, pl. XII.

Neiman, E. L., and G. R. Manglitz. 1972. The biology and ecology of the bromegrass seed midge in Nebraska. Univ. Neb. Coll. Agric. Res. Bull. 252: 1–23.

Neiswander, C. R. 1962. The phlox midge, *Hyperdiplosis phlox* Greene, and its insect enemies. Ohio Agric. Expt. Stn. Res. Bull. 915: 1–19.

Nijveldt, W. C. 1967. *Neolasioptera martelli* sp. n., a gall midge injuring *Agave* in Mexico. Entomol. Ber. 27: 125–127.

——. 1969. Gall Midges of Economic Importance. VIII. Miscellaneous. Crosby Lockwood & Son, London. 221 pp., 12 pls.

——. 1984. The host plant range of *Janetiella siskiyou* (Felt, 1917) (Diptera: Cecidomyiidae). Entomol. Ber. 44: 187.

——. 1985. Nieuwe galmuggen voor de Nederlandse fauna (IX) (Diptera: Cecidomyiidae). Entomol. Ber. 45: 85–88.

——. 1987. On four new Palaearctic species of the genus *Cecidomyia* (Diptera, Cecidomyiidae). Tijdschr. Entomol. 130: 211–223.

Nijveldt, W. C., and J. Yukawa. 1982. A taxonomic study on *Salix*-inhabiting gall midges in Japan (Diptera, Cecidomyiidae). Bull. Kitakyushu Mus. Nat. Hist. 4: 23–56.

Noble, W. B. 1931. Two wild grasses as hosts of the Hessian fly *Phytophaga destructor*. J. Agric. Res. 42: 589–592.

Noonan, G. R. 1986. Distribution of insects in the Northern Hemisphere: continental drift and epicontinental seas. Bull. Entomol. Soc. Am. 32: 80–84.

Ohno, K., and J. Yukawa. 1984. Description of a new gall midge (Diptera, Cecidomyiidae) causing leaf galls on *Camellia japonica* L., with notes on its bionomics. Kontyû 52: 427–434.

Osgood, E. A., and R. J. Gagné. 1978. Biology and taxonomy of two gall midges (Diptera: Cecidomyiidae) found in galls on balsam fir needles with description of a new species of *Paradiplosis*. Ann. Entomol. Soc. Am. 71: 85–91.

Osten Sacken, C. R. 1862. V. On the North American Cecidomyidae, pp. 173–205, 2 pls. *In* H. Loew, Monographs of the Diptera of North America. Part I. Smithson. Inst. Misc. Coll. 6.

——. 1863. *Lasioptera* reared from a gall of the golden-rod. Proc. Entomol. Soc. Phila. (1861–1863) 1: 368–370.

——. 1869. Biological notes on Diptera (galls on *Solidago*). Trans. Am. Entomol. Soc. 2: 299–303.

——. 1870. Biological notes on Diptera. (Article 2nd.). Trans. Am. Entomol. Soc. 3: 51–54.

——. 1871. Biological notes on Diptera. (Article 3d.). Trans. Am. Entomol. Soc. 3: 345–347.

——. 1875. Three new galls of Cecidomyiae. Can. Entomol. 7: 201–202.

——. 1878. Catalogue of the described Diptera of North America. Smithson. Inst. Misc. Coll. 16: 1–276.

Otter, G. W. 1938. On the morphology of the larvae of three species of Cecidomyiidae (Diptera) from knapweed (*Centaurea*) flowers. Trans. R. Entomol. Soc. Lond. 87: 39–68.

Painter, R. H. 1935. The biology of some dipterous gall-makers from Texas. J. Kans. Entomol. Soc. 8: 81–95, pls. I–II.

——. 1951. Insect Resistance in Crop Plants. Macmillan, New York. ix and 520 pp.

Panelius, S. 1965. A revision of the European gall midges of the subfamily Porricondylinae (Diptera: Itonididae). Acta Zool. Fenn. 113: 1–157.

Pantone, D. J., and S. M. Brown. 1985. New host records for *Schizomyia macrofila* (Felt) (Diptera: Cecidomyiidae). Pan-Pac. Entomol. 61: 153.

Pantone, D. J., S. M. Brown, and C. Womersley. 1985. Biological control of fiddleneck. Calif. Agric. July–Aug. 1985: 4–5.

Parnell, J. R. 1963. Three gall midges (Diptera: Cecidomyiidae) and their parasites found in the pods of broom (*Sarothamnus scoparius* (L.) Wimmer. Trans. R. Entomol. Soc. Lond. 115: 261–275.

——. 1964. Investigations on the biology and larval morphology of the insects associated with the galls of *Asphondylia sarothamni* H. Loew (Diptera: Cecidomyiidae) on broom (*Sarothamnus scoparius* [L.] Wimmer). Trans. R. Entomol. Soc. Lond. 116: 255–273.

——. 1971. A revision of the Nearctic Porricondylinae (Diptera: Cecidomyiidae) based largely on an examination of the Felt types. Misc. Pubs. Entomol. Soc. Am. 7: 273–348.

Pecora, P., A. Rizza, and M. Stazi. In prep. Life history and host specificity of *Dasineura capitigena* (Diptera: Cecidomyiidae), a gall midge attacking leafy spurge (*Euphorbia esula* L. "complex").

Peschken, D. P., A. McClay, J. L. Derby, and R. DeClerck. In press. *Cystiphora sonchi* (Diptera: Cecidomyiidae), a new biological control agent established on the weed perennial sow-thistle (*Sonchus arvensis*) (Compositae) in Canada. Can. Entomol.

Pettersson, J. 1976. Ethology of *Dasyneura brassicae* Winn. (Dipt., Cecidomyiidae). I. Laboratory studies of olfactory reactions to the host-plant. Symp. Biol. Hung. 16: 203–208.

Pierce, W. D., and A. W. Morrill. 1914. Notes on the entomology of the Arizona wild cotton. Proc. Entomol. Soc. Wash. 16: 14–29.

Pinto, J. D., and S. I. Frommer. 1980. A survey of the arthropods on jojoba *Simmondsia chinensis*. Environ. Entomol. 9: 137–143.

Pitcher, R. S. 1955a. A comparison of a group of gall midges of the genus *Thomasiniana* Strand, E. 1916 and a description of a new species from wild blackberry. Bull. Entomol. Res. 46: 27–38.

——. 1955b. The larval instars of some gall midges of the genus *Thomasiniana* Strand, E., 1916 (Diptera: Cecidomyiidae). Proc. R. Entomol. Soc. Lond. (A) 30: 97–102.

——. 1957. The abrasion of the sternal spatula of the larva of *Dasyneura tetensi* (Rübs.) (Diptera: Cecidomyidae) during the post-feeding phase. Proc. R. Entomol. Soc. Lond. (A) 32: 83–88.

Plakidas, J. D. 1982. Note on gall inhabitants of *Asphondylia helianthiglobulus* Osten Sacken (Diptera: Cecidomyiidae) in western Pennsylvania. J. N.Y. Entomol. Soc. 90: 2–4.

——. 1988. The newly discovered spring crown gall of *Asphondylia rudbeckiaeconspicua* (Diptera: Cecidomyiidae) on *Rudbeckia laciniata* (Asteraceae) in Pennsylvania. Proc. Entomol. Soc. Wash. 90: 393.

Price, P. W., G. W. Fernandes, and G. L. Waring. 1987. Adaptive nature of insect galls. Environ. Entomol. 16: 15–24.

Principi, M. M. 1959. Ricerce di morfologia e di etologia su di un Dittero Cecidomyiide galligeno, la *Putoniella marsupialis* F. Loew, vivente su piante del gen. *Prunus*. Boll. Ist. Entomol. Univ. Bologna 23: 35–68.

Puzanowa-Malysheva, E. W. 1935. *Asphondylia prunorum* Wachtl (Diptera, Cecidomyidae) und deren Pilzgallen am Pflaumenbaum. Z. Angew. Entomol. 21: 443–462.

Rainwater, C. F. 1934. Insects and a mite of potential economic importance found on wild cotton in Florida. J. Econ. Entomol. 27: 756–761.

Ranasinghe, M. A. S. K. 1977. Bionomics of a subterranean gall midge (Diptera: Cecidomyiidae) from *Artemisia ludoviciana*. Great Basin Nat. 37: 429–442.

Readshaw, J. L. 1966. The ecology of the swede midge, *Contarinia nasturtii* (Kieff.) (Diptera, Cecidomyiidae). I. Life-history and influence of temperature and moisture on development. Bull. Entomol. Res. 56: 685–700.

Redfern, M. 1975. The life history and morphology of the early stages of the yew gall midge, *Taxomyia taxi* (Inchbald) (Diptera: Cecidomyiidae). J. Nat. Hist. 9: 513–533.

Redfern, M., and R. A. D. Cameron. 1978. Population dynamics of the yew gall midge, *Taxomyia taxi* (Inchbald) (Diptera: Cecidomyiidae). Ecol. Entomol. 3: 251–263.

Reeks, W. A. 1960. Observations on the life history, distribution, and abundance of two species of *Cecidomyia* (Diptera, Cecidomyiidae) on jack pine in Manitoba and Saskatchewan. Can. Entomol. 92: 154–160.

Refai, F. Y., B. S. Miller, E. T. Jones, and J. E. Wolfe. 1956. The feeding mechanism of Hessian fly larvae. J. Econ. Entomol. 49: 182–184.

Richter-Vollert, I. 1964. Untersuchungen zur Morphologie und Ökologie von *Asphondylia sarothamni* H. Lw. Zoologica 112: iii and 1–54.

Robinson, W. H. 1971. Old and new biologies of *Megaselia* species (Dipt. Phoridae). Stud. Entomol. 14: 321–348.

Rogers, C. E. 1972a. Midges associated with guar in Texas and Oklahoma. J. Econ. Entomol. 65: 1203–1208.

——. 1972b. Economic impact of midges on guar seed production in the Texas Rolling Plains. J. Econ. Entomol. 65: 1689–1694.

——. 1973. Bionomics of a gall midge, *Asphondylia prosopidis* Cockerell, attacking mesquite buds. Cecidol. Indica 8: 131–142.

——. 1974. Natural biological control of midges attacking guar. Cecidol. Indica 9:69–81.

Rogers, C. E., T. E. Thompson, and R. J. Gagné. 1979. Cecidomyiidae of *Helianthus*: taxonomy, hosts and distribution. Ann. Entomol. Soc. Am. 72: 109–113.

Rohfritsch, O. 1974. Etude de la galle du *Lobopteromyia ramachandrani* Mani et de son développement sur l'*Acacia ferruginea* D.C. Marcellia 38: 67–75.

——. 1980. Relations hôte-parasite au début de la cécidogenèse du *Hartigiola annulipes* Hartig sur le hêtre. Bull. Soc. Bot. Fr. Actual. Bot. 127: 199–207.

——. 1987a. Gall development and feeding strategy of *Didymomyia reaumuriana* F.L.W. (Cecidomyiidae) on *Tilia* (*T. cordata* Mill. x *T. platyphyllos* Scop.). Phytophaga 1: 1–20.

——. 1987b. Different food supply strategies in midge-induced plant galls, pp. 195–200. *In* V. Labeyrie, G. Fabres, and D. Lachaise (eds.), Proceedings of the 6th International Symposium on Insect-Plant Relationships (Pau, 1986). Dr. W. Junk, Boston. Series Entomologica Vol. 41.

Rohfritsch, O., and J. D. Shorthouse. 1982. Insect galls, pp. 131–151. *In* G. Kahl, and J. S. Schell (eds.), Molecular Biology of Plant Tumors. Academic Press, New York. 615 pp.

Rohfritsch, O., and V. Skuhravý. 1970. Etude du développement de la galle de *Thecodiplosis brachyntera* (Schwägr.) sur *Pinus mugo* sbsp. *mughus* Scop. Marcellia 36: 175–186.

Roskam, J. C. 1977. Biosystematics of insects living in female birch catkins. I. Gall midges of the genus *Semudobia* Kieffer (Diptera, Cecidomyiidae). Tijdschr. Entomol. 120: 153–197.

——. 1979. Biosystematics of insects living in female birch catkins. II. Inquiline and predaceous gall midges belonging to various genera. Neth. J. Zool. 29: 283–251.

——. 1985. Evolutionary patterns in gall midge—host plant associations (Diptera, Cecidomyiidae). Tijdschr. Entomol. 128: 193–213.

——. 1986. Biosystematics of insects living in female birch catkins. IV. Egg-larval

parasitoids of the genera *Platygaster* Latreille and *Metaclisis* Förster (Hymenoptera, Platygastridae). Tijdschr. Entomol. 129: 125–140.
Roskam, J. C., and G. A. Van Uffelen. 1981. Biosystematics of insects living in female birch catkins. III. Plant-insect relation between white birches, *Betula* L., section *excelsae* (Koch) and gall midges of the genus *Semudobia* Kieffer (Diptera, Cecidomyiidae). Neth. J. Zool. 31: 533–553.
Rübsaamen, E. H. 1911. Ueber deutsche Gallmücken und Gallen. Wiss. Insektenbiol. 7: 120–125, 162–172, 278–282, 350–353, 390–394.
Rübsaamen, E. H., and H. Hedicke. 1926–1939. Die Zoocecidien, durch Tiere erzeugte Pflanzengallen Deutschlands und ihre Bewohner. Band II: Die Cecidomyiden (Gallmücken) und ihre Cecidien. Zoologica 77: 1–350, 42 pls.
Russo, R. A. 1979. Plant Galls of the California Region. Boxwood Press, Pacific Grove, California. xi and 203 pp.
Sanui, T., and J. Yukawa. 1985. A new gall midge of the genus *Resseliella* (Diptera: Cecidomyiidae) inhabiting resin of the Japanese cedar, *Cryptomeria japonica* (Taxodiaceae). Appl. Entomol. Zool. 20: 27–33.
Say, T. 1817. Some account of the insect known by the name of Hessian fly, and of a parasitic insect that feeds on it. J. Acad. Nat. Sci. Phila. 1: 45–48.
Schread, J. C. 1959. Pod gall of honey locust. Conn. Agric. Expt. Sta. New Haven Circ. 206: 1–4.
——. 1964. Dogwood club gall. Conn. Agric. Expt. Stn. Circ. 225: 1–6.
Schulz, J. T. 1973. Damage to cultivated sunflower by *Contarinia schulzi*. J. Econ. Entomol. 66: 282.
Seitner, M. 1908. Die Fichtensamengallmücke (*Plemeliella abietina*). Centralbl. Gesam. Forstwes. 34: 189–190.
Shorthouse, J. D., and R. J. West. 1986. Role of the inquiline, *Dasineura balsamicola* (Diptera: Cecidomyiidae), in the balsam fir needle gall. Proc. Entomol. Soc. Ont. 117: 1–7.
Sivinski, J., and M. Stowe. 1981. A kleptoparasitic cecidomyiid and other flies associated with spiders. Psyche 87: 337–348.
Skuhravá, M. 1973. Monographie der Gallmückengattung *Clinodiplosis* Kieffer, 1894. Stud. CSAV Praha 1973 (17): 1–80.
——. 1986. Family Cecidomyiidae, pp. 72–297. *In* A. Soos, Catalogue of Palaearctic Diptera. 4. Sciaridae: Anisopodidae. Akademiai Kiado, Budapest.
Skuhravá, M., and V. Skuhravý. 1981. Die Gallmücken (Cecidomyiidae, Diptera) des Schilfes (*Phragmites communis* Trin.). Stud. CSAV 1981 (3): 1–152.
——. 1986. Outbreak of two gall midges, *Harrisomyia* n. gen. *vitrina* (Kieffer) and *Drisina glutinosa* Giard (Cecidomyiidae, Diptera) on maple, *Acer pseudoplatanus* L. in Czechoslovakia, with descriptions of the two genera and species. J. Appl. Entomol. 101: 256–274.
Skuhravý, V. 1973. "Needle blight" and "needle droop" on *Pinus sylvestris* L. in Europe and *P. resinosa* Ait. in North America (Diptera: Cecidomyiidae). Z. Angew. Entomol. 72: 421–428.
——. 1981. Insects in reed stands as a source of bird food during the winter season. *In* V. Skuhravý (ed.), Invertebrates and vertebrates attacking common reed stands (*Phragmites communis*) in Czechoslovakia. Stud. CSAV Praha 1981 (1): 93–99.
Slepyan, E. I. 1960. On the differences in the structure of the alimentary tract in the

gall-midge larvae (Diptera, Itonididae) as related to their life-habits. Zool. Zh. 39: 1362–1369.

Slifer, E. H., and S. S. Sekhon. 1971. Circumfila and other sense organs on the antenna of the sorghum midge (Diptera, Cecidomyiidae). J. Morphol. 133: 281–301.

Smith, C. C. 1952. The life-history and galls of a spruce gall midge, *Phytophaga piceae* Felt (Diptera: Cecidomyiidae). Can. Entomol. 84: 272–275.

Smith, E. L. 1968. Biosystematics and morphology of Symphyta. I. Stem-galling *Euura* of the California region, and a new female genitalic nomenclature. Ann. Entomol. Soc. Am. 61: 1389–1407.

——. 1970. Biosystematics and morphology of Symphyta. II. Biology of gall-making nematine sawflies in the California region. Ann. Entomol. Soc. Am. 63: 36–51.

Smith, F. F., and H. J. Fisher. 1930. The boxwood leaf miner. Bull. Pa. Dept. Agric. 13: 1–14.

Smith, F. F., and R. E. Webb. 1976. The rose midge. Am. Rose Annu. 1976: 57–73.

Smith, F. F., R. E. Webb, G. W. Argus, J. A. Dickerson, and H. W. Everett. 1983. Willow beaked-gall midge, *Mayetiola rigidae* (Osten Sacken), (Diptera: Cecidomyiidae): differential susceptibility of willows. Envir. Entomol. 12: 1175–1184.

Smith, J. B. 1890. The insects injuriously affecting cranberries. N.J. Agric. Coll. Expt. Stn. Spec. Bull. K.: 1–43.

Snow, W. A., and H. Mills. 1900. The destructive *Diplosis* of the Monterey pine. Entomol. News 11: 489–494.

Sobhian, R., and L. A. Andres. 1978. The response of the skeletonweed gall midge, *Cystiphora schmidti* (Diptera: Cecidomyiidae), and gall mite, *Aceria chondrillae* (Eriophyidae) to North American strains of rush skeletonweed (*Chondrilla juncea*). Env. Entomol. 7: 506–508.

Solinas, M. 1965a. Studi sui Ditteri Cecidomyiidi. I. *Contarinia medicaginis* Kieffer. Boll. Ist. Entomol. Univ. Bol. 27: 249–300.

——. 1965b. Studi sui Ditteri Cecidomyiidi. II. *Jaapiella medicaginis* Rübsaamen e *Anabremia inquilina* sp. n. Entomologica 1: 211–281.

——. 1965c. Studi sui Ditteri Cecidomyiidi. III. *Diplosiola bursaria* n.g. Entomologica 2: 55–92.

——. 1967. Osservazioni biologiche condotte in puglia sulla *Prolasioptera berlesiana* Paoli, con particulare riferimento ai rapporti simbiotici col *Dacus oleae* Gmel. e con la *Sphaeropsis dalmatica* (Thüm.) Gigantc. Entomologica 3: 129–176.

——. 1968. Morfologia, anatomia e organizzazione funzionale del capo della larva matura de *Phaenobremia aphidimyza* (Rondani). Entomologica 4: 7–44.

——. 1971. Adattamenti morfologici nel capo delle larve di ditteri cecidomiidi a regime dietetico specializzato. Atti dell'VIII Congr. Naz. Ital. Entomol. Firenze 1969: 70–92.

Solinas, M., and M. Bucci. 1982. An ecological investigation into flower galls on *Diplotaxis muralis* D. C. caused by the gall midge *Paragephyraulus diplotaxis* Solinas. Entomologica 17: 5–22.

Solinas, M., and G. Nuzzaci. 1987. Antennal sensilla of *Mycodiplosis erysiphes* Ruebs. (Cecidomyiidae, Diptera). Boll. Ist. Entomol. Univ. Studi Bologna 41: 173–194.

Solinas, M., and P. Pecora. 1984. The midge complex (Diptera, Cecidomyiidae) on *Euphorbia* spp. I. Entomologica 19: 168–213.

Soné, K. 1985. Gregariousness and development of larvae of the pine needle gall midge, *Thecodiplosis japonensis* Uchida et Inouye (Diptera: Cecidomyiidae), in a gall. Appl. Entomol. Zool. 20: 431–438.

——. 1986a. Mortality factors before gall formation by the pine needle gall midge, *Thecodiplosis japonensis* Uchida et Inouye (Diptera: Cecidomyiidae). J. Jpn. For. Soc. 68: 32–34.

——. 1986b. Ecology of host-parasitoid community on the pine needle gall midge, *Thecodiplodis japonensis* Uchida et Inouye (Diptera, Cecidomyiidae). J. Appl. Entomol. 102: 517–527.

Specker, D. R., and W. T. Johnson. 1988. Biology and immature stages of the rhododendron gall midge, *Clinodiplosis rhododendri* Felt (Diptera: Cecidomyiidae). Proc. Entomol. Soc. Wash. 90: 343–355.

Spence, A. P. 1969. Biology of the gall midge *Rhopalomyia hirtipes* O. S. (Diptera: Cecidomyiidae) on *Solidago juncea* Ait. Cornell Exp. Stn. Mem. 409: 1–59.

Stark, A., and T. Wetzel. 1987. Fliegen der Gattung *Platypalpus* (Diptera, Empididae)—bisher wenig beachtete Prädatoren im Getreidebestand. L. Appl. Entomol. 103: 1–14.

Stebbins, F. A. 1910. Insect galls of Springfield, Massachusetts, and vicinity. Springfield Mus. Nat. Hist. Bull. 2: 1–139.

Stegmaier, C. E. 1973. The corkscrew 3-awn, *Aristida gyrans* (Gramineae), and its insect associates in South Florida. Fla. Entomol. 56: 135–139.

Steinhauer, J. R. 1983. Notes on the life history and control of the juniper midge, *Contarinia juniperina* (Diptera: Cecidomyiidae), in Pennsylvania. Melsheimer Entomol. Ser. 33: 9–14.

Stelter, H. 1982. Untersuchungen über Gallmücken XXIII. Die Gallmücken an *Salix repens* (Diptera, Cecidomyiidae). Entomol. Abhandl. 45: 15–33.

Steyskal, G. C., W. L. Murphy, and E. M. Hoover, eds. 1987. Insects and mites: techniques for collection and preservation. U.S.D.A. Agric. Res. Serv. Misc. Pub. No. 1443: (i–iv) 1–103.

Stokes, B. M. 1953. The host plant range of the swede midge (*Contarinia nasturtii* Kieffer) with special reference to types of plant damage. Tidjschr. Plantenziekten 59: 82–90.

Struble, D. B., and E. A. Osgood. 1976. Predation on larvae of the balsam gall midge, *Dasineura balsamicola* (Diptera: Cecidomyiidae), within galls in Maine. Can. Entomol. 108: 1443–1444.

Stuart, J. J., and J. H. Hatchett. 1987. Morphogenesis and cytology of the salivary gland of the Hessian fly, *Mayetiola destructor* (Diptera: Cecidomyiidae). Ann. Entomol. Soc. Am. 80: 475–482.

Sunose, T. 1978. Studies on extended diapause in *Hasegawaia sasacola* Monzen (Diptera, Cecidomyiidae) and its parasites. Kontyû 46: 400–415.

——. 1979. Two colour types of the *Euonymus* gall midge, *Masakimyia pustulae* Yukawa & Sunose (Diptera: Cecidomyiidae). Kontyû 47: 352–357.

——. 1983a. Redescription of *Asphondylia morivorella* (Naito), comb. n. (Diptera: Cecidomyiidae), with notes on its bionomics. Appl. Entomol. Zool. 18: 22–29.

——. 1983b. Studies on the popupation dynamics of the *Euonymus* gall midge *Masakimyia pustulae* Yukawa et Sunose (Diptera: Cecidomyiidae) under different environmental conditions. Kontyû 51: 29–40.

——. 1984. Parasitoid complex of the *Euonymus* gall midge *Masakimyia pustulae* Yukawa et Sunose (Diptera: Cecidomyiidae) in Japan. Kontyû 52: 557–564.

——. 1985. Geographical distribution of two gall types of *Masakimyia pustulae* Yukawa & Sunose (Diptera: Cecidomyiidae). Kontyû 53: 677–689.

Sunose, T., and J. Yukawa. 1979. Interrelationship between the leaf longevity of the evergreen spindle tree, *Euonymus japonicus* Thunb. and the *Euonymus* gall midge, *Masakimyia pustulae* Yukawa and Sunose (Diptera, Cecidomyiidae) in different environments. Jpn. J. Ecol. 29: 29–34.

Sylvén, E. 1970. Field movement of radioactively labelled adults of *Dasyneura brassicae* Winn. (Dipt., Cecidomyiidae). Entomol. Scand. 1: 161–187.

——. 1975. Study of relationships between habits and external structures in Oligotrophidi larvae (Diptera, Cecidomyiidae). Zool. Scr. 4: 55–92.

——. 1979. Gall midges (Diptera, Cecidomyiidae) as plant taxonomists. Symb. Bot. Ups. 22: 62–69.

Sylvén, E., and U. Carlbäcker. 1981. Allometric relationshops concerning antenna and wing in Oligotrophini (Diptera: Cecidomyiidae). Acta Entomol. Fenn. 42: 90–102.

Sylvén, E., and G. Svenson. 1975. Relationship between density of *Ceuthorrhynchus assimilis* Payk. (Col.) and damage by *Dasineura brassicae* Winn. (Cec.) in a cage experiment in summer turnip rape. Statens Vaxtskyddsanst. Medd. 16: 53–71.

Takasu, K., and J. Yukawa. 1984. Two-year life history of the *Neolitsea* leaf gall midge, *Pseudasphondylia neolitseae* Yukawa (Diptera, Cecidomyiidae). Kontyû 52: 596–604.

Teetes, G. L. 1985. Insect resistant sorghums in pest management. Insect Sci. Applic. 6: 443–451.

Tilden, J. W. 1951. Observations on *Rhopalomyia californica* Felt (Diptera: Itonididae). Pan-Pac. Entomol. 27: 20–22.

Tokuhisa, E., S. Nagai, and J. Yukawa. 1979. Jumping behaviour of the larvae of the Japanese cedar gall midge, *Contarinia inouyei* Mani (Diptera: Cecidomyiidae). Kontyû 47: 599–605.

Tripp, H. A. 1955. Descriptions and habits of Cecidomyiidae (Diptera) from white spruce cones. Can. Entomol. 87: 253–263.

Tscharnke, T. 1987. Growth regulation of *Phragmites australis* by the gall midge *Giraudiella inclusa*, pp. 201–205. *In* V. Labeyrie, G. Fabres, and D. Lachaise (eds.), Proceedings of the 6th International Symposium on Insect-Plant Relation ships (Pau, 1986). Dr. W. Junk, Boston. Series Entomologica Vol. 41.

Tsuda, K. 1982. Midge galls produced on the leaves of *Fagus*-species in Kyushu. Satsuma 31(88): 117–128.

Urban, J., and M. Skuhravá. 1982. Bejlomorka *Resseliella dizygomyzae* (Barnes 1933) (Cecidomyidae, Diptera) na vrbě košíkářské (*Salix viminalis* L.) v ČSSR. Lesnictví 28: 927–939.

Valley, K. 1981. Boxwood leafminer, *Monarthropalpus buxi* (Laboulbène). Diptera: Cecidomyiidae. Regul. Hort. Entomol. Circ. 7(1): 19–20.

Vallotton, R. 1969. Contribution à la biologie de la cécidomyie du pois *Contarinia pisi* Winn. (Diptera, Cecidomyiidae) avec étude particulière du phénomène de la diapause. Bull. Soc. Entomol. Suisse 42: 241–293.

Vincinaux, C., and J. Biernaux. 1974. Premières observations biologiques sur la cécidomyie du chou (*Contarinia nasturtii* Kief.), déprédatrice des choux de Bruxelles. Int. Symp. Fytofarm Fytiat 1974, 25th (2): 1213–1224.

Walsh, B. D. 1864. On the insects, coleopterous, hymenopterous, and dipterous, inhabiting the galls of certain species of willow. Part 1st. Diptera. Proc. Entomol. Soc. Phila. 3: 543–644.

Wangberg, J. K. 1981. Observation on the bionomics of *Rhopalomyia utahensis* Felt (Diptera, Cecidomyiidae), its gall and insect associates. Cecidol. Int. 2: 17–24.

Watson, W. Y. 1973. *Dasineura aceris* (Shimer), midge on silver maple (Diptera: Cecidomyiidae). Entomol. Soc. Ont. 104: 31–33.

Watts, J. G., and A. C. Bellotti. 1967. Some new and little-known insects of economic importance on range grasses. J. Econ. Entomol. 60: 961–963.

Webster, F. M. 1912. The alfalfa gall midge. (*Asphondylia miki* Wachtl). U.S. Dep. Agric. Bur. Entomol. Circ. No. 147: 1–4.

Webster, J. A., and H. L. Kohls. 1972. Occurrence of a cecidomyiid, *Clinidiplosis* sp., on grain sorghum in Michigan. J. Econ. Entomol. 65: 923.

Wehrle, L. P. 1946. The cucurbit midge, *Itonida citrulli*. J. Econ. Entomol. 39: 415–416.

Weis, A. E. 1982a. Resource utilization patterns in a community of gall-attacking parasitoids. Environ. Entomol. 11: 809–815.

——. 1982b. Use of a symbiotic fungus by the gall maker *Asteromyia carbonifera* to inhibit attack by the parasitoid *Torymus capite*. Ecology 63: 1602–1605.

——. 1982c. Ecological and evolutionary consequences of variation in clutch size in *Asteromyia carbonifera* (O. S.) (Diptera: Cecidomyiidae). Dissert. Abstr. Int. (B) 42(7): 2682.

——. 1984. Apical dominance asserted over lateral buds by the gall of *Rhabdophaga strobiloides* (Diptera: Cecidomyiidae). Can. Entomol. 116: 1277–1279.

Weis, A. E., and A. Kapelinski. 1984. Manipulation of host plant development by the gall-midge *Rhabdophaga strobiloides*. Ecol. Entomol. 9: 457–465.

Weiss, H. B., and E. West. 1921. Notes on the insects on the spreading dogbane, *Apocynum androsaemifolium* L., with a description of a new dogbane midge, by Dr. E. P. Felt. Can. Entomol. 53: 146–152.

Wellhouse, W. H. 1922. The insect fauna of the genus *Crataegus*. Cornell Univ. Agric. Expt. Stn. Mem. 56: 1041–1136.

Wells, B. W. 1914. Some unreported cecidia from Connecticut. Ohio Nat. 14: 289–298.

——. 1915. A survey of the zoocecidia on species of *Hicoria* caused by parasites belonging to the Eriophyidae and the Itonididae (Cecidomyiidae). Ohio J. Sci. 16: 37–57, pls. I–II.

——. 1916. The comparative morphology of the zoocecidia of *Celtis occidentalis*. Ohio J. Sci. 16: 249–290, pls. XII–XIX.

——. 1921a. Evolution of Zoocecidia. Bot. Gaz. 71: 358–357, pls. XXI–XXII.

——. 1921b. New United States Zoocecidia. Psyche 28: 35–45.

Wellso, S. G. and R. D. Freed. 1982. Positive association of the wheat midge (Diptera: Cecidomyiidae) with glume blotch. J. Econ. Entomol. 75: 885–887.

West, R. J., and J. D. Shorthouse. 1982. Morphology of the balsam fir needle gall induced by the midge *Paradiplosis tumifex* (Diptera: Cecidomyiidae). Can. J. Bot. 60: 131–140.

Wheeler, A. G., Jr., and E. R. Hoebeke. 1985. The insect fauna of ninebark, *Physocarpus opulifolius* (Rosaceae). Proc. Entomol. Soc. Wash. 87: 356–370.

Wheeler, A. G., Jr., J. R. Steinhauer, J. F. Stimmel, K. R. Valley, and T. J. Henry.

1981. Insects and Mites of Pennsylvania Junipers. Pennsylvania Dept. Agric., Harrisburg. iv and 32 pp.

Wheeler, W. M. 1889. On two new species of cecidomyid flies producing galls on *Antennaria plantaginifolia.* Proc. Wisc. Nat. Hist. Soc. 1888: 209–215.

White, M. J. D. 1950. Cytological studies on gall midges (Cecidomyidae). Univ. Tex. Pub. 5007: 1–80.

Whitney, H. S., and J. A. Baranyay. 1968. An undescribed gall midge leaf spot of balsam poplar. Phytopathology 58: 262–263.

Williams, I. L., and C. W. Fatzinger. 1977. A new cone midge, *Cecidomyia* spp. (Diptera: Cecidomyiidae) affecting slash pine cones. 14th South. For. Tree Improv. Conf. Proc. 1977: 104–107.

Wilson, L. F. 1966. Life history, habits, and damage of the boxelder leaf gall midge, *Contarinia negundifolia* Felt (Diptera: Cecidomyiidae) ın Michigan. Can. Entomol. 98: 777–784.

——. 1968a. Life history and habits of *Rhabdophaga* sp. (Diptera: Cecidomyiidae), a gall midge attacking willow in Michigan. Can. Entomol. 100: 184–189.

——. 1968b. Life history and habits of the willow beaked gall midge, *Mayetiola rigidae* (Diptera: Cecidomyiidae), in Michigan. Can. Entomol. 100: 202–206.

——. 1968c. Life history and habits of the pine cone willow gall midge, *Rhabdophaga strobiloides* (Diptera: Cecidomyiidae), in Michigan. Can. Entomol. 100: 430–433.

——. 1968d. Life history, habits, and damage of a gall midge, *Oligotrophus papyriferae* (Diptera: Cecidomyiidae), injurious to paper birch in Michigan. Can. Entomol. 100: 663–669.

Wilson, L. F., and G. C. Heaton. 1987. Life history, damage, and gall development of the gall midge, *Neolasioptera brevis* (Diptera: Cecidomyiidae), injurious to honeylocust in Michigan. Great Lakes Entomol. 20: 111–118.

Wilson, L. F., F. J. Sapio, and G. A. Simmons. 1987. Biology, injury, and control of the European needle-bending midge (Diptera: Cecidomyiidae) on Scotch pine in Michigan. Great Lakes Entomol.

Wünsch, A. 1979. Gallenerzeugende Insekten Nordkolumbiens, Speziell Asphondyliidi und Lasiopteridi (Diptera, Cecidomyiidae) aus dem Küstenbereich um Santa Marta. Offsettdruck Koppenhöfer, Waiblingen. 238 pp.

Wyatt, I. J. 1967. Pupal paedogenesis in the Cecidomyiidae (Diptera). 3. A reclassification of the Heteropezini. Trans. R. Entomol. Soc. Lond. 119: 71–98.

Young, A. M. 1985. Studies of cecidomyiid midges (Diptera: Cecidomyiidae) as cocoa pollinators (*Theobroma cacao* L.) in Central America. Proc. Entomol. Soc. Wash. 87: 49–79.

Yukawa, J. 1971. A revision of the Japanese gall midges (Diptera: Cecidomyiidae). Mem. Fac. Agric. Kagoshima Univ. 8: 1–203.

——. 1980. Life histories and host plants of the genus *Asphondylia* (Diptera: Cecidomyiidae) in Japan. Proc. Assoc. Plant Prot. Kyushu 26: 125–127.

——. 1983. Arthropod community centred upon the *Neolitsea* leaf gall midge, *Pseudaphondylia neolitseae* Yukawa (Diptera, Cecidomyiidae) and its host plant, *Neolitsea sericea* (Blume) Koidz. (Lauraceae). Mem. Fac. Agric. Kagoshima Univ. 19: 89–96.

Yukawa, J., and K. Miyamoto. 1979. Redescription of *Asphondylia sphaera* Monzen (Diptera, Cecidomyiidae), with notes on its bionomics. Mem. Fac. Agric. Kagoshima Univ. 15: 99–106.

Yukawa, J., S. Ohsako, and H. Ikenaga. 1981. Parasite complex of the Japanese species of the genus *Asphondylia* (Diptera, Cecidomyiidae), including the soybean pod gall midge. Proc. Assoc. Pl. Prot. Kyushu 27: 113–115.

Yukawa, J., and T. Sunose. 1976. Description of a new gall midge (Diptera, Cecidomyiidae) on *Euonymus*, with notes on its bionomics. Kontyû 44: 159–168.

Yukawa, J., K. Takahashi, and N. Ohsaki. 1976. Population behaviour of the *Neolitsea* leaf gall midge, *Pseudasphondylia neolitseae* Yukawa (Diptera, Cecidomyiidae). Kontyû 44: 358–365.

Yukawa, J., and K. Tsuda. 1986. Leaf longevity of *Quercus glauca* Thunb., with reference to the influence of gall formation by *Contarinia* sp. (Diptera: Cecidomyiidae) on the early mortality of fresh leaves. Mem. Fac. Agric. Kagoshima Univ. 22: 73–77.

——. 1987. A new gall midge (Diptera, Cecidomyiidae) causing conical leaf galls on *Celtis* (Ulmaceae) in Japan. Kontyû 55: 123–131.

Index

Cecidomyiid names in italics are junior synonyms or emendations. Numbers in boldface type indicate pages where the main mention occurs. Numbers in italics refer to pages with illustrations. Headings followed by page numbers punctuated with semicolons (e.g., "Asphondylia sp.") indicate that a different species of the taxon is mentioned on the page(s) listed between semicolons. The color plates can be found between pages 100 and 101.

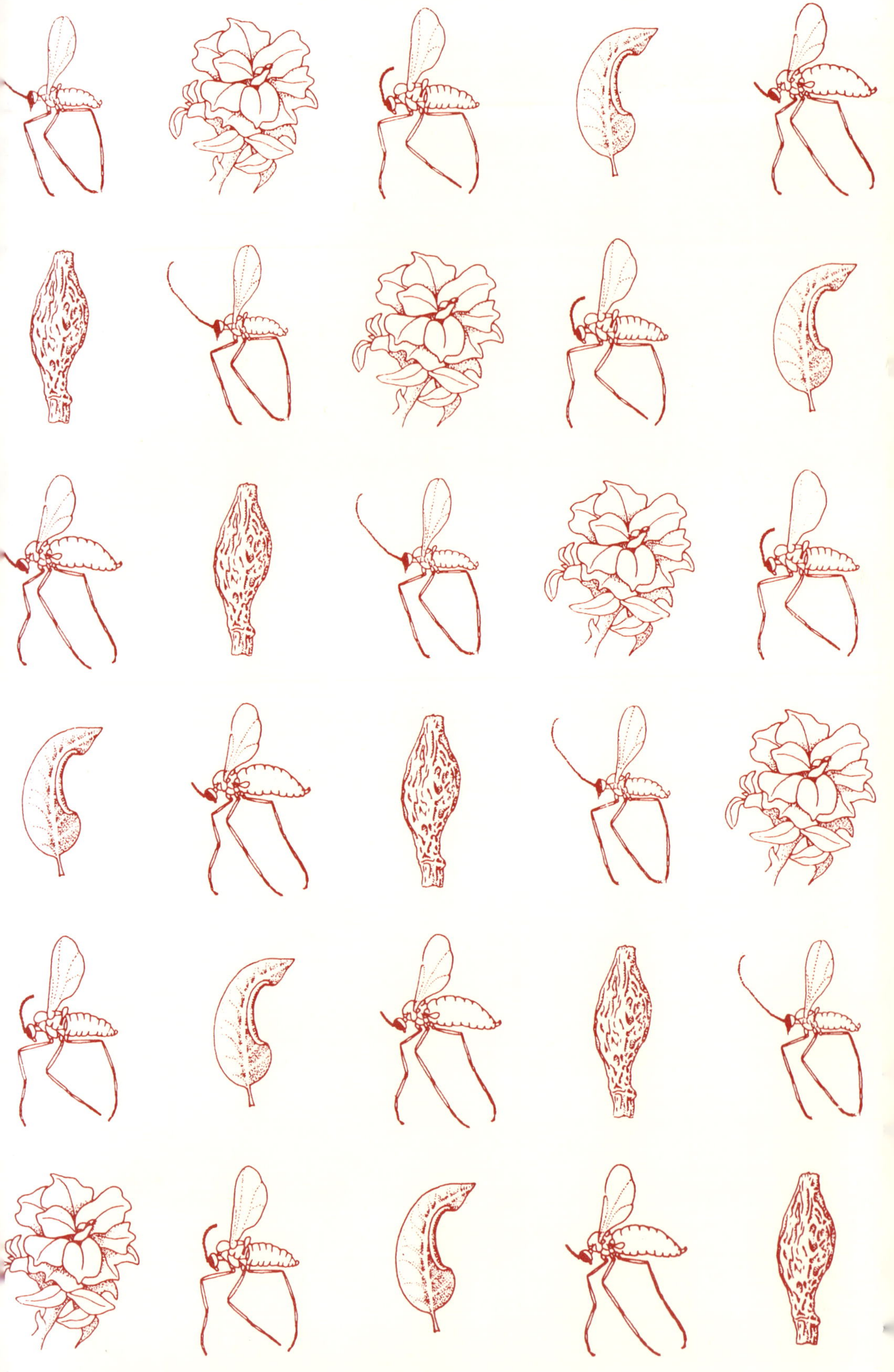